Nutrition and the Adult
Macronutrients

Human Nutrition
A COMPREHENSIVE TREATISE

General Editors:
Roslyn B. Alfin-Slater, University of California, Los Angeles
David Kritchevsky, The Wistar Institute, Philadelphia

Nutrition and the Adult

Macronutrients

Edited by

Roslyn B. Alfin-Slater

University of California
Los Angeles, California

and

David Kritchevsky

The Wistar Institute
Philadelphia, Pennsylvania

PLENUM PRESS · NEW YORK AND LONDON

Library of Congress Cataloging in Publication Data

Main entry under title:

Nutrition and the adult: macronutrients.

  (Human nutrition; v. 3A)
  Includes bibliographies and index.
  1. Nutrition. I. Alfin-Slater, Roslyn, 1916-   II. Kritchevsky, David, 1920-
III. Series. [DNLM: 1. Nutrition—In adulthood. QU145.3 H9182 v. 3]
QP141.H78 vol. 3A   612'.3'08s [612'.3]   79-25119
ISBN 0-306-40287-4

Printed in the United States of America

*This volume is dedicated to all the
scientists whose work has advanced
our understanding of
how to apply knowledge
to benefit human nutrition.*

Contributors

Lilla Aftergood • Environmental and Nutritional Sciences, School of Public Health, University of California, Los Angeles, California 90024

Roslyn B. Alfin-Slater • Environmental and Nutritional Sciences, School of Public Health, University of California, Los Angeles, California 90024

P. L. Baldwin • Department of Animal Science, University of California, Davis, California 95616

Elsworth R. Buskirk • Laboratory for Human Performance Research, Intercollege Research Programs, The Pennsylvania State University, University Park, Pennsylvania 16802

Steven D. Clarke • Department of Food Science and Nutrition, Ohio State University, Columbus, Ohio 43210

Andrew L. Clifford • Department of Nutrition, University of California, Davis, California 95616

Keith A. Crist • Department of Animal Science, University of California, Davis, California 95616

Susanne K. Czarnecki • The Wistar Institute of Anatomy and Biology, Philadelphia, Pennsylvania 19104

David Kritchevsky • The Wistar Institute of Anatomy and Biology, Philadelphia, Pennsylvania 19104

Ian Macdonald • Department of Physiology, Guy's Hospital Medical School, London SE1 9RT, United Kingdom

James F. Mead • Department of Biological Chemistry, Schools of Medicine and of Public Health, and Laboratory of Nuclear Medicine and Radiation Biology, University of California, Los Angeles, California 90024

Jose Mendez • Laboratory for Human Performance Research, Intercollege Research Programs, The Pennsylvania State University, University Park, Pennsylvania 16802

Rose Mirenda • Department of Home Economics, Hunter College, City University of New York, New York, New York 10021

Dale R. Romsos • Department of Food Science and Human Nutrition, Michigan State University, East Lansing, Michigan 48824

Judith S. Stern • Department of Nutrition, University of California, Davis, California 95616

Jon A. Story • The Wistar Institute of Anatomy and Biology, Philadelphia, Pennsylvania 19104

Foreword

The science of nutrition has advanced beyond expectation since Antoine Lavoisier as early as the 18th century showed that oxygen was necessary to change nutrients in foods to compounds which would become a part of the human body. He was also the first to measure metabolism and to show that oxidation within the body produces heat and energy. In the two hundred years that have elapsed, the essentiality of nitrogen-containing nutrients and of proteins for growth and maintenance of tissue has been established; the necessity for carbohydrates and certain types of fat for health has been documented; vitamins necessary to prevent deficiency diseases have been identified and isolated; and the requirement of many mineral elements for health has been demonstrated.

Further investigations have defined the role of these nutrients in metabolic processes and quantitated their requirements at various stages of development. Additional studies have involved their use in the possible prevention of, and therapy for, disease conditions.

This series of books was designed for the researcher or advanced student of nutritional science. The first volume is concerned with prenatal and postnatal nutrient requirements; the second volume with nutrient requirements for growth and development; the third with nutritional requirements of the adult; and the fourth with the role of nutrition in disease states. Our objectives were to review and evaluate that which is known and to point out those areas in which uncertainties and/or a lack of knowledge still exists with the hope of encouraging further research into the intricacies of human nutrition.

Roslyn B. Alfin-Slater
David Kritchevsky

Preface

Nutrition has been defined as the study of nutrients in foods from the time they are ingested to the time their metabolic end products are excreted. Included in the study of these processes are a host of catabolic and anabolic reactions following digestion and absorption, the manner in which these nutrients interact with each other, and the systems in which they participate. Furthermore, nutrition implies a study of nutrient requirements for all segments of a healthy population.

Whether the requirements for nutrients are met requires nutritional assessment. For infants, requirements are based on nutrients present in human milk. For children, anthropometric measures—height, weight, arm circumference—are compared with standards to see whether adequate weight gain and growth are maintained. For adults, there are no simple determinants to estimate adequate nutrition other than the maintenance of "normal" weight and "health."

These volumes, dealing with nutrition of the adult, were divided so that the macronutrients would be considered separately from the micronutrients. Volume 3A starts with a discussion of the recommended dietary allowances and how they are formulated. Thereafter, the digestion, metabolism, and functions of the various macronutrients are presented. Volume 3B is concerned with the micronutrients and also with the interactions of the various nutrients with common drugs. Volume 3B ends with a discussion of nutrition for the older adult and some of the problems which arise with aging. All of the authors in these volumes are experts in their respective areas.

It is hoped that calling attention to the known and unknown in nutrition will act as a stimulus for further research. It is only when we can relate functions and metabolic reactions of nutrients with external symptoms that we will be able to properly assess nutritional status and requirements and eventually arrive at an optimum rather than an adequate diet.

We wish to thank Miss Jane T. Kolimaga for her expert assistance in the preparation of the indices.

Roslyn B. Alfin-Slater
David Kritchevsky

Los Angeles and Philadelphia

Contents

Chapter 2

Energy: Caloric Requirements

Elsworth R. Buskirk and Jose Mendez

Chapter 3

Suppliers of Energy: Carbohydrates

Ian Macdonald

Chapter 4

Suppliers of Energy: Fat

Roslyn B. Alfin-Slater and Lilla Aftergood

Chapter 5

Suppliers of Energy: Carbohydrate–Fat Interrelationships

Dale R. Romsos and Steven D. Clarke

Chapter 6

Energetics and the Demands for Maintenance

Keith A. Crist, R. L. Baldwin, and Judith S. Stern

Chapter 7
Nutrients with Special Functions: Proteins and Amino Acids in Tissue Maintenance
Andrew J. Clifford

Chapter 8
Nutrients with Special Functions: Essential Fatty Acids
James F. Mead

Nutrient Requirements: What They Are and Bases for Recommendations

Roslyn B. Alfin-Slater and Rose Mirenda

1. Introduction

Biblical references to the essential nature of food for the survival of mankind offer evidence of our persistent concern with nourishment and the quality of life which we enjoy. Food, as consumed by the average, healthy individual, satisfies not only hunger but also a number of other needs, i.e., physiological, psychological, and social. In addition, the specific requirements for nutrients depend on the individual biochemistry, on the ability to digest food properly, and on the efficiency of nutrient absorption and metabolism. It becomes evident that although food is essential to life, there are differences in the amount of nutrients individuals require for survival and well-being.

The present state of knowledge about nutrition is the result of an accumulation of information from a variety of records passed on from early Greek and Roman times to the latest research reports of the renowned nutrition scientists. Through trial and error and sometimes serendipity, the nutritional scientist has successfully supplemented the body of historical knowledge with the latest scientific observations to supply facts of such a high degree of certainty that recommendations can be made about the daily nutrient needs of man. It is within this frame of reference and through the utilization of the latest available scientific information that dietary standards and recommendations have been and are proposed in the United States and in other countries.

Unfortunately, in the field of nutrition as in other sciences, information is never complete. The unknown facts about nutrition provide the impetus for continued experimentation. Meanwhile, the accepted scientific data are

Roslyn B. Alfin-Slater • Environmental and Nutritional Sciences, School of Public Health, University of California, Los Angeles, California 90024. *Rose Mirenda* • Department of Home Economics, Hunter College, City University of New York, New York, New York 10021.

interpreted and applied to the local population in its unique environment. Hence, dietary recommendations based on the same scientific information may differ in different areas of the world.

Although the validity of applying incomplete information is open to question, to delay until absolute certainty is available would be an interminable wait. Therefore, when information is not complete or not available, value judgments are made by nutrition scientists who are experts in their particular fields and who are selected by their peers and mandated to formulate dietary recommendations which will be useful in providing guidelines. These allowances are intended to provide adequate nutrient intake for the great majority of healthy persons in a specific national environment. The Food and Nutrition Board of the National Research Council, National Academy of Sciences, serves this function in the United States.

One of the difficulties in formulating recommendations is to adequately define "health." The most comprehensive definition is that offered by the World Health Organization of the United Nations, in which health is defined as "the state of complete physical, mental and social well-being and not merely the absence of disease or infirmity" (Anonymous, 1970). The question remains as to the definition of the standards with which to measure this state of well-being. It is probable that only a minority of persons would be classified as consistently and flawlessly healthy by WHO standards. However, this definition has been interpreted differently in various parts of the world based on both scientific and philosophic criteria familiar to the particular health evaluator.

There is no single set of criteria that describes optimum health. Nonetheless, health assessment continues to be a routine task of medical practitioners whose decisions about patient health care and diet recommendations are influenced by their medical expertise, established standards of nutritional assessment (Christakis, 1973; Foman, 1976), and possibly, too, by societal dicta.

Nutritionists agree that many dietary patterns can meet the nutritional needs of various populations, since food habits, meal patterns, and food values vary within countries as well as on an international basis. In addition, the different nutritional needs of individuals influenced by genetic and environmental factors as well as by sex, age, growth rate, reproductive state, and size must also be considered. To date, no single universally acceptable food plan has been devised which will provide an adequate diet for everyone, although efforts have been made and continue to be made in the United States (Page and Phipard, 1957; Phipard and Page, 1962) and in other nations (Ahlstrom and Rasanen, 1973; Taylor and Pye, 1966; Taylor and Riddle, 1971) to adapt indigenous foods to the nutritional needs of unique populations in particular areas.

In order for food to fulfill its functions, that is, to provide energy, to promote tissue growth and maintenance, and to regulate body processes, it must provide the nutrients contained therein in amounts quantitatively and qualitatively sufficient to meet human nutrient needs. The six classes of nutrients—carbohydrate, fat, protein, minerals, vitamins, and water—provided in the usual diet are able to supply the required amounts of nutrients if the diet is properly selected. Consumption of a mixed diet composed of a variety of foods from both animal and vegetable sources and taken in quantities consistent with ideal weight for age and sex is recommended and easily obtained.

2. Techniques Used to Determine Nutritional Status

Ideally, nutritional status may be determined by a review of data collected during dietary, biochemical, and clinical evaluations of individuals or groups. However, the number of assessments included in any nutrition survey is limited by the goals and resources available. One or more of the three accepted examination procedures may be included in any nutrition survey, with the knowledge that each of these methods provides a measure of different states of nutrition and that these measures may not correlate with each other. Despite these limitations, and the additional problem of inherent human variability, nutrition surveys continue to be conducted with the purpose of collecting data useful in identifying the specific factors affecting the gross nutritional status of population groups and in defining the magnitude and distribution of possible malnutrition within the groups.

2.1. Dietary Surveys

Becker *et al.* (1960) have indicated that the need for dietary surveys arises from epidemiologic concerns for factors which affect health. Generally, dietary surveys evaluate current food intake. A review of the actual food consumption record of an individual will reveal not only the food habits and dietary patterns, but also the levels of nutrient intake. Current methods of collecting food intake information for epidemiologic studies are classified into two broad categories: (1) food records in estimated, measured, or weighed amounts and (2) the dietary history. Explanations of the various techniques and the advantages of each have been given by Bigwood (1939) and others (Cantoni *et al.*, 1959; Young and Trulson, 1960).

Food records provide intake data concerning specific meals recorded concurrently by means of weights, household measures, or estimates of quantity, whereas when a dietary history is recorded, the data collected are in terms of individual recall, for example, estimates of quantity of foods eaten at meals and/or snacks for one, three, or seven days and frequency of occurrence. The selection of the appropriate dietary survey method is determined by the data analysis techniques and survey goals planned by the director of the dietary or the nutrition survey.

Guides for the calculation and evaluation of dietary intakes in the United States have been developed by the Interdepartmental Committee on Nutrition for National Defense (ICNND, 1963). The nutrient composition of foods is also available in various handbooks published by the United States Department of Agriculture (Adams, 1975; Adams and Richardson, 1977; Davenport, 1964; Merrill *et al.,* 1966; Watt and Merrill, 1963).

When making any nutritional evaluation, it must be realized that these are numerical values derived from tables of food composition which, in themselves, present limitations in evaluating dietary intakes. Lack of information about the level of differences in nutrient content of similar raw foods and commercially prepared and processed foods and of losses incurred during storage, preparation, and cooking present problems in assessing nutrient intake values and in relating these to biochemical and clinical data collected in a nutrition survey.

Although the nutrient intake information per se may not be clearly indicative of a specific level of deficiency, it does provide valuable information about food availability and consumption practices of subjects and is generally interpreted in terms of population groups. Data are evaluated in the light of dietary standards such as those developed by the Interdepartmental Committee on Nutrition for National Defense (ICNND, 1963), and more recently by the Nutrition Section of the United States Department of Health, Education and Welfare (USDHEW, 1972). It was anticipated that these dietary standards would provide a more uniform manner for the collecting and evaluating of nutrition survey data and also facilitate in the development of practical measures to raise the nutritional status of malnourished people. Since the first discussion of recommended allowances for the United States by Stiebling (1933) until the latest edition of Recommended Dietary Allowances (1979), as well as the FAO/WHO standards (1973), the purposes of establishing these standards have not changed. It is generally accepted that these dietary standards represent expert value judgments based on experimental data but are limited by the variables imposed by human metabolic studies. The data lack the exactitude that would make the standards ideal for the evaluation of the nutritional status of individuals. In practice, dietary standards are most useful in the evaluation and planning of diets for civilian and military population groups, as well as for providing a basis for recommendations concerning national agriculture, import, and export policies.

2.2. Biochemical Evaluations

Biochemical analyses of human blood, tissue, and excreta are helpful in evaluating current nutritional status, since they indicate the nutrients and metabolites to be found in, or missing from, the body. The collection of data during the biochemical evaluation of nutritional status may be achieved by analysis of a sample of blood, hair, or urine and/or by biopsy of liver or bone, the most quantitative being analysis of blood and urine. Often, the simplicity, sensitivity, and practicality of the analytical method will determine which of the macro and/or micro techniques will be selected for use during a survey (Baker and Frank, 1968, Lowry, 1952; Lowry and Bessey, 1945; Sauberlich *et al.*, 1973). Some of the more traditional procedures of blood and urine analysis have been modified and complemented by new highly sensitive analytical methodologies (Leveille, 1972; Sauberlich *et al.*, 1972).

During the National Nutrition Survey (USDHEW, 1972), blood biochemical determinations used in evaluating nutritional status included hematocrit, hemoglobin, total serum protein, serum albumin, serum globulin, serum cholesterol, plasma amino acid ratio, plasma vitamin A and carotene, transketolase, serum ascorbic acid, total serum iron and iron-binding capacity, and serum and whole blood folic acid. Urine was analyzed for its content of albumin, creatine, creatinine, glucose, hydroxyproline, iodine, *N*-methylnicotinamide, riboflavin, thiamine, and urea nitrogen. When evaluated against a set of standards, the data collected from these biochemical and additional chemical and dietary investigations provided a general overview of the nutritional status of the population groups included in the Ten State Nutrition Survey. Other biochemical tests, including determinations of vitamin E and vitamin K, have provided additional data in other studies.

Interpretation of the biochemical assessment values of nutrients or metabolites in the blood or urine is difficult. Human variables present complications, as do the changing levels of nutrients and metabolites which occur throughout the day and the effort of the body to maintain homeostasis. In an effort to provide a broader basis for interpretation and possible correlation of the biochemical data, suggested guides to the interpretation of data have provided four categories for classification of blood, urinary excretion, and nutrient intake data: deficient, low, acceptable, and high.

2.3. Clinical Evaluations

Clinical examinations and assessments are important adjuncts to dietary and biochemical information as an indication of nutritional status. However, it is important to remember that clinical assessment values relate to long-term nutritional history and hence, do not necessarily correlate with other evaluative data of more recent nutritional status. The influence of clinical, dietary, and biochemical factors, as these affect nutritional deficiencies, has been postulated by Krehl (1964).

Because the observation and recognition of deficiency symptoms is the most subjective aspect in the determination of nutritional status, trained physicians are provided with well-defined standards of physical evaluation to guide them in the assessment of the health of various parts of the body (USDHEW, 1972; WHO/ECMANS, 1963). Clinical observations of the tissues of the eyes, hair, lower extremities, mucous membranes, skin, teeth, thyroid gland, and tongue are included in the physical examination. Due consideration is given to the fact that some changes are specific for a single nutrient, others are not, and still others may be caused by inadequacy of several nutrients. Hence, although the presence of deformities, lesions, or scars of earlier deficiencies may be visible, these and current aberrations in body tissues are not considered as diagnostic unless there is strong correlation with other evaluative data of nutriture.

2.4. Anthropometric Evaluations

Nutritional anthropometry has been considered one of the important methods of assessing nutriture because it provides body measurements which are indicative of body fat stores and, by inference, of nutriture. In 1956, the Committee on Nutritional Anthropometry published a list of the minimum number of measurements considered essential to indicate thickness of subcutaneous fat and skeletal build (Brozek, 1956). These were designated as the measurements of height, weight, and skinfold thickness. Since that time, efforts to standardize measurements have resulted in the formulation of several height–weight tables by the USDHEW (1953), State University of Iowa (1943a,b), and Wetzel (1940). These tables are age-related for children. The Metropolitan Insurance Company developed similar tables for adults (1960). Jelliffe (1969) has used tables to measure growth in children which are age independent in order to evaluate those children whose actual ages are unknown.

It has been suggested that, in children in whom malnutrition is or has been prevalent, the cumulative anthropometric indices should include height, weight, arm length, midarm muscle, and fat measurements as indicators of mass, linear growth, and protein and calorie reserves. Scales and tape measures facilitate the collection of height and weight data from which relationships to growth can be drawn, but these measurements alone are no more definitive of nutritional status than is a simple visual appraisal. Also, the propriety of applying the standard United States height–weight measurements to evaluate population groups of unlike or mixed genetic and nutritional backgrounds has been questioned (Cravioto *et al.,* 1966).

Measurements of skinfold thickness provide additional information to data collected during a nutrition survey. Skinfold measurement is obtained by use of a standard skinfold caliper and serves to estimate total body fat. Subcutaneous adipose tissue constitues approximately one-half of body stores; hence, through the use of standardized measurement techniques on three anatomical sites, a trained technician can obtain a quick estimate of total body fat by methods developed by Keys and Brozek (1953) and Brozek *et al.* (1963). Because of the reported high correlation between pinch caliper and X-ray measurements of skin and subcutaneous fat, it has been suggested that mass radiography can also be used as a means of assessing obesity (Garn, 1956). Most of these methodologies lend themselves to field surveys more readily than do other sophisticated laboratory techniques which require costly equipment, highly trained personnel, and more time and subject cooperation than is usually available. Advanced laboratory techniques are used to determine body density and lean body mass (Behnke *et al.,* 1942) by water (Rathbun and Pace, 1945; Young *et al.,* 1963), helium (Siri, 1953), and air (Gnaedinger *et al.,* 1963) displacement; body cell mass by simultaneous multiple-isotope dilution

(Moore and Boyden, 1963; Moore *et al.* 1963), and total-body potassium content (Forbes *et al.,* 1961).

3. Standards of Nutrient Intake

3.1. History and Definition

During the past three decades, nine editions of the *Recommended Dietary Allowances* of the Food and Nutrition Board have been published (1943, 1945, 1948, 1953, 1958, 1964, 1968, 1974, and 1979). These standards indicate levels of intake of essential nutrients considered to be adequate to meet the known nutritional needs of practically all healthy persons in the United States. The recommended dietary allowances (RDA) are not absolute, but rather are value judgments of nutritional needs made by nutrition specialists on the basis of available scientific knowledge. They are not to be confused with either basic nutrient requirements essential to meet physiologic needs of most people or the USRDA devised by the Food and Drug Administration for food labeling purposes. Misunderstanding of the term RDA and the objectives of formulating the recommended allowances has created confusion in the interpretation and application of these dietary standards.

The original objective of the RDAs was to serve as a guide for advising ''on nutrition problems in connection with national defense'' (Roberts, 1958). Currently, these standards serve as a guide to assess group food consumption records for nutrient intake, to evaluate the adequacy of national food supplies, to establish standards for public assistance programs, to plan and acquire food supplies for groups of people, to develop nutrition education programs, to develop new food products, to establish guidelines for nutrition labeling, and to regulate nutritional quality. The recommendations are also used to plan diets and evaluate nutrient intake.

When the first RDAs were formulated, it was planned that all of the allowances except that for energy and those for nutrients known to be toxic in excess should be sufficiently in excess of mean nutritional requirements to meet the needs of almost all of the population; thus, the concept of incorporating a ''margin of safety'' into the allowance was introduced. Recommended allowances were given above the known nutrient requirements which would maintain normal function and health of the general population. This amount above the requirement provides leeway for unpredictable higher needs caused by genetic differences in a healthy population. It should be emphasized that, to date, there is no means of predicting higher or lower nutrient requirements for individuals except for energy. Knowledge of human nutritional needs remains incomplete. Of the 45 known essential nutrients, there are RDAs for only 17. While research into nutrient needs and interrelationships continues, the Food and Nutrition Board of the National Research Council, National Academy of Science, suggests that the RDAs be consumed daily in a varied and acceptable diet to ensure the possibility of meeting unrecognized nutritional needs of healthy persons. Therapeutic nutritional needs are not covered by the RDAs, nor are nutrient losses that occur during processing, storage, and preparation of food.

3.2. Establishing the Allowances

There is a paucity of data that define human nutrient needs. Data are required for a population which is known to differ because of the influence of genetic makeup, physiologic state, activity, environmental conditions, age, weight, and sex. The difficulty of making a recommendation for every possible human variation is apparent. Hence, the available information has been tabulated by age–weight–sex groups, based on the knowledge that most nutrient requirements vary with age, body size, and sex and that there are increased needs to sustain rapid growth, pregnancy, and lactation.

After information about human nutrient needs has been collected, the members of the Food and Nutrition Board must agree on the criterion for judging minimum intake necessary to maintain normal function and health in infants, in children, and in adults. For infants, the criterion may be the maintenance of a satisfactory growth rate; for adults, the maintenance of body weight and acceptable levels of nutrients in body tissues or the prevention of specific deficiency symptoms. Considerable judgment is involved in deciding between the higher level required to maintain body stores and the lower level necessary to prevent the occurrence of deficiency symptoms caused by a lack of specific nutrients. Further considerations in setting allowances include those which take into account differences in requirements arising from inherent human variability and factors that influence nutrient utilization, including the chemical forms in which nutrients occur (active or precursor) and interrelationships among nutrients and of nutrients with other food components, drugs, and disease states.

Despite the numberous variables to be considered and the paucity of information about human requirements, the Committee on Dietary Allowances specifies recommended allowances which, in its judgment, represent adequate levels of nutrient intake for practically all healthy individuals in this country. In addition to utilizing the data on hand, in those cases where only limited information is available from human experiments, data on the requirements of other mammals are considered, as is information about the minimum amounts of the nutrients needed if these are known. Furthermore, information from food analyses and dietary surveys may serve as bases for estimating nutrient requirements.

The calculation of allowances is, except for energy, based on known average requirements in a statistically normal distribution. To this mean figure, two standard deviations are added. Thus, these levels of allowances, if consumed in a mixed diet of indigenous foods, are considered adequate to meet the nutrient needs of all but approximately 2.5% of the healthy population in the United States. Furthermore, except for energy, these RDAs supply nutrients in amounts 30–40% above the average reported in biologic measurements of groups of normal healthy subjects. This safety margin above the average requirement, which varies with each nutrient, provides a cushion for individual variations and fluctuations in daily intake. A low intake on one day is usually balanced by a high intake on another in healthy persons, so that he intake usually meets the RDA over a five- to eight-day period.

4. Energy

The allowance for energy is estimated differently from allowances for the other nutrients. Calorie allowances have been established which will provide sufficient energy to support a normal rate of growth or ideal body weight at the lowest levels consonant with good health of average healthy persons in each age group. Energy needs are based on energy expenditure for metabolic processes, growth, lactation, maintenance of body temperature, and to support physical activity.

Calories consumed in excess of needs are stored as fat or reserve energy. In order to preserve health and to prevent excessive weight gain, the continued storage of excess fat is discouraged. By close observation of fluctuations in body weight, individuals can adapt the recommended allowance for energy to their own needs. The allowances for energy represent the average needs of population groups, and not recommended intakes for individuals.

Energy allowances are inclusive of physiologically available or metabolizable energy yield of foods consumed, but exclusive of food energy lost in feces, urine, and as plate waste. The terms "kilocalories" or "kilojoules per gram" are used to describe food energy values and allowances. Energy values in kilocalories for food carbohydrate, protein, and fat are calculated on the basis of the Atwater conversion factors (Merrill and Watt, 1955) (4, 9, and 4 kcal/g of food protein, fat, and carbohydrate, respectively). Expressed in kilojoules per gram, the energy values for protein and carbohydrate are 17 and for fat, 38. Because it has been estimated that Americans consume 76 kcal/day of alcohol (ethanol) on a per capita basis (RDA, 1968), figures for the calculation of food energy contributed by ethyl alcohol must also be considered. Ethanol contributes 7 kcal or 30 kJ/g consumed.

Adjustments of energy intakes must be made for variations in physical activity, body size, and special energy demands of pregnancy and lactation. The effects of climate changes are limited, because clothing selection is adjusted to the weather. As physical activity increases, energy needs also increase. Unfortunately, in the United States, physical activity is limited to light or sedentary tasks, energy expenditures are low, and excessive amounts of fat accumulate. The recommendation is to increase activity levels while reducing energy intake. Increasing activity will expend more energy, reduce the reserves of adipose tissue, and, at the same time, limit obesity and its role in degenerative diseases. There are problems in limiting food intake below the 1800–2000 kcal level, because such reductions may limit the ingestion of the essential nutrients in foods. Diets lower in calories should be restricted in alcohol, fat, and sugar content to ensure the ingestion of required nutrients.

Persons who vary in body size from the average require more or less energy in direct proportion to the degree and direction of the difference observed. Ideal or desirable weight is the criterion used to determine adjusted calorie needs. Adjustments usually fall within 300 kcal of allowances for 97.5% of the population.

With increasing age, basal metabolic rate and physical activity generally decrease. In adults, a 2% decrease per decade in resting metabolic rate has been estimated (Durnin and Passmore, 1967). Individual differences and lack of reliable scientific data make it difficult to estimate the degree of reduction of physical activity with advancing age. A 10% reduction in energy allowances has been proposed for mature adults over 50 years of age, unless activity levels remain high, in which case decreases in energy intake should be lowered proportionately.

Although there appears to be little need to adjust energy allowances to compensate for climate changes, heavy physical activity performed in a mean temperature below 14°C or above 30°C requires increments of approximately 5% in the cold and 0.5% for every degree of temperature above 30°C.

Pregnancy and lactation also impose additional energy requirements to compensate for the building of new tissue and production of milk. The desired weight gain of 11 kg during pregnancy can be provided by an added allowance of 300 kcal/day. Women who have gained 11–12.5 kg during pregnancy will have stored 2–4 kg of adipose tissue that may be used to provide some of the energy needs for lactation. In addition, energy allowances should be increased by 500 kcal/day during the first three months of lactation, to allow for adjustment to completion of the reproductive cycle. Thereafter, increments of decrements in allowances should be adjusted to ideal weight for height.

Energy allowances for infants, children, and adolescents have been set to reflect energy needs determined by age, body size, and growth. Relatively high levels of energy intake per unit of body weight are recommended at infancy. The levels are reduced gradually through ten years of age and more so for adolescent females than for adolescent males. It has been recommended that energy allowances for very active children be adjusted individually upward to maintain growth and metabolic needs. On the other hand, energy allowances of the more sedentary may need to be decreased to levels which will support weight gain for growth without imposing obesity.

Carbohydrate and fat contribute almost equally to the American diet, while protein intake has been recorded as being fairly constant at 11–12% of dietary energy (Frind, 1972). Alcohol may contribute from 5 to 10% of total dietary energy among those who drink alcoholic beverages in moderation, while excesses are known to be toxic and to limit the consumption of other foods (Hartroft, 1967).

In the light of the close association between dietary fat and coronary heart disease and atherosclerosis, the American Heart Association has recommended that the proportion of energy derived from fat not exceed 35%, with less than 10% of total calories in the form of saturated fatty acids and up to 10% polyunsaturated fatty acids. A decrease in fat energy intake and a constant protein intake would increase carbohydrate energy contribution to the diet. This has been demonstrated as harmful to those susceptible individuals who may develop hypertriglyceridemia in the presence of a high carbohydrate intake, sucrose in particular (Anderson *et al.,* 1963).

5. Fats

Fats provide more than satiety and energy to the diet. They serve as carriers of two essential fatty acids, linoleic acid and arachidonic acid, which occur in vegetable and animal sources, respectively. Essential fatty acids are involved in fat transport and in the prevention of dermatitis. Lowered serum cholesterol levels have resulted from increased intake of diets high in polyunsaturated fats, inclusive of essential fatty acids which are known to be precursors of prostaglandins. An intake of approximately 1–2% of the total calories is probably necessary to meet minimum requirements of essential fatty acids (Alfin-Slater and Aftergood, 1968).

6. Protein

Protein is an essential structural element of all living matter. As such, it is involved in every biological process of the organism and is necessary for growth and maintenance. Body proteins are synthesized from food proteins which also provide nitrogen for synthesis of other tissue components. A limited number of amino acids serve as building blocks for the various proteins, and all are manufactured by cells by the same process. A dynamic state of destruction and resynthesis of both protein and other, nonprotein nitrogenous compounds facilitates the regeneration of cells and their constituents. Losses are incurred by the elimination of nitrogen in body secretions, excretions, and sloughed off skin, nails, and hair; therefore, replacement of amino acids and nitrogen is required on a continuous basis for tissue maintenance past the growth years. Proteins, and therefore amino acids, habitually consumed in excess of RDAs by healthy persons exceed the needs to synthesize body tissues. These amino acids are not stored but are deaminized. The nitrogen component is eliminated, while the remainder of the amino acid is utilized as energy or converted to carbohydrate and fat. Under these circumstances, proteins (amino acids) serve as expensive sources of energy from the viewpoint of metabolism and cost (Holt *et al.,* 1962).

Histidine, isoleucine, leucine, lysine, methionine, phenylalanine, threonine, tryptophan, and valine are the known "essential" amino acids, indispensable to tissue growth and maintenance in humans. Rose (1957) defined an essential amino acid as "one which cannot be synthesized by the animal organism out of material ordinarily available, at a speed commensurate with the demands for normal growth." Arginine, originally labeled an essential amino acid, is not required by the human infant (Holt and Snyderman, 1955). The protein requirements of humans extend beyond the essential amino acids to include nonspecific requirements for nitrogen, which can be provided by other amino acids and inorganic nitrogen sources to structure body amino acids and other nitrogenous organic substances.

Human protein needs are known to be influenced by dietary calorie and protein intake, amino acid composition of the protein, and the nutritional and physiologic state of the individual. Protein requirements have been estimated on the basis of data from nitrogen equilibrium, protein depletion–repletion, and growth promotion studies. Nitrogen losses have been calculated, converted to protein losses, and then to the minimum protein requirement in humans. A margin of safety and an additional correction for 75% efficiency of utilization of mixed protein in the American diet are superimposed on the minimum protein requirement.

Because sufficient protein must be available to provide for satisfactory growth and maintenance, the requirements and allowances represent a greater percentage of body weight for infants, children, and adolescents than for adults. Allowances for children and adolescents have been estimated from growth rates (Hathaway, 1957) and body composition data (Foman, 1974) and based on an assumed efficiency of dietary protein utilization similar to that for maintenance in the adult.

Increased protein is recommended during pregnancy from the second month on because of the observed decrease in efficiency of protein utilization during gestation and the improved reproductive performance when protein intake is generous and caloric intake is adequate (NRC/CMN, 1973). Increased protein above maintenance is also recommended, based on the assumption of a 70% efficiency of utilization of mixed dietary protein consumed and, too, a yield of 10 to 15 g of milk protein/day.

7. Vitamins

Vitamins, those "accessory factors" as described by Frederick G. Hopkins, have become as much a household word as they are required dietary components. The relatively small quantities of these organic compounds which are required for metabolism and physiologic well-being do not reflect the essential nature of their involvement in fundamental biochemical life processes. Vitamins have distinct chemical compositions and serve unique body functions. While acting as catalysts or components of enzyme systems, only minimal amounts of vitamins are used up. Replacement is facilitated if an adequate mixed diet is consumed, because food is the source of single or multiple vitamins. No one food, however, is known to supply all of the identified vitamins.

It is customary to classify vitamins by physical properties as either fat-soluble or water-soluble. The fat-soluble vitamins are carried by lipid in foods, and, as such, their absorption is directly affected by the same factors which influence the digestion and absorption of consumed fats. These vitamins tend to be stored in the body in varying quantities, and, hence, there is little danger of precipitating deficiency even if these are not consumed in recommended amounts on a daily basis. Vitamins A, D, E, and K are the fat-soluble vitamins which have been identified, and for all, except vitamin K, recommended allowances have been formulated. For vitamin K, an adequate and safe intake has been suggested (RDA, 1979).

All of the other vitamins have been classified as water-soluble and, as such, are not stored in the body in appreciable amounts but are excreted in the urine. This necessitates daily dietary intake in order to maintain desirable tissue levels and to sustain normal biochemical and physiological body processes. Recommended allowances of the following water-soluble vitamins have been established: ascorbic acid, folacin, niacin, riboflavin, thiamine, vitamin B_6, and vitamin B_{12}. Other water-soluble vitamins known to affect the well-being of humans are pantothenic acid and biotin; however, insufficient evidence exists for establishing recommended allowances of these two vitamins.

7.1. Fat-Soluble Vitamins

7.1.1. Vitamin A

Although the primary metabolic role of vitamin A is not clearly defined, it is known to be active in maintaining the health of epithelial cells and the stability of membranes, visual purple and dim-light vision, growth, and reproduction. In humans, deficiency of vitamin A has been related specifically to biological defects in the pigments of the eye (Wald, 1973).

Data collected from studies conducted to determine human requirements among adult males have indicated differences between levels of intake required to maintain normal blood levels and scotopic vision (Hume and Krebs, 1949), body equilibrium, taste and smell thresholds, and lesion-free skin (Hodges and Kolder, 1971). It was concluded that the 500–600 μg of retinol is a minimum requirement for adults. A margin of safety was added to allow for reserves and for the fact that in the American diet, vitamin A is contributed almost equally by plant and animal food sources (Roberts, 1958).

Evaluation of vitamin A status is difficult, because blood plasma levels may reflect recent intake of carotene or a defect in the protein carrier system or even a zinc deficiency rather than liver reserves. For infants, recommendations are usually based on the content of the nutrient in human milk. Other allowances for children and adolescents have been extrapolated from those for infants and male adults and are based on body weight with an additional amount to meet growth needs. Allowances for women are lower than for men to compensate for smaller body size; however, during the latter two-thirds of pregnancy and during lactation, the recommended amounts are increased to satisfy needs of fetal growth and to provide for vitamin A secreted in milk.

Because both vitamin A and β-carotene can be stored in the body, cases of acute hypervitaminosis A have been reported. Caution must be advised concerning intake. Consumption of large doses of preformed vitamin A can be toxic, while continued and excessive indulgence in provitamin-A-rich foods have been known to cause yellow coloration of the skin, which is reversible when ingestion of the carotenoids is discontinued (RDA, 1974, 1979).

7.1.2. Vitamin D

Several years after vitamin A had been identified as a growth factor, Mellanby proved that vitamin A in cod-liver oil could be destroyed while some other constituent provided protection against rickets. McCollum (1957) and his associates discovered the unidentified nutrient and named it vitamin D. By current definitions, vitamin D can be considered both a vitamin and a hormone (DeLuca and Schnoes, 1976). It is important as a regulator of calcium and phosphorus metabolism in human nutrition, because it influences intestinal absorption of calcium, bone mineralization, and collagen synthesis. Vitamin D is a collective term used to include closely related steroid alcohols that demonstrate vitamin D activity. For its role in nutrition, vitamin D is identified in two forms: the synthetic vitamin D_2 (ergocalciferol) and the naturally occurring vitamin D_3 (cholecalciferol).

Hypovitaminosis D in children results in rickets which, unless treated early, will produce permanent bone malformation. Premature infants and the elderly who lack exposure to the sun, persons consuming low-fat diets, and vegans are most susceptible to deficiences. Adults lacking adequate vitamin D intake develop osteomalacia or adult rickets, with characteristic accumulation of uncalcified bone tissue in the costochondral joints.

No minimum requirement for vitamin D has been established because of the existence of more than one form of vitamin D (natural and synthetic) and, too, because other factors may influence body needs. Among these are degree of exposure to sunlight, skin color (Loomis, 1967), level of dietary intake, and unique personal metabolic needs.

Reports from Britain (DHSS/PCN, 1970) describe the toxic effect of excessive intake of vitamin D on infants and children. Hypervitaminosis D is dangerous and provides no benefit. This fat-soluble vitamin can be stored for extended periods of time, metabolizes slowly, and is excreted gradually by way of the bile. Toxicity symptoms appear in adults as well as children who consume massive doses over prolonged periods of time. Fortunately, excessive doses of vitamin D are not available from natural sources.

7.1.3. Vitamin E

Today, vitamin E is described as the vitamin in search of a disease, but, as early as 1922, it was named the "fertility vitamin," because its absence from the diet of experimental animals produced sterility, premature delivery, fetal death, and failure of the pregnant rat to deliver her young (McCollum, 1957). Similar results have not been observed in humans, but it is agreed that vitamin E is essential for higher animals, including humans. Unrelated and puzzling deficiency symptoms have been observed in different species of animals (Dam, 1962), and some investigators have tried unsuccessfully to extrapolate these to humans. Recently, deficiences have been observed in individuals with prolonged fat malabsorption and vitamin E absorption problems (Binder and Spiro, 1967). Low-birth-weight newborn infants and experimentally depleted adult males have shown increased red blood cell hemolysis in the presence of low serum tocopherol levels (Hassan *et al.*, 1966; Horwitt, 1962). Vitamin E deficiency is probably not an important nutritional problem for those American adults consuming an adequate mixed diet.

Eight substances (four tocopherols and four tocotrienols) comprise the vitamin E group. All have identical physiological properties but α-tocopherol is the most biologically active, and vitamin E is produced commercially in this form. Vitamin E acts as an antioxidant to protect other metabolites, such as preformed vitamin A and unsaturated fatty acids, from oxidation. Similarly, tocopherols act as antioxidants in natural fats to prevent rancidity.

Vitamin E nutriture is determined by plasma tocopherol concentrations and correlates to a high degree in normal persons with the level of plasma total lipids. Variation in dietary intake or clinical conditions can affect these plasma levels directly (Horwitt, 1962). Data indicate that the dietary requirement for vitamin E increases as the intake of polyunsaturated fatty acids (PUFAs) increases. Foods high in PUFAs are usually also rich sources of vitamin E. Because of the lack of evidence concerning the exact mechanism by which vitamin E functions and the fact that vitamin E intake by Americans is inadequate, it is assumed the ratio of PUFAs to vitamin E activity is adequate. The minimum vitamin E requirement is unknown, but it is assumed that consumption of lesser amounts of PUFAs will require less vitamin E.

During pregnancy and lactation, additional vitamin E should be taken to permit adequate deposits in the fetus and in the milk.

Although toxicity of hypervitaminosis E has not been reported in humans there is evidence of this condition in animals (Alfin-Slater *et al.*, 1972). Extended and excessive use of large doses of this fat-soluble vitamin is discouraged, especially among anemic children, since it depresses response to parenteral iron (Melhorn and Gross, 1969).

7.1.4. Vitamin K

The detection of the difference between the diets of hemorrhagic chickens fed ether-extracted fish meal and nonbleeding chickens fed unextracted meal eventually led to the isolation of vitamin K, the antihemorrhagic nutrient, by Dam in 1935. He isolated this fat-soluble vitamin, whose primary function is to catalyze the synthesis of prothrombin in the normal liver (Ferguson, 1946). Vitamin K can alleviate hypoprothrombinemia only when the liver is capable of producing prothrombin, the precursor enzyme essential to the clotting of blood. Vitamin K, which exists in nature as phylloquinones (K_1) and menaquinones (K_2) and also as the synthetic menadione (K_3), functions to regulate the synthesis of other plasma clotting factors, as well as that of prothrombin, but the control site of these activities is still unknown (Owen, 1971).

Defective blood coagulation in animals and low concentrations of blood-clotting factors prior to the introduction of bacteria through food into the intestinal tract of the newborn infant are the only available evidence of deficiencies of this vitamin (AAP/CN, 1971; Sutherland *et al.,* 1967). Vitamin K_2 is synthesized by bacteria in the intestinal tract. This amount plus the vitamin K derived from leafy green and yellow plants in the diet supply sufficient amounts to satisfy adult dietary requirements. It is recognized that antibiotics and anticoagulant drugs act as vitamin K antagonists and increase the requirement of this vitamin.

Kernicterus, a form of vitamin K toxicity in low-birth-weight infants, has been attributed to excessive use of synthetic vitamin K (Crosse *et al.,* 1955).

7.2. Water-Soluble Vitamins

7.2.1. Ascorbic Acid (Vitamin C)

Between 1753, when Lind published the first edition of his *A Treatise on the Scurvy,* and 1932, when Waugh and King isolated the antiscorbutic factor and named it ascorbic acid, several other scientists were exploring the nature of this nutrient and the identification of those species lacking the metabolic means to produce ascorbic acid (see McCollum, 1957). Humans, some primates, and guinea pigs must rely on external sources to avoid scurvy and its clinical symptoms, including weakening of collagenous structures, capillary hemorrhaging, and a decrease in many cellular and serum enzymes (RDA, 1974, 1979).

Ascorbic acid is absorbed in the upper intestinal tract and passes from the portal blood into the tissues to the point of saturation. Although extensive storage occurs in the adrenal cortex, excess ascorbic acid is excreted in the urine, thus reducing the possibility of toxicity. Recent studies have indicated that massive doses of ascorbic acid may lead to formation of kidney stones or diarrhea and may interfere with fertility, fetal development, and activity of other drugs (Rheed and Schrauzer, 1971).

Recommended allowances of ascorbic acid for adults have been established to include a safety margin above the minimum requirements to prevent scurvy, to protect against estimated losses in carbon dioxide expired from the catabolism of the vitamin and by urinary excretion, and to provide an intake required to achieve the most desirable levels of tissue saturation. Continued stress (Baker, 1967), certain drugs, and possibly an ascorbic acid antagonist (Lan and Sealock, 1944; Wooley and Krampitz, 1943) may increase the need for this vitamin.

7.2.2. Thiamine

Almost two decades before Eijkmann, in 1906, made known his observation of the beriberi syndrome in birds fed a polished rice diet, Takaki, the

Director-General of the Japanese Navy, demonstrated that the incidence of "Shipboard beriberi" could be decreased among sailors by the addition of meat and whole grains to the diet (Takaki, 1887). Eijkmann cured his afflicted chicks with rice bran. Later, Gryns postulated the existence of the curative substance in whole grains. In 1926, Jansen and Donath isolated the antineuritic vitamin, and ten years later, R. R. Williams synthesized thiamine (McCollum, 1957).

It has been established that thiamine is active in several metabolic reactions, but its principal role is as a part of the coenzyme thiamine pyrophosphate (Lohmann and Schuster, 1937). As cocarboxylase, it is a coenzyme for energy metabolism, facilitating the oxidative decarboxylation of α-keto acids during carbohydrate and fat metabolism, for transketolase activity in all cells, and for condensation of glycolate and α-ketoglutarate (Liang, 1963).

Several tests are available to test thiamine adequacy or to detect deficiency in humans. These tests are based on the different biochemical functions of thiamine pyrophosphate in energy production, in transketolase activity, or in membrane function and nerve conduction. The measurement of whole blood or transketolase activity is considered the most reliable evaluation measure of adequacy in humans (McCormick and Wright, 1970).

The duodenum is the primary absorption site for thiamine, which, in high concentrations, is absorbed by passive diffusion, and in low concentrations, by an active process which has not been clearly explained (Rindi and Ventura, 1972). Multiple factors may affect thiamine deficiency in chronic alcoholics, including low intake, malabsorption in conjunction with folate deficiency (Thomson *et al.*, 1971), a phosphorylating defect of thiamine, and a deficiency of apotransketolase (Sauberlich, 1967). Two groups of antithiamine factors which occur in food have been identified as altering the structure and reducing the biological activity of thiamine (Shimazono and Katsura, 1965).

Balance studies between dietary intake and urinary output of thiamine, as well as transketolase assessments, have been conducted to investigate minimum needs and optimum intake. It has been determined that thiamine requirements are related to energy intake and that fat "spares" thiamine. Information about the thiamine requirements of infants and children is limited.

7.2.3. Riboflavin

During the years 1917 and 1920, Emmett and his co-worker (McCollum, 1957) reported the existence of separate factors in the "water-soluble vitamin B," one of which, later named riboflavin, was identified as being heat stable and as having retained some of its growth potential after the antineuritic properties of thiamine had been destroyed by heat. In 1935, the synthesis of the yellow enzyme was accomplished, and riboflavin was identified as a five-carbon sugar occurring in many foods and essential to many species because of its role as a coenzyme in a group of flavoproteins involved in tissue respiration and the oxidation–reduction reactions of protein and energy metabolism (Sebrell and Harris, 1972).

Urinary excretion tests and studies of the relationship of dietary intake and the appearance of symptoms characteristic of riboflavin deficiency have served as bases for assessment of riboflavin nutriture (Horwitt *et al.*, 1950). Although some signs of ariboflavinosis have been produced in adults of both sexes (Sebrell and Butler, 1938), no lethal defiency symptoms have been identified.

Riboflavin is absorbed in the upper gastrointestinal tract. Its absorption is regulated by a transport system which determines the upper level of uptake. Excesses are excreted in the urine, since there is no storage of this water-soluble vitamin.

Allowances are calculated on the basis of protein and energy needs and metabolic body size because of the close interrelation of these factors. Data from urinary excretion, dietary intake, and tests of erythrocyte riboflavin levels indicate that tissue reserves cannot be maintained on riboflavin intake levels which fall much below 0.6 mg/1000 kcal for people of all ages. Additional riboflavin is recommended during pregnancy to meet energy requirements and during lactation to equal that secreted in milk daily and to ensure tissue saturation (RDA, 1974, 1979).

Several riboflavin antagonists have been identified and are used primarily in experimental studies. Galactoflavin is one of these. A few foods, such as the Jamaican ackee nut, are known to contain antagonists (Fox and Miller, 1960; Lane and Alfrey, 1970).

7.2.4. Niacin

Casals, a Spanish physician, was the first to describe pellagra in the early 18th century. Goethe, having crossed the Brenner Pass in 1786, commented on the appearance of the population in Italy and associated the displeasing brownish color of their skin to the corn in their diet. But it was Goldberger, in 1917, who proved conclusively that the incidence of pellagra was related to the quality of the diet. The pellagrogenic diet of the poor, which consisted of "3 Ms"—meal (corn), meat (fatback), and molasses—plus poverty was identified as the cause of the "3 Ds"—dermatitis on sun-exposed skin, dementia, and diarrhea. In 1937, Elvehjem and his co-workers described the efficacy of nicotinic acid in curing black tongue in dogs and pellagra in humans (McCollum, 1957).

Niacin is the generic term accepted to include nicotinic acid and its amide, niacinamide. Because several forms of niacin are available to the body, and tryptophan can be converted to niacin at the ratio of 60 mg tryptophan to 1 mg niacin (Krehl *et al.*, 1945; Sarett and Goldsmith, 1947), the term "niacinamide equivalent" is used to describe the contribution to dietary intake of all active forms of niacin (AIN/CN, 1975; IUNS/CN, 1970).

Niacin is vital to oxidation of all living cells, where, as a functional group of the coenzymes nicotinamide adenine dinucleotide (NAD) and nicotinamide adenine dinucleotide phosphate (NADP), it is consumed with the release of energy from carbohydrate, fat, and protein and the synthesis of protein, fat, and pentoses for nucleic acid formation (Chaykin, 1967).

Absorption of niacin and utilization of its metabolites appear to be regulated at the cellular (Gholson, 1966) and systemic (Dietrich *et al.*, 1968) levels. Excesses of nicotinamide are methylated before excretion in the urine, along with other known niacin metabolites (Lee *et al.*, 1969).

Recommended allowances for niacin are based on minimum requirements observed to prevent the deficiency disease pellagra and on data from balance and dietary studies (Goldsmith *et al.*, 1952; Horwitt *et al.*, 1956). Minimum niacin requirements for infants, children, and pregnant and lactating women have not been determined. There appear to be differences in tryptophan conversions in these population groups. For infants up to six months, allowances are usually based on consideration of the tryptophan and niacin content of breast milk.

Several niacin antagonists are known (Kodicek, 1966; Wooley, 1945) to bind niacin, forming compounds resistant to enzymes. Maize is thought to contain an antiniacin substance which increases the requirement for niacin, but the condition may be alleviated by adding tryptophan or niacin to the diet (Gontzea and Sutzescu, 1968).

A high intake of leucine has been reported to cause black tongue in dogs and an increased excretion of niacin metabolites in humans, conditions which were corrected by the addition of niacin to the diet. Addition of trytophan had no corrective effect (Raghuramulu *et al.*, 1965).

7.2.5. *Folacin*

The term folacin is a generic one that comprises folic acid and its derivatives. Investigation leading to the discovery of this hematopoietic vitamin began with Wills, who observed megaloblastic anemia in pregnant women in 1931 and produced it in monkeys fed the same diet as the gravid women (Wills, 1933). The active principle was isolated and its structure determined, and the name folic acid was proposed by Mitchell *et al.* (1941) and pteroylglutamic acid by Waller *et al.* (1948).

Folic acid is the bacterial growth factor in spinach leaves and liver and also is the nutritionally essential precursor of folates that serve as coenzymes for one-carbon transfer reactions in almost all living systems (FAO/WHO, 1970).

It has been determined that pteroylglutamic acid is the form essential to most mammals. Folates, however, are synthesized in plants and by bacteria in such a wide variety of forms (Baugh and Knumdieck, 1971) as to make it difficult to determine species requirements.

Complete assessment of folacin nutriture may include several techniques, since no one method can measure all folic acid active compounds. Among the assay methods available are column chromatographic, enzymatic, microbiological, and radioisotopic techniques (Baker and Sobotka, 1962; McCormick and Wright, 1971).

Tetrahydrofolic acid is the active form of the vitamin. In conjunction with its derivatives, and sometimes with vitamin B_{12}, it transfers single-carbon units for the synthesis of purines and pyrimidines, the initiation of a variety of methylation reactions, and the interconversion of certain amino acids. Absorption of dietary folacin by active transport and diffusion occurs in the upper and lower intestinal tract, after conjugase action. Some folacin is stored in the liver. A large number of folacin antagonists have been identified (Robinson, 1966). Some drugs used as anticonvulsants, as antitumor agents, as immunosuppressants, and in the treatment of psoriasis (Ellegaard *et al.*, 1972) act as folacin antagonists.

On the basis of assessment of the amounts necessary to relieve symptoms in subjects with folacin deficiency, and results of time depletion studies, average daily requirements have been established. These were based on hematological and deficiency recovery responses to doses of crystalline folic acid (Herbert, 1962).

Little information is available concerning folacin requirements of infants and children. Estimates of needs based on folic acid absorbed from human and cow's milk have served as a basis for the recommended allowances.

7.2.6. Vitamin B_6

Investigations of the multiple factors in the vitamin B complex led to the recognition of vitamin B_6 by Gÿorgy in 1934. Its isolation and synthesis followed shortly thereafter, and the product was named pyridoxine (McCollum, 1957). Snell revealed the existence of pyridoxamine and pyridoxal as parts of the vitamin B_6 complex found in animal tissue, whereas pyridoxine is found in plants (Snell, 1945). The dietary needs of humans for vitamin B_6 can be met by any of the three forms of this vitamin.

Vitamin B_6 is water soluble and can be measured by animal bioassays, chemical procedures, or microbiological assays. Comprehensive reviews of the chemistry and roles of vitamin B_6 indicate that this coenzyme may activate several vitamin-B-dependent enzymes and be involved in atherosclerosis, dental caries, antibody and renal calculi formation, the metabolism of the central nervous system, and hormone activity. Vitamin B_6 also has been implicated in lipid metabolism, but the vitamin primarily is associated with nitrogen and amino acid metabolism (Sauberlich and Canham, 1973).

Human requirements of vitamin B_6 have been based on amounts required to treat deficiency symptoms, most of which are associated with abnormal amino acid and protein metabolism; observation of the level of urinary excretion of vitamin B_6 or its metabolite; measurement of the excretion of end products of the tryptophan load test; and measurement of transaminase levels in the blood. Nowhere has vitamin B_6 deficiency been observed in humans (RDA, 1974, 1979).

It has been reported that vitamin B_6 requirements are elevated by the intake of steroid contraceptive pills (Price *et al.*, 1967) and other drugs and chemicals which act as vitamin B_6 antagonists (Brown, 1972). For some individuals, due to inborn errors of metabolism, supplements of the vitamin are essential. Age may also affect an increased requirement of the vitamin (Gershoff, 1976).

Pyridoxine is often included in vitamin-B-complex medication and has been used in higher doses when antagonistic drugs, such as isoniazid, are used in the treatment of disease (Biehl and Vilter, 1954). Toxicity is very low because vitamin B_6 is readily metabolized in man (Sebrell and Harris, 1972).

7.2.7. *Vitamin B_{12}*

Cyanocobalamin (vitamin B_{12}), referred to as the "animal protein factor" found in cow manure and effective in stimulating "growth in chicks," was discovered initially by Minot and Murphy (1926) as the substance in liver used in the treatment of pernicious anemia. Castle (Castle and Minot, 1936) named it the "extrinsic factor" since it came from sources outside the body. The active principle in liver was isolated later by two teams of scientists working independently in laboratories in England (Smith and Parker, 1948) and in the United States (Rickles *et al.,* 1948). Hodgkin *et al.* (1957) explained the chemical structure of the red vitamin, which was synthesized a quarter of a century after its isolation (Maugh, 1973).

In nutritional and pharmacological literature, the name "vitamin B_{12}" is inclusive of all cobalt-containing corrinoids, but chemists refer to it as cyanocobalamin. Only two active forms of vitamin B_{12} have been identified as functional in human metabolism. Although the vitamin is essential to all cells, it is particularly important to bone marrow for the development of mature red cells (Beck, 1968), to the nervous system to prevent demyelination and progressive neuropathy (Frenkel, 1973), and to the gastroenteric tract to forestall structural changes in the epithelial tissue and facilitate B_{12} absorption (Herbert, 1972).

Absorption of vitamin B_{12} occurs by simple diffusion or by means of the intrinsic factor system involving the glycoprotein, calcium ion, and mucosal cells of the ileum for transportation into the blood, where it is delivered to hematopoietic tissues and to the liver by two binding proteins (Herbert, 1975).

Estimations of requirements of vitamin B_{12} in humans have been made by studying minimum doses affecting and maintaining normal hematologic response; comparing serum and liver levels of B_{12} in normal and deficient individuals; and measuring turnover rates and body stores by using radioactive B_{12} tracer doses. Requirements for adults range from 0.1 to 1 μg (FAO/WHO, 1970; Herbert, 1968).

Serum and human milk levels are identical at 0.3 μg (FAO/WHO, 1970), and no deficiency has been observed in breast-fed infants. During the latter half of pregnancy, when fetal needs are greatest, and during lactation, when there is a loss of B_{12} in the breast milk, needs increase; hence, an increment is recommended. Inborn errors of metabolism responsive to therapeutic doses of B_{12} are known to occur (Nyhan, 1975), but the indiscriminate use of the nontoxic red vitamin is discouraged.

Microorganisms synthesize vitamin B_{12}; hence, foods of animal origin serve as excellent-to-good sources. Feces are rich in vitamin B_{12} and may provide a supply to those who consume unwashed foods contaminated by fecal material, Certain legumes whose root nodules have been contaminated by microorganisms will contribute vitamin B_{12} to the diet when the root nodules are ingested. Otherwise, plants are devoid of the vitamin, creating a problem for true vegans who, over long periods of time, may develop deficiency symptoms (Herbert, 1973).

7.2.8. Pantothenic Acid

Pantothenic acid, a water-soluble vitamin isolated by Williams and his co-workers (1938) and synthesized by Stiller and his associates (1940), is a component of coenzyme A (CoA) and, as such, is involved in energy-related functions in humans, most other animals, and some microorganisms. CoA and acyl carrier protein (ACP) are the two known active forms. De Vries *et al.* (1950) identified pantothenic acid as a component of CoA. Later, Pugh and Wakil (1965) described ACP.

Pantothenic acid as a component of CoA is associated with the release of energy through the metabolism of carbohydrates, fats, and proteins, gluconeogenesis and the synthesis of fatty acids, as well as other basic metabolic reactions (Lipmann *et al.*, 1947).

In the past, pantothenic acid has been assayed colorimetrically (Crokaert, 1949), but today the best method for determining the vitamin value in natural materials is by microbiological techniques (AOAC, 1960) and in pharmaceutical products by gas chromatography (Tarli *et al.*, 1971).

As its name implies, the occurrence of pantothenic acid in nature is extensive, thereby limiting the possibility of developing a dietary deficiency. Different types of deficiency symptoms have been observed in various species of animals, but no evidence of dietary deficiency has been reported in humans. Bean (Bean *et al.*, 1955) and Hodges (Hodges *et al.*, 1959) and their co-workers induced clinical deficiency within one month by treating human subjects with a diet deficient in pantothenic acid and administering its antagonists.

There is a dearth of evidence on which to base requirements of pantothenic acid. The average healthy adult, consuming a diet of mixed foods, has not shown deficiency symptoms.

The omnipresence of pantothenic acid in food facilitates its inclusion in the diet. Heating of food at temperatures above boiling may cause loss. Toxicity due to excessive intake of pantothenic acid has not been reported.

7.2.9. Biotin

In the early 1900s, numberous investigations were undertaken to study the effects of feeding raw egg white. Whereas Osborn and Mendel reported that 18% egg white as the sole dietary protein supported growth in young rats,

Bateman, in 1916, and Boas, in 1927, found it injurious to the health of rats (McCollum, 1957). Parsons and her co-workers (1937) reported the occurrence of similar symptoms when rats were fed a diet including raw egg white. Two years later, Gÿorgy (1939) designated vitamin H, present in liver and yeast, as protective against egg white injury in rats. It was identical to the potent crystalline substance which Kogl and Tonnis (1936) had isolated in boiled duck egg yolk and was found to be a growth factor for yeast strain M. Later, Harris and his associates (1945) were able to synthesize biotin, the "anti-egg white injury factor."

By definition, biotin is a vitamin because it performs very specific and essential functions in the body. However, it is synthesized by bacteria in the gastro-intestinal tract, and some appears to be absorbed. It has been reported that levels of biotin excretion may exceed the intake by three to six times (Oppel, 1942). At least five active forms of biotin have been identified and their activity reviewed (Mistry and Dakshinamurti, 1964). Biotin is an extremely active enzyme in two general types of carboxylation reactions, one energy dependent and the other involving only the exchange of carboxyl groups.

Biotin plays a role in the release of energy from carbohydrates and in the deamination of amino acids, but there is a poverty of evidence of the role of biotin in carbohydrate and protein metabolism (Bridgens, 1967). As a functional component and the prosthetic group of acetyl-CoA carboxylase, biotin is active in fatty acid biosynthesis (Wakil and Gibson, 1960).

Measurement of yeast and bacterial growth is used to assay biotin values in biological materials. Few microorganisms are specifically biotin dependent. Baker and his co-workers (1962) found *Ochromonas danica* specific and useful for biotin assays in blood, serum urine, and animal tissues.

Deficiency symptoms have appeared in humans under experimental and unusual dietary conditions in which raw egg white was fed in large quantities (Baugh *et al.*, 1968; Scott, 1958; Sydenstricker *et al.*, 1942; Williams, 1943). The glycoprotein in the raw egg white, avidin, which is destroyed during cooking, acts as a biotin antagonist. Dietary intake, reported to be 100—300 μg (Oppel, 1942), plus gastrointestinal synthesis appear to provide adequate quantities of biotin to support the physiologic functions in which it is involved. Human blood levels range from 12 to 55 μg/100 ml through the life cycle, with slightly lower levels during pregnancy (Bhagavan and Coursin, 1967).

There is no evidence of natural biotin deficiency, and requirements have not been defined.

7.2.10. Choline

In 1844 and 1846, Gobley isolated lecithin from egg yolk, which on hydrolysis yielded a nitrogenous base, identified by Strecker in 1868 as choline. Thudichum, in 1884, and Trier, in 1913, identified kephalin as a part of the nitrogenous base. These initial investigations into phospholipids did not arouse special attention among nutrition investigators until later, when it was to be determined whether these compounds were dietary essentials or synthesized in the body (McCollum, 1957).

Some investigators question the classification of choline as a vitamin, although it has been shown to be essential for several animal species (Best and Lucas, 1962) and is known to be important in metabolism as a source of labile methyl groups (Griffith et al., 1971). Biotin occurs in free form in biological tissues and is a component of sphingomyelin and acetylcholine. It is known to be synthesized within the body. No deficiency due to dietary restriction has been reported in humans. In rats, evidence of choline deficiency and fatty liver (Best and Huntsman, 1932) and kidney degeneration (Griffith and Wade, 1939) has been reported. Meneghello and Neimeyer (1950) have shown choline to be effective in the treatment of fatty liver in human protein-calorie malnutrition, but these results have not been confirmed.

Folic acid, methionine, and vitamin B_{12} in the diet are known to influence choline requirements in animals, as do growth, energy intake and output, and the quality and quantity of fat consumed (Best and Lucas, 1962; Griffith et al., 1971).

Because of the widespread occurrence of both choline and methionine (its methyl donor) in foodstuffs of animal and plant origin, dietary deficiency is unlikely to occur when an otherwise adequate diet is consumed (Griffith et al., 1971). Such diets provide 400–900 mg of choline/day. Body synthesis supplements dietary intake to meet body needs.

8. Mineral Elements

Mineral elements are normally occurring inorganic chemical elements. Approximately 100 have been identified, and of these, half have been found in the human body. Minerals are indestructible; hence, they remain as ash. This characteristic facilitates their analysis in biological tissues, where these are present in large amounts (AOAC, 1960) as wet ash or dry ash. Required in the diet at levels of 100 mg/day or more are the macronutrients—calcium, chlorine, magnesium, phosphorus, potassium, sodium, and sulfur. The trace mineral elements, or micronutrients, are needed in lesser amounts, no more than a few milligrams per day. Chromium, cobalt, copper, fluorine, iodine, iron, magnesium, molybdenum, selenium, and zinc are among the trace elements which have been identified in the body. As new laboratory techniques have been developed, more minute traces of mineral elements, formerly considered contaminants, have been found to serve some function in the human body or are actually needed by animals (Laitinen, 1972); for example, cadmium, nickel, silicon, tin, and vanadium may be essential to humans and are needed by animals. For the following mineral elements or contaminants that form a part of the basic composition of the body, there is no known function at present: aluminum, arsenic, barium, beryllium, boron, bromine, gold, lead, lithium, mercury, rubidium, and strontium (Anderson, 1977; Sherman, 1952).

The distribution of mineral elements in the body is ubiquitous, but uneven. Some tissues contain traces, but quantity is no clue to the importance of the mineral element. These are important to the structure of every tissue and cell in the body. Calcium, phosphorus, and magnesium constitute the greater proportion of hard skeletal and tooth tissue, while soft tissues contain more potassium. Mineral elements may be present as free ions or as inorganic or organic compounds. They serve as structural components of enzymes, vitamins, hormones, hemoglobin, and other protein-containing tissues. Permeability of cell membranes, osmotic pressure, and acid–base equilibrium are regulated by the presence of mineral elements in body fluids. Nerve response to stimuli and muscle contractility require the presence of mineral elements.

Characteristically, these essential elements are absorbed from the gastrointestinal tract and excreted in the urine or feces and sweat. The processes are regulated by body homeostatic mechanisms. Thus, under normal conditions, overloading and/or toxicity caused by excessive intake may be avoided.

8.1. Calcium

A normal adult human body contains about 1.2 kg of calcium, all but 1% of which is a component of bones and teeth. Calcium is constantly being mobilized in and out of bone tissue. The small amount present in body fluids is ionized calcium in blood serum or is bound to serum proteins. It is essential in the activation of several enzymes, the control of nerve irritability and muscle contractility, blood coagulation and myocardial function, and the maintenance of the integrity of all membranes and intracellular cement substances. Homeostasis is maintained by hormonal intervention and the presence of vitamin D (Tanaka *et al.*, 1973).

When needs are increased, as during pregnancy, lactation, growth, or when intake is decreased, absorption by active transport is increased (Willis, 1973). Some dietary factors, such as lactase and certain amino acids, may enhance absorption of dietary calcium, whereas others, such as oxalates and phytates, decrease the absorption of calcium. If calcium intake is adequate, the formation of insoluble oxalate and phytate salts is not considered of practical importance (Hegsted, 1973). High protein intake has been demonstrated to increase calcium excretion and may initiate osteoporosis (Johnson *et al.*, 1970; Margen *et al.*, 1974) in a fairly constant relationship to intake (Hegsted, 1973). Adaptation to lower calcium intakes occurs in adults (Davidson and Passmore, 1970). There is no evidence that, in normal adults, high calcium intake per se is cause for high levels of this mineral element in the serum or urine (Hegsted, 1973). Low intakes of calcium have been implicated in the loss of bone or osteoporosis (Garn, 1970; Nordin, 1962). The understanding of calcium metabolism and tissue mineralization is still incomplete.

8.2. Phosphorus

Phosphorus is a universal cell component, constituting 1% of body weight. It is so readily available in nature as to preclude deficiency in humans under normal conditions (Hegsted, 1973). Human milk provides adequate phosphorus for the full-term infant, but additional phosphorus has been shown to be required by premature infants in order to meet their more rapid growth needs (Van Sydow, 1946). Inversely, hypocalcemic tetany has resulted from consumption of the highly concentrated phosphorus in cow's milk during early infancy (Harrison, 1972; Mizrahi *et al.,* 1968), because the parathyroid gland is relatively inactive and is incapable of maintaining calcium homeostasis.

Calcium and phosphorus form an insoluble salt to prove rigidity to bone and tooth structure, provided vitamin D is present and there is an appropriate calcium:phosphorus ratio. Organic phosphates, as components of ATP and ADP, control the energy release for cellular activity resulting from the metabolism of carbohydrate, fat, and protein. The absorption of glucose is phosphorus dependent, as are the control of all cell permeability in which phospholipids are involved and all cell protein synthesis in which RNA and DNA control cell reproduction. Phosphorus-containing compounds also serve as buffers to control the acid–base balance of the body and assist in the utilization of the vitamin B complex.

The amount of phosphorus absorbed from the gastrointestinal tract is affected by the availability of the mineral element. Some will be lost when phosphorus forms insoluble salts with iron, magnesium, and other elements. Then, phosphorus is excreted in the feces as inorganic and organic phosphates, in conjunction with desquamated intestinal wall cells. Clinical signs of phosphorus deficiency have been observed in situations in which inadequacy of available phosphorus has resulted from excessive and prolonged consumption of phosphate-precipitating agents (Lotz *et al.,* 1968), intravenous hyperalimentation with low-phosphate amino acid and glucose solutions, or starvation. Repletion with dietary phosphorus, abundant in foods, and cessation of deprivation conditions will correct the deficiency symptoms.

Recommendations for phosphorus are usually the same as for calcium, except during infancy, where the recommended ratio of calcium to phosphorus is higher.

8.3. Magnesium

Until 1915, when Denis first determined the magnesium content of blood plasma, few investigations of this mineral element had been attempted. In 1931, McCollum and Orent later described magnesium deficiency in rats (McCollum, 1957). More recently, magnesium depletion in humans has been reported by Shils (1964) and by Dunn and Walser (1966).

Approximately 20–28 g of magnesium are found in the body, 55% of which is deposited on the surface of the skeletal tissue as phosphate and carbonates, half as much in muscles and organ tissues, and a small amount in extracellular fluids (Shils, 1973). The element is involved in phosphate-transfer systems and changes in energy states. Magnesium is also associated with amino acid synthesis, lipid metabolism, thiamine utilization, neuromuscular transmission, and other enzymatic reactions (Aikawa, 1963: Wacker and Parisi, 1968). It bears a resemblance to calcium, phosphorus, and sodium in its involvement in body chemistry, and there is evidence that a deficiency of magnesium will affect the metabolism of these ions (Shils, 1969).

Magnesium absorption is affected by several physiological factors in the intestine, including transit time, level of magnesium concentration, water absorption, and the amounts of calcium, lactose, magnesium, and phosphate in the diet (Walser, 1967). Urinary excretion is limited during severe restriction of dietary intake and increases when there is magnesium supplementation. Renal threshold determines the serum level. It is postulated that reabsorption occurs as for calcium and sodium and that their clearance is interdependent (Walser, 1967).

Deficiency symptoms have been observed in humans in the presence of childhood malnutrition (Caddell, 1969), magnesium dietary depletion as in parenteral feeding, malabsorption, chronic alcoholism, renal tubular dysfunction, genetic interference with renal and/or intestinal conservation, large lactation losses, hyperparathyroidism (Shils, 1969, 1973), and the intake of certain diuretics (Martin *et al.,* 1952). The exact role of magnesium deficiency in the disease state has not been clearly elucidated because of the variety of factors which affect the homeostatic mechanism (Seelig, 1971).

Recommendations for magnesium are based on studies of magnesium content of the American diet, balance studies (Seelig, 1971; Shils, 1973), and the rarity of deficiency occurring in normal adults. Allowances for infants are usually based on the magnesium content and consumption of human milk. Data about needs of children and adolescents are not available; hence, estimates are usually made to allow for rapid bone growth and other increased needs based on age and sex.

8.4. Sodium

Sodium as a component of sea salt has been involved in the physiological processes of cells since the beginning of time. Salt, formerly considered a treasure, is now accepted as a commonplace dietary item. Approximately 2700–3000 mEq of sodium are found in the adult male body. About 50% of this is present in extracellular fluid and the remainder in bone, where half is held as a part of the body labile sodium pool.

Sodium, as the chief cation in extracellular fluid, acts with potassium, the principal cation in intracellular fluid, to regulate and maintain osmotic equilibrium and body fluid balance. In combination with chloride and bicarbonate, sodium plays an active role in maintaining acid–base balance in extracellular fluid. In addition, sodium is a major factor affecting the permeability of cell walls and, subsequently, muscle contraction and nerve transmission.

There is evidence which indicates that sodium deficiency retards growth in animals (Forbes, 1962); however, deficiency is unlikely to occur in humans, because sodium occurs naturally in foods of animal and plant sources. More sodium is usually added during processing of food. Sodium intakes in this country vary. Dahl (1958) has estimated an average daily consumption of 6–18 g of table salt, which is equivalent to 2–7 g of sodium.

Dahl (1958) has theorized that salt intake is an induced rather than innate response and bears no relationship to requirement. It appears that the adult daily requirement for sodium may be no more than 0.5 g, but Americans consume much more, sometimes with resulting hypertension (Dahl, 1972), depending upon individual susceptibility. Acute sodium chloride toxicity may occur with daily intakes of 35 g or more, but some persons show clinical evidence of chronic toxicity at lower levels.

Under normal conditions of health and dietary intake, there is renal regulation and homeostatic control of sodium concentration in body fluids and tissues by aldosterone. High salt intake will provoke thirst, and any excess is excreted in urine and sweat. Excessive losses of sodium or lack of intake may cause temporary depletion associated with water loss (Rogers *et al.*, 1963). Evidence suggests an increased need for sodium during pregnancy due to the increase in body mass (Pike and Smiciklas, 1972; Pitkin *et al.*, 1972). The requirement of pregnant women has been estimated to be 25 g of sodium/ day.

8.5. Potassium

Potassium, the key cation in intracellular fluid, is also present in a small amount in extracellular fluid under normal conditions. An adult male has in his body 3200 MEq of potassium. It works in consortium with sodium to maintain body fluid equilibrium and acid–base balance. Potassium is found in all cells and reflects cell mass. This cation influences muscle activity, particularly of cardiac muscle, and is involved in the transmission of nerve impulses. It acts as a catalyst in carbohydrate and protein metabolism and in the synthesis of glycogen and protein. There is evidence which indicates that potassium exerts a protective effect in the event of chronic sodium chloride toxicity (Addison, 1928).

It has been estimated that healthy adults need 2.5 g of potassium daily (RDA, 1974, 1979). The usual daily intake in the United States has been reported to be 50–150 MEq day (Wilde, 1962), which far exceeds estimated daily losses of potassium. Because potassium is widely distributed in foods, dietary deficiency is unlikely to occur in the average healthy person.

Severe losses of potassium due to starvation, decreased absorption, increased excretion, and vomiting may precipitate clinical signs of hypokalemia. The same homeostatic mechanism which controls sodium balance affects potassium equilibrium; however, the ability to conserve potassium during deficiency is less efficient than for sodium, and, hence, hypokalemia may develop.

With a daily intake of 25 g of potassium chloride, acute toxicity has been observed (Blum, 1970). Treatment with diuretics, high sodium intake, and certain disease states may result in hyperkalemia. This condition is unlikely to occur with the consumption of an adequate diet of mixed foods.

8.6. Chlorine

Chlorine exists in the body as the anion chloride, which is associated with the cation sodium, primarily in extracellular fluid, but to a limited extent within the cell with the cation potassium. It comprises 0.15% of body weight, but is active is both digestion and metabolism. It passes freely through cell membranes to help maintain acid–base balance by means of the chloride–bicarbonate shift in red cells and, with sodium, regulates osmotic pressure in the body. It is a constituent of gastric juice and, as such, functions as an activator of amylases (Cottove and Hagben, 1962).

Chloride is readily and almost completely absorbed in the gastrointestinal tract. Its excretion via the urine parallels intake.

The daily turnover of 85–250 mEq in adults is more than adequately met by recorded dietary intakes of 3–9 g (RDA, 1974, 1979). Factors affecting sodium loss will similarly affect chloride output, and a deficiency may occur with chloride loss due to vomiting. For all practical purposes, chloride intake is generally more than adequate.

8.7. Iron

In 1684, Robert Boyle, the first chemist to analyze blood, observed the brick-red color of blood but did not suspect the presence of iron compounds. More than six decades later, Merghini proved that blood contained iron. Since that time, numerous investigators have studied the properties of blood and the presence, utilization, and functions of body iron (McCollum, 1957). The entire body of a well-nourished adult contains approximately 4 g of iron, which consists of two fractions: the major portion present in hemoglobin, myoglobin, and certain oxidative enzymes, and the lesser amount constituting a mobilizable iron reserve. There is practically no ionic iron present. About one-fourth of body iron is stored as iron–protein complexes, ferritin and hemosiderin, in the cells of bone marrow, liver, and spleen. Lesser quantities are stored in other cells (Macara *et al.*, 1972). Negative iron balance precedes depletion of stores, which results in reduced hemoglobin concentration.

Iron is absorbed at about 5–10% levels, primarily from the upper intestinal tract, as needed. Absorption is aided by ascorbic acid and gastric acidity. Iron is absorbed more readily in the reduced ferrous form. Most iron-dependent metabolic systems are dependent upon the interconversion of ferrous and ferric iron (Frieden, 1973). The main function of iron is to assist in the oxidative processes of the body, with its compounds serving as the carriers of oxygen to cells. There is no efficient mechanism for regulating the excretion of iron; however, iron metabolism is controlled at the absorption site. Cook *et al.* (1972) have reviewed the variety of factors that affect availability of iron for absorption. When body iron is adequate, the intestinal mucosa regulates iron absorption to maintain body iron at a constant content. In a deficiency state, iron absorption increases above normal, but even this may not be adequate to prevent anemia in iron-deficient persons. Deficiency most frequently occurs in infants, preschool children, and women in their childbearing years (US-DHEW, 1972). In addition to the function of iron in oxygen transport and cellular respiration, data indicate that iron-containing enzymes catalyze certain reactions in the synthesis of collagen (Prockop, 1971).

RDA for iron have been based on balance studies and results of nutritional and epidemiological surveys and include a sufficient margin of safety to cover average physiological requirements among almost all healthy persons in the general population.

8.8. Copper

Copper was demonstrated to be indispensable for mammalian utilization of iron by Hart and his associates in 1928 (see McCollum, 1957) and reaffirmed later by Elvehjem (1935), when they found copper to be essential to the formation of hemoglobin. Copper is also a component of proteins and enzymes associated with cellular oxidation and other metabolic reactions (Frieden *et al.*, 1965). In addition, copper plays a role in connective tissue metabolism (Carnes, 1971). The exact mechanism of copper in enzyme systems has not been clearly defined (O'Dell, 1976).

The adult body contains approximately 75–150 mg of copper (Cartwright and Wintrobe, 1964a,b). However, some mild to severe symptoms of copper deficiencies have been reported in Peruvian and American infants. Hypocupremia has been observed in protein–calorie malnutrition, accompanied by anemia, neutropenia, and bone disease (Cardano *et al.*, 1968), and similar symptoms have developed in premature infants fed only modified cow's milk or receiving prolonged parenteral alimentation (Al-Rashid and Spangler, 1971). Hypocupremia may be unrelated to dietary intake, but rather may be of genetic origin, in which event there is defective formation of ceruloplasmin. Abnormalities of copper deficiency in animals and humans have been described by Underwood (1971).

Requirements for copper have been determined by balance studies. Infants at birth, women taking contraceptive drugs, and persons with certain infectious diseases demonstrate elevated serum copper concentrations (Sass-Kortsak, 1965). Dietary deficiencies and body balance are maintained by the 2 mg daily provided by a mixed diet (Schroeder *et al.*, 1966b). Human milk meets the needs of breast-fed infants.

8.9. Iodine

Typical of an essential trace mineral element whose function is far more important than its quantitative presence would indicate, iodine represents 0.0004% of the body weight, but is a structural component of thyroxine and other compounds produced by the thyroid gland. In 1896, Baumann discovered that the thyroid gland was rich in iodine, and it became recognized as an essential micronutrient for animals and humans (McCollum, 1957).

After absorption from the gastrointestinal tract, 30% is removed by the thyroid gland to produce thyroxine and triiodothyronine (T_3), which are stored in the thyroid as thyroglobulin. The function of thyroxine, and, hence, iodine, is to regulate energy metabolism. During a period of iodine deficiency or prolonged low level of iodine availability, the thyroid gland increases its secretory activity in an attempt to compensate. Endemic goiter results and is likely to remain, despite subsequent increased iodine intake (RDA, 1974, 1979). Additional factors which may contribute to simple nontoxic goiter if taken in large amounts include goitrogenic compounds found in foods, such as broccoli, brussels sprouts, cabbage, cauliflower, kale, kohlrabi, rutabaga, turnips, and rapeseed meal. Dietary iodine does not counteract the antithyroid effect of these compounds. Thiocyanate and dialkylsulfides in foods are mildly goitrogenic but can be counteracted by adequate dietary intake of iodine (Van Etten and Wolff, 1973). The increased use of processed and convenience foods, which are manufactured with noniodized salt, further decreases iodine intake, as does the limited use of iodized table salt in the United States, despite the fact that iodization programs have proven themselves to be successful goiter-preventive measures. Iodized salt provides 76 μg of iodine/g of salt (RDA, 1974, 1979).

The iodine requirement is based on the amount necessary for the prevention of goiter and on data about circulating thyroxine levels in the blood of adults. Breast-fed infants will receive an adequate amount of iodine if the diet of the lactating mother is adequate. No demonstrable adverse effects have been reported for adult intakes of between 50 and 1000 μg daily.

8.10. Fluorine

Fluorides are omnipresent, abundant in the earth's crust and occurring in low to high concentrations in all foodstuffs and water supplies. The nutritional significance of fluoride is still being investigated. It is an important dietary constituent of humans, but it has not been proven to be an essential trace element in the diet of certain other species. Messer and Singer (1976) question the attributed effects of fluorides on rodent growth rates, reproduction, and hematopoiesis and on *in vitro* biological calcification, as well as the essential status of fluorides.

In humans, the stomach is the major site of fluoride absorption (Carlson *et al.*, 1960). Whereas absorption of fluoride in food ranges from 50 to 80%, almost all of the soluble flourides in food are absorbed (Hodge and Smith, 1965). Plasma fluoride levels are effectively regulated by the homeostatic action of the kidneys and skeleton. Absorbed fluoride is deposited in mineralized tissues, and a fluoride pool is maintainted to protect against dietary fluoride inadequacy. With age, skeletal content of fluoride increases, but soft-tissue content does not fluctuate. Excesses of fluoride are excreted in the urine, with minor amounts being lost through sweat and feces (Armstrong and Singer, 1970). Total body content varies with individual consumption patterns.

Fluorine has demonstrated its beneficial effects in the prevention of dental caries through the life cycle, so long as fluoride is consumed or is applied topically (Scherp, 1971), so that it may reduce the solubility and increase the size and strength of the apatite crystals in tooth enamel. A high fluoride intake during the tooth formation period provides even more protection agains caries (Jenkins, 1971). When high intake is continued throughout life, it may reduce the development of senile osteoporosis (Leone, 1970). Bernstein *et al.* (1966) have reported decreased arterial calcification in high-versus low-fluoride areas. Less severe periodontal disease has been observed in those communities where fluoridated water is consumed (Messer *et al.*, 1973).

Requirements for fluoride are unknown, although a deficiency during infancy and childhood will negate the critical protection against dental caries in primary teeth (Horowitz, 1973). For maximum protection, daily supplementation to intake levels of 0.25 to 0.5 mg fluoride during the first year and 0.5 mg until three years is recommended (Brudevold and McCann, 1966). Adults may require more fluoride than children since, 4 ppm fluorine reduces the prevalence of osteoporosis (Leone, 1970) in adults but causes mottling of enamel during tooth formation in children without other side effects.

Acute and chronic toxicity effects of excessive fluoride intake have been described (Roholm, 1937). Although acute toxicity in humans is rare, chronic toxicity is more common, occurring in areas with high-fluoride-bearing waters. Artificially fluoridated water containing 0.7–1.5 ppm fluorine is not likely to cause toxicity, allergic reactions, or other ailments which have been attributed to it (*J. Allergy,* 1971). Water containing 2 ppm fluorine or more will cause various degrees of mottled enamel if consumed during the tooth formation period.

8.11. Zinc

Trace amounts of zinc are present in all living matter. Approximately 263 mg are distributed throughout the body. Zinc is an essential component of several enzyme systems and influences the structural configuration of certain nonenzyme organic ligands (Parisi and Vallee, 1969). Zinc is involved in most major metabolic reactions and in nucleic acid, carbohydrate, and protein metabolism (Underwood, 1971). It is essential for growth and normal sexual development in children.

A deficiency of zinc can be detected in humans by testing for a decreased serum alkaline phosphatase level (Arakawa *et al.*, 1976). All body systems can be negatively affected by a zinc deficiency. Increased sensitivity to insulin (Sandstead *et al.*, 1967), dwarfism and retarded gonadal maturation (Prasad *et al.*, 1963), impaired healing of leg ulcers (Haeger *et al.*, 1972) and burns (Larson, 1974), loss of taste and smell (Henkin *et al.*, 1975), dermatitis in infants (Arakawa *et al.*, 1976), and correlation of fetal size with zinc concentrations in amniotic fluid (Favier *et al.*, 1972) are clinical symptoms reported in human zinc deficiency.

In Egypt (Prasad *et al.*, 1963) and Iran (Halstead *et al.*, 1972), studies have revealed zinc deficiencies among men and women. The consumption of diets high in cereals, which are rich in phytate and fiber, is the primary cause, inhibiting the absorption of zinc by forming insoluble chelates (Reinhold *et al.*, 1976).

In the United States, where soils are low in zinc content, animals may develop zinc deficiency unless supplementary zinc is added to their feed (RDA, 1974, 1979). Marginal zinc deficiency has been reported among apparently healthy preschool children in Denver (Hambidge *et al.*, 1972). The diets of some American women and children have been found to be marginally deficient in zinc (Sandstead, 1973).

Based on data from zinc balance studies and radioisotope studies of turnover, an intake of 15 mg/day seems to be sufficient for all persons over ten years of age. An increment of 5 mg is recommended during pregnancy and a 10-mg increment during lactation. Data indicate lower requirements for children. Since no data are available about infant zinc requirements, the zinc concentration in human milk is used as a guideline.

8.12. Chromium

The existence of a glucose tolerance factor (GTF) was postulated in 1957 by Schwartz and Mertz. Two years later, trivalent chromium was identified as the active ingredient (Schwartz and Mertz, 1959). Within the decade, the nature in which the GTF functioned to facilitate the reaction of insulin was described (Mertz, 1969). Chromium exists in most organic matter, in more than one ionic state, and occurs in a variety of complexes in foods. Mertz *et al.* (1974) have classified chromium compounds in food into two categories, according to their level of insulin-potentiating activity: low (simple compounds) or outstanding (others whose metabolic behavior is similar to that of trivalent chromium in yeast). Depending on the form and availability of food chromium, approximately 10–25% of trivalent chromium and 1% of inorganic chromium is absorbed (Levine *et al.*, 1968; Mertz and Roginski, 1971).

Less than 6 mg of chromium is found in the adult body. High concentrations occur in hair, kidneys, spleen, and testes. The plasma chromium is about 3 ppb (Hambidge, 1974). Chromium is absorbed from the intestinal tract and excreted in the urine; however, the rate of excretion does not reflect intake, or deficiencies would be rampant (Mertz and Roginski, 1971). Schroeder *et al.* (1962) and Levander (1975) associated a decline in body chromium with age. Marginal deficiencies have been related to old age (Levine *et al.*, 1968), pregnancy (Hopkins *et al.*, 1968), and protein–calorie malnutrition (Gurson and Saner, 1971).

Niacin is known to impart high biological activity to chromium (Polansky, 1974), but the exact mechanism of trivalent chromium activity as a cofactor in insulin utilization and glucose metabolism remains to be defined. Chromium therapy has been effective in some cases of impaired glucose tolerance in protein–calorie-deficient children and in diabetics (Sandstead *et al.*, 1967).

8.13. Cobalt

Cobalt was identified as a dietary essential component of vitamin B_{12} (Rickles *et al.,* 1948) 11 years after it was shown to be the cure for coast disease in cattle and sheep (Filmer and Underwood, 1937). The body is incapable of utilizing the simple element, cobalt; hence, it must be supplied in the diet in its physiologically active form as cobalamin or vitamin B_{12}. Cobalt is easily absorbed from the intestinal tract but serves no known physiological function other than as a part of vitamin B_{12}. Humans do not synthesize this vitamin, so that it must be absorbed from food via the intestinal tract (Schroeder *et al.,* 1967).

Murphy *et al.* (1971) found dietary cobalt intake of institutionalized children in the United States to exceed that of adults (Tipton *et al.,* 1966); it was two to five times the cobalt concentration of an adequate intake for cattle and sheep (Rickles *et al.,* 1948). Cobalt is absorbed from the intestine and is excreted in the urine in direct relation to intake (Valberg, 1971). Small amounts are lost by way of feces (Schroeder *et al.,* 1967) and sweat (Consolazio *et al.,* 1964). The average adult male body content of cobalt is 1.1 mg (Yamagata *et al.,* 1962). Because cobalt is only essential as a component of vitamin B_{12} neither requirements nor allowances have been defined.

The relationship between the metabolic activity of cobalt and other mineral elements is under investigation. It has been reported that the absorption of both cobalt and iron is increased during iron deficiency in humans; however, the reason for this reaction is unknown (Valberg, 1971). Cobalt is known to stimulate production of erythropoietin and has been used on a limited basis (20–30 mg cobalt/day) as a nonspecific erythropoietic stimulant in humans. The treatment is toxic to infants (Sederholm *et al.,* 1968). The body has a high tolerance for cobalt, but toxicity has been known to occur in men who were heavy beer drinkers. The combination of concentrated doses of cobalt and alcohol in beer precipitated cardiomyopathy (*Nutr. Rev.,* 1968). The practice of adding cobalt to beer to produce a foam is no longer condoned. An inverse correlation has been observed between the cobalt–iodine content of soil and the incidence of goiter in humans and farm animals (Kovalsky, 1970).

8.14. Manganese

Manganese was recognized as an essential element for rat metabolism in 1931, first by Orent and McCollum and later by several other investigators working with other species of animals (McCollum, 1957). Continued investigations have demonstrated the involvement of manganese in bone structure, reproduction, and nervous system activity. Its specific role in human metabolism is not clearly defined (Cotzias, 1958; RDA, 1974).

The average adult male body contains a fairly constant 10–20 mg of manganese, stored chiefly in the kidney and the liver (Schroeder *et al.,* 1966a). Homeostasis is maintained by regulation of intestinal absorption and excretion as a bile complex in the feces (Bertinchamps *et al.,* 1966).

Manganese is active in human metabolism as a component of several manganese-containing metalloproteins (Cotzias and Greenough, 1958) and as a cofactor activating non-manganese-containing enzymes (Leach, 1976). It is involved in enzymatic reactions of carbohydrate metabolism, as a component of pyruvate carboxylase (Scrutton *et al.*, 1966), in mucopolysaccharide synthesis (Leach, 1971), in lipid metabolism as a lipotropic agent interacting with choline (Underwood, 1971), and in the metabolism of biogenic amines and brain function (Cotzias *et al.*, 1977).

Cotzias (1958) has reported manganese toxicity in miners inhaling dust ore, but it is unlikely to happen as a result of dietary intake.

8.15. Molybdenum

Molybdenum is assumed to be an essential mineral element for man, because it is a component of the enzymes aldehyde oxidase and xanthine oxidase (Rickert and Westerfeld, 1953). Tissues of all animal species are low in molybdenum content, but no clinical deficiency signs have been observed in humans, although they have been experimentally produced in animals (Underwood, 1971).

This mineral is absorbed from the intestine as molybdate and is excreted in the urine. An estimated daily intake of from 45 to 500 mg of molybdenum is considered adequate to meet human requirements (Schroeder *et al.*, 1970).

Molybdenum is related to copper and sulfur metabolism. It has been suggested, too, that this mineral possesses cariostatic properties, because children living in areas where the soil content of this mineral is high have been reported to have fewer than average dental caries (Davidson *et al.*, 1975). Excess molybdenum intake by cattle causes teart. Experimentally, molybdenum toxicity effects have been counteracted by administration of copper or methionine (Miller and Engel, 1960). High molybdenum intake also appears to interfere with the utilization of dietary copper (O'Dell, 1972).

8.16. Selenium

The essential nature of selenium in the metabolism of humans has not been clearly defined, but inference has been drawn since 1957, when the relevance of selenium to nutrition was accepted because the mineral prevented liver necrosis in the rat (Schwarz and Foltz, 1957). Since that time, investigators have reported the effects of deficiencies in various animals, but not in humans (NRC/CAN, 1971).

Selenium is present in all tissues and is highly concentrated in the kidney, liver, pancreas, spleen, and testes. Significantly, different human blood levels of selenium have been reported (Allaway, 1969). An adequate mixed American diet is assumed to meet the human requirement (Morris and Levander, 1970), though the amount needed is not known. There appears to be better absorption of dissolved than solid sodium selenite. A homeostatic mechanism is not believed to exist in humans, since only minute amounts of test doses have been absorbed (Thomson, 1974). This may mitigate against human toxicity due to dietary selenium.

Selenium is believed to be essential to humans. The mineral is a component of glutathione peroxidase and, as a selenoenzyme, protects against lipid peroxide-induced damage (Rotruck *et al.*, 1973). Other selenoproteins have been identified and are known to function in electron-transfer reactions (Stadtman, 1974). Selenium may also be involved in the hepatic microsomal cytochrome system (Burk *et al.*, 1974).

In some of the infrequent human nutrition studies, low selenium blood levels have been found in children with kwashiorkor in Guatemala (Burk *et al.*, 1967) and Thailand (Levine and Olson, 1970). Selenium deficiency has been suggested as a possible carcinogenic factor (Shapiro, 1972). Trace amounts have been reported to be protective against human cancer (Scott, 1973), but additional research is needed in this area.

Excess dietary selenium intake has been implicated in the higher incidence of dental caries among children who live in the seleniferous areas of Oregon (Hadjimarkos, 1956, Tank and Storvick, 1960).

8.17. Nickel, Silicon, Tin, and Vanadium

Whether nickel, silicon, vanadium, and tin are essential for humans is a matter of conjecture. These minerals appear in the human body, but their exact roles are not clear. Investigations continue to be made in an effort to discover whether these mineral elements are essential nutrients for humans (Nielsen and Standstead, 1974). Nickel (Nielsen, 1971), silicon (Carlisle, 1974), tin (Schwarz, 1971), and vanadium (Hopkins and Mohr, 1971) have each been acknowledged as essential to the growth of animals fed a purified diet and raised in isolation systems free of mineral contaminants. Although deficiencies of these elements are unlikely to occur in humans on earth, these data bear consideration in the light of projected space travel.

Traces of nickel are found in human tissues. Approximately 0.3–0.6 mg of nickel is consumed daily. It may play a role in cell membrane structure and in activating liver arginase (Orten and Neuhaus, 1975).

Human blood serum carries approximately 1 mg of silicon/100 ml of blood. Silicon deficiencies in humans have not been reported. There is adequate silicon in plant and animal foodstuffs to meet the needs for growth and development. Among workers in the stone-cutting industry, the inhalation of silicon dust carries an increased concentration of the element to their lungs, oftentimes precipitating silicosis.

Although tin is essential for growth in rats, the implications for human nutrition are not clear. The human diet contains approximately 3.5–17 mg of tin/day, which is far below the 250 mg/day upper limit permissible in canned food (Davidson *et al.*, 1975). Tin is poorly absorbed and is excreted primarily via the feces.

Vanadium has been proven to be essential for the growth of chicks (Hopkins and Mohr, 1971) and rats (Schwarz and Milne, 1971). It is assumed, therefore, that vanadium may also be necessary for humans. Vanadium is present in human tissues. The American diet contains about ten times the 0.1–0.3 mg/day considered to be the possible requirement.

As research progresses and as methodology becomes more sophisticated, quantitative recommendations for trace minerals will undoubtedly become available. In the meantime, protection against deficiencies continues to be the ingestion of sufficient quantities of varied foodstuffs representative of the different food groups.

9. References

AAP/CN (American Academy of Pediatrics, Committee on Nutrition), 1971, Vitamin K supplementation for infants receiving milk substitute infant formulas and for those with fat malabsorption, *Pediatrics* **48:**483.

Adams, C. F., 1975, *Nutritive Value of American Foods in Common Units*, U.S. Dep. Agric. Agric. Handb. No. 456.

Adams, C. F., and Richardson, M., 1977, *Nutritive Value of Foods*, U.S. Dep. Agric. Home Gard. Bull. No. 72.

Addison, W. L. T., 1928, The uses of sodium chloride, potassium chloride, sodium bromide and potassium bromide in cases of arterial hypertension which are amenable to potassium chloride, *Can. Med. Assoc. J.* **18:**281.

Ahlstrom, A., and Rasanen, L., 1973, Review of food grouping systems in nutrition education, *J. Nutr. Educ.* **5:**13.

Aikawa, J. K., 1963, *The Role of Magnesium in Biologic Processes*, C. C. Thomas, Springfield, Ill.

AIN/CN (American Institute of Nutrition, Committee on Nomenclature), 1975, Nomenclature Policy: Generic descriptors and trivial names for vitamins and related compounds, *J. Nutr.* **105:**135.

Alfin-Slater, R. B., and Aftergood, L., 1968, Essential fatty acids reinvestigated, *Physiol. Rev.* **48:**758.

Alfin-Slater, R. B., Aftergood, L., and Kishineff, S., 1972, Investigations on hypervitaminosis E in rats, in: *Summaria, IX International Congress of Nutrition*, Mexico City, p. 191.

Allaway, W. H., 1969, Control of environmental levels of selenium, in: *Trace Substances in Environmental Health* (D. D. Hemphill, ed.), pp. 181–206, University of Missouri, Columbia.

Al-Rashid, R. A., and Spangler, J., 1971, Neonatal copper deficiency, *N. Engl. J. Med.* **285:**841.

Anderson, C. E., 1977, Minerals, in: *Nutritional Support of Medical Practice* (H. A. Schneider, C. E. Anderson, and D. B. Coursin, eds.), pp. 57–72, Harper and Row, New York.

Anderson, J. T., Grande, F., Matsumoto, Y., and Keys, A., 1963, Glucose, sucrose and lactose in the diet and blood lipids in man, *J. Nutr.* **79:**349.

Anonymous, 1970, On a definition of health (editorial), *Calif. Med.* **112**(4):63–64.

AOAC (Association of Official Agricultural Chemists), 1960, *Official Methods of Analysis*, Collegiate Press, G. Banta, Menasha, Wisc.

Arakawa, T., Tamura, T., Igarshi, Y., Suzuki, H., and Sandstead, H., 1976, Zinc deficiency in two infants during "total parenteral alimentation for diarrhea," *Am. J. Clin. Nutr.* **29:**197.

Armstrong, W. D., and Singer, L., 1970, Distribution of fluorides. 2. Distribution in body fluids and soft tissues, in: *Fluorides and Human Health*, Chapter 4, pp. 94–110, Monograph 59, World Health Organization, Geneva.

Baker, E. M., 1967, Vitamin C requirements in stress, *Am. J. Clin. Nutr.* **20:**583.

Baker, H., and Frank, O., 1968, *Clinical Vitaminology*, Wiley–InterScience, New York.

Baker, H., and Sobotka, H., 1962, Microbiological assay methods for vitamins, in: *Advances in Clinical Chemistry* (H. Sobotka and C. P. Stewart, eds.), pp. 173–235, Academic Press, New York.

Baker, H., Frank, O., Matovich, V. B., Pasher, I., Aaronson, S., Hutner, S. H., and Sobotka, H., 1962, A new assay method for biotin in blood, serum, urine and tissues, *Anal. Biochem.* **3**:31.

Baugh, C. M., and Knumdieck C. L., 1971, Naturally occurring folates, *Ann. N. Y. Acad. Sci.* **186**:7.

Baugh, C. M., Malone, J. M., and Butterworth, C. E., 1968, Human biotin deficiency. A case history of biotin deficiency induced by raw egg consumption in a cirrhotic patient, *Am. J. Clin. Nutr.* **21**:73.

Bean, W. B., Hodges, R. E., and Daum, K., 1955, Pantothenic acid deficiency induced in human subjects, *J. Clin. Invest.* **34**:1073.

Beck, W. S., 1968, Deoxyribonucleotide synthesis and the role of vitamin B_{12} in erythropoiesis, *Vitam. Horm. (N.Y.)* **26**:413.

Becker, B. G., Indik, B. P., and Beeuwkes, A. M., 1960, *Dietary Intake Methodologies–A Review,* University of Michigan, Ann Arbor.

Behnke, A. R., Feen, B. G., and Welham, W. C., 1942, The specific gravity of healthy men, *J. Am. Med. Assoc.* **118**:495.

Bernstein, D. S., Sandowsky, N., Hegsted, M., Guri, C. D., and Stare, F. J., 1966, Prevalence of osteoporosis in high and low fluoride areas in North Dakota, *J. Am. Med. Assoc.* **198**:85.

Bertinchamps, A. J., Miller, S. T., and Cotzias, G. C., 1966, Interdependence of routes excreting manganese, *Am. J. Physiol.* **211**:217.

Best, C. H., and Huntsman, M. E., 1932, The effects of the components of lecithin upon deposition of fat in the liver, *J. Physiol. (London)* **75**:405.

Best, C. H., and Lucas, C. C., 1962, Choline malnutrition, in: *Clinical Nutrition,* 2nd ed. (N. Jolliffe, ed.), pp. 277–260, Harper, New York.

Bhagavan, H. N., and Coursin, D. B., 1967, Biotin content of blood in normal infants and adults, *Am. J. Clin. Nutr.* **20**:903.

Biehl, J. P., and Vilter, R. W., 1954, Effects of isoniazid on pyridoxine metabolism, *J. Am. Med. Assoc.* **156**:1549.

Bigwood, E. J., 1939, *Guiding Principles for Studies on the Nutrition of Populations,* League of Nations Organization, Geneva.

Binder, H. J., and Spiro, H. M., 1967, Tocopherol deficiency in man, *Am. J. Clin. Nutr.* **20**:594.

Blum, L., 1970, Recherches sur le rôle des sels alcalins dans la pathogenie des oedèmes. L'action diurétique du chlorure de potassium, *Presse Méd.* **28**:685.

Bridgens, W. F., 1967, Present knowledge of biotin, *Nutr. Rev.* **25**:65.

Brown, R. R., 1972, Normal and Pathological conditions which may alter the human requirement for vitamin B_6, *J. Agric. Food Chem.* **20**:498.

Brozek, J. (ed.), *Body Measurements and Human Nutrition,* 1956, National Research Council, Committee on Nutritional Anthropometry, Wayne University Press, Detroit.

Brozek, J., Kehelberg, J. K., Taylor, H. L., and Keys, A., 1963, Skinfold distributions in middle-aged American men: A contribution to norms of leanness-fatness, *Ann. N.Y. Acad. Sci.* **110**:492.

Brudevold, F., and McCann, H. G., 1966, Fluoride and caries control—Mechanism of action, in: *The Science of Nutrition and Its Application in Clinical Dentistry* (A. E. Nizel, ed.), pp. 331–347, W. B. Saunders, Philadelphia.

Burk, R. F., Pearson, W. N., Wood, R. P.,II, and Viteri, F., 1967, Blood selenium levels and *in vitro* red blood cell uptake of ^{75}Se in kwashiorkor, *Am. J. Clin. Nutr.* **20**:723.

Burk, R. F., Mackinnon, A. M., and Simon, F. R., 1974, Selenium and hepatic microsomal hemoproteins, *Biochem. Biophys. Res. Commun.* **56**:431.

Caddell, J. L., 1969, Magnesium deficiency in protein–calorie malnutrition: A follow up study, *Ann. N.Y. Acad. Sci.* **162**:874.

Cantoni, M., Paffenbarger, R. S., and Krueger, D. E., 1959, Methods of dietary assessment in current epidemiologic studies of cardiovascular diseases, Report presented at American Public Health Association meeting, Atlantic City, N.J. (Mimeo).

Cardano, A., Baertl, J. M., and Graham, G. G., 1968, Copper deficiency in infancy, *Pediatrics* **34**:324.

Carlisle, E. M., 1974, Silicon as an essential element. *Fed. Proc. Fed. Am. Soc. Exp. Biol.* **33:**1758.

Carlson, C. H., Armstrong, W. D., and Singer L., 1960, Distribution and excretion of radiofluoride in the human, *Proc. Soc. Exp. Biol. Med.* **104:**235.

Carnes, W. H., 1971, Role of copper in connective tissue metabolism, *Fed. Proc. Fed. Am. Soc. Exp. Biol.* **30:**995.

Cartwright, G. E., and Wintrobe, M. M., 1964a, Copper metabolism in normal subjects, *Am. J. Clin. Nutr.* **14:**224.

Cartwright, G. E., and Wintrobe, M. M., 1964b, The question of copper deficiency in man, *Am. J. Clin. Nutr.* **15:**94.

Castle, W. B., and Minot, G. R., 1936, *Pathological Physiology and Clinical Description of Anemias* (H. A. Christian, ed.), Oxford University Press, New York.

Chaykin, G., 1967, Nicotinamide coenzymes, *Annu. Rev. Biochem.* **36:**149.

Christakis, G., 1973, Nutritional assessment in health programs, *Am. J. Public Health,* Part II, Vol. 63, 82 pp.

Consolazio, C. F., Nelson, R. A., Matoush, L. O., Hughes, R. C., and Urone, P., 1964, *The Trace Mineral Losses in Sweat,* U.S. Army Med. Res. Nutr. Lab. Rep. No. 284, pp. 1–13.

Cook, J. D., Layrisse, M., Martinez-Torres, C., Walker, R., Monsen, E., and Finch, C. A., 1972, Food iron absorption measured by an extrinsic tag, *J. Clin. Invest.* **51:**805.

Cottove, E., and Hagben, C. A. M., 1962, Chloride, in: *Mineral Metabolism,* Vol. 2, Part B (C. L. Comar and F. Bronner, eds.), pp. 109–173, Academic Press, New York.

Cotzias, G. C., 1958, Manganese in health and disease, *Physiol. Rev.* **38:**503.

Cotzias, G. C., and Greenough, J. J., 1958, The high specificity of the manganese pathway through the body, *J. Clin. Invest.* **37:**1298.

Cotzias, G. C., Tang, L. C., Miller, S. T., Sladic-Simig, D., and Hurley, L. S., 1977, A mutation influencing the transportation of manganese, L-dopa and L-tryptophan, *Science* **176:**410.

Cravioto, J., De Licardie, E. R., and Birch, H. G., 1966, Nutrition, growth and neurointegrative development: An experimental and ecologic study, *Pediatrics* **38:**319.

Crokaert, R., 1949, Du dosage chimique de l'acide pantothénique, *Bull. Soc. Chim. Biol.* **31:**903.

Crosse, V. M., Meyer, T. C., and Gerrard, J. W., 1955, Kernicterus and prematurity, *Arch. Dis. Child.* **30:**501.

Dahl, L. K., 1958, Salt intake and salt need, *N. Engl. J. Med.* **258:**1152; 1205.

Dahl, L. K., 1972, Salt and hypertension, *Am. J. Clin. Nutr.* **25:**231.

Dam, H., 1935, Antihaemorrhagic vitamin of chicks: Occurrence and chemical nature, *Nature* **135:**652.

Dam, H., 1962, Interrelations between vitamin E and polyunsaturated fatty acids in animals, *Vitam. Horm.* **20:**527.

Davenport, E., 1964, *Calculating the Nutritive Value of Diets,* U.S. Dep. Agric. Agric. Res. Serv. Report No. A R S 62-10-1.

Davidson, L. S. P., and Passmore, R., 1970, *Human Nutrition and Dietetics,* Williams and Wilkins, Baltimore.

Davidson, S., Passmore, R., Brock, J. F., and Truswell, A. S., 1975, *Human Nutrition and Dietetics,* 6th ed., Churchill Livingston, New York.

DeLuca, H. F., and Schnoes, H. K., 1976, Metabolism and mechanism of action of vitamin D, *Annu. Rev. Biochem.* **45:**631.

De Vries, W. H., Grovier, W. M., Evans, J. S., Gregory, J. D., Novelli, G. D., Soodak, M., and Lipmann, F., 1950, Purification of coenzyme A from fermentation sources and its further partial identification, *J. Am. Chem. Soc.* **72:**4838.

DHSS/PCN (Department of Health and Social Security, Panel on Child Nutrition), 1970, *Interim Report on Vitamin D,* Report on Public Health and Medical Subjects No. 123, H. M., Stationery Office, London.

Dietrich, L. S., Martinez, L., and Franklin, L., 1968, Role of the liver in systemic pyridine nucleotide metabolism, *Naturwissenschaften* **55:**231.

Dunn, M. J., and Walser, M., 1966, Magnesium depletion in normal man, *Metabolism* **15:**884.

Durnin, J. V. G. A., and Passmore, R., 1967, *Energy, Work and Leisure,* Heinemann Educational Books, London.

Ellegaard, J., Ehmann, V., and Henriksen, L., 1972, Deficient folate activity during treatment of psoriasis with methotrexate diagnosed by determination of serine synthesis in lymphocytes, *Br. J. Dermatol.* **87:**248.

Elvehjem, C. A., 1935, The biological significance of copper and its relation to iron metabolism, *Physiol. Rev.* **15:**471.

FAO/WHO (Food and Agriculture Organization World Health Organization), 1970, *Requirements of Ascorbic Acid, Vitamin D, Vitamin B_{12}, Folate and Iron,* W.H.O. Tech. Rep. Ser. No. 452.

FAO/WHO (Food and Agriculture Organization World Health Organization), 1973, *Energy and Protein Requirement,* W.H.O. Tech. Rep. Ser. No. 522.

Favier, M., Yacoub, M., Racinet, C., Marka, C., Chabert, P., and Benbassa, B., 1972, Lésions métalliques dans le liquide amniotique au cours du troisième trimestre de la gestation. Relation significative entre la concentration en zinc et le poids foetal, *Rev. Fr. Gynécol. Obstét.* **67:**707.

Ferguson, J. H., 1946, Blood coagulation, thrombosis and hemorrhagic disorders, *Annu. Rev. Physiol.* **8:**231.

Filmer, J. F., and Underwood, E. J., 1937, Enzootic marasmus: Further data concerning the potency of cobalt as a curative and prophylactic agent, *Aust. Vet. J.* **13:**57.

Fomon, S. J., 1974, *Infant Nutrition,* 2nd ed., W. B. Saunders, Philadelphia.

Fomon, S. J., 1976, *Nutritional Disorders of Children,* U.S. Dep. Health Educ. Welfare Publ. No. (HSA) 76–5612.

Forbes, G. B., 1962, Sodium, in: *Mineral Metabolism,* Vol. 2, Part B (C. L. Comar and F. Bronner, eds.), pp. 1–72, Academic Press, New York.

Forbes, G. B., Gallup, J., and Hursh, J. B., 1961, Estimation of total body fat from potassium-40 content, *Science* **133:**101.

Fox, H. C., and Miller, D. S., 1960, Ackee toxin: A riboflavin antimetabolite? *Nature* **186:**561.

Frenkel, E. P., 1973, Abnormal fatty acid metabolism in peripheral nerves of patients with pernicious anemia, *J. Clin. Invest.* **52:**1237.

Frieden, E., 1973, The ferrous to ferric cycles in iron metabolism, *Nutr. Rev.* **31:**41.

Frieden, E., Osaki, S., and Kobayashi, H., 1965, Copper proteins and oxygen. Correlations between structure and function of the copper oxidases, *J. Gen. Physiol. (Suppl.)* **49:**213–252.

Frind, B., 1972, Nutritional review, in: *National Food Situation,* No. 142, pp. 25–28, U.S. Dep. of Agric., Hyattsville, Md.

Garn, S. M., 1956, Comparison of pinch-caliper and X-ray measurements of skin plus subcutaneous fat, *Science* **24:**178.

Gershoff, S. N., 1976, Vitamin B_6, in: *Present Knowledge of Nutrition,* 4th ed., pp. 149–161, Nutrition Foundation, New York.

Gholson, R. K., 1966, The pyridine nucleotide cycle, *Nature* **212:**933.

Gnaedinger, R. H., Reineke, E. P., Pearson, A. M., Van Huss, W. D., Wessel, J. A., and Montoye, H. J., 1963, Determination of body density by air displacement, helium dilution and underwater weighing, *Ann. N.Y. Acad. Sci.* **110:**96.

Goldsmith, G. A., Starett, H. P., Register, U. D., and Gibbens, J., 1952, Studies of niacin requirements in man. I. Experimental pellagra in subjects on corn diets low in niacin and tryptophan, *J. Clin. Invest.* **31:**533.

Gontzea, I., and Sutzescu, P., 1968, *Natural Antinutritive Substances in Foodstuffs and Forages,* S. Karger, Basel.

Griffith, W. H. and Wade, N. J., 1939, Choline metabolism. I. The occurrence and prevention of hemorrhagic degeneration in young rats on a low choline diet, *J. Biol. Chem.* **131:**567.

Griffith, W. H., Nye, J. F., Hartroft, W. S., and Porta, E. A., 1971, Choline, in: *The Vitamins,* Vol. III, 2nd ed. (W. H. Sebrell and R. S. Harris, eds.), pp. 1–154, Academic Press, New York.

Gurson, C. T., and Saner, G., 1971, Effect of chromium on glucose utilization in marasmic protein–calorie malnutrition, *Am. J. Clin. Nutr.* **24:**1313.

Gÿorgy, P., 1939, The curative factor (vitamin H) for egg white injury, with particular reference to its presence in different foodstuffs and in yeast, *J. Biol. Chem.* **131:**733.

Hadjimarkos, D. M., 1956, Geographic variations of dental caries in Oregon, *J. Pediatr.* **48:**195.

Haeger, K. Lanner, E., and Magnusson, P. O., 1972, Oral zinc sulfate in the treatment of venous leg ulcers, *VASA* **1:**62.

Halstead, J. A., Ronaghey, H. A., Abadi, P., Haghshenass, M., Amirhakeml, G. H., Barakat, R. M., and Reinhold, J. C., 1972, Zinc deficiency in man, *Am. J. Med.* **53:**277.

Hambidge, K. M., 1974, Chromium nutrition in man, *Am. J. Clin. Nutr.* **27:**505.

Hambidge, K. M., Hambidge, C., Jacobs, M., and Baum, J. D., 1972, Low levels of zinc in hair, anorexia, poor growth and hypogeusia in children, *Pediatrics* **6:**868.

Harris, H. C., Wolf, D. E., Mozingo, R., Arth, G. E., Anderson, R. C., Easton, N. R., and Folkers, K., 1945, Biotin. V. Synthesis of di-biotin, di-allobiotin and di-epi-allobiotin, *J. Am. Chem. Soc.* **67:**2096.

Harrison, H. E., 1972, Calcium metabolism, in: *Pediatrics,* 15th ed. (H. L. Barnett and A. H. Einhorn, eds.), pp. 195–213, Appleton-Century-Crofts, New York.

Hartroft, W. S. (chm.), 1967, Nutrition Society Symposium: Alcohol, metabolism and liver disease, *Fed Proc. Fed Am. Soc. Exp. Biol.* **26:**1432.

Hassan, H., Hashim, S. A., Van Itallie, T. B., and Sebrell, W. H., 1966, Syndrome in premature infants associated with low plasma vitamin E levels and high polyunsaturated fatty acid diet, *Am. J. Clin. Nutr.* **19:**147.

Hathaway, M. L., 1957, *Heights and Weights of Children in the United States,* U.S. Dept. Agric. Home. Econ. Res. Rep. No. 2.

Hegsted, D. M., 1973, Calcium and phosphorus, in: *Modern Nutrition in Health and Disease,* 5th ed. (R. S. Goodhart and M. E. Shils, eds.), pp. 268–286, Lea and Febiger, Philadelphia.

Henkin, R. I., Patten, B. M., Re, P. K., and Bronzert, D. A., 1975, A syndrome of acute zinc loss, *Arch. Neurol.* **32:**745.

Herbert, V., 1962, Experimental nutritional folate deficiency in man, *Trans. Assoc. Am. Physicians* **75:**307.

Herbert, V., 1968, Nutritional requirements for vitamin B_{12} and folic acid, *Am. J. Clin. Nutr.* **21:**743.

Herbert, V., 1972, Detection of malabsorption of vitamin B_{12} due to gastric or intestinal dysfunction, *Semin. Nucl. Med.* **2:**220.

Herbert, V., 1973, Folic acid and vitamin B_{12}, in: *Modern Nutrition in Health and Disease,* 5th ed. (R. S. Goodhart and M. E. Shils, eds.), pp. 221–244, Lea and Febiger, Philadelphia.

Herbert, V., 1975, Megaloblastic anemia, in: *Textbook of Medicine,* 14th ed. (P. B. Beeson and W. McDermott, eds.), p. 1404–1413, W. B. Saunders, Philadelphia.

Hodge, H. C., and Smith, F. A., 1965, Biological properties of inorganic fluorides, in: *Fluorine Chemistry,* Vol. 4 (H. Simons, ed.) pp. 2–376, Academic Press, New York.

Hodges, R. E., and Kolder, H., 1971, Experimental vitamin A deficiency in human volunteers, in: *Summary of Proceedings, Workshop on Biochemical and Clinical Criteria for Determining Human Vitamin A Nutriture,* pp. 10–16, National Academy of Sciences, Washington, D. C.

Hodges, R. E., Bean, W. B., Ohlson, M. A., and Bleiler, R., 1959, Human pantothenic acid deficiency produced by omega-methyl pantothenic acid, *J. Clin. Invest.* **38:**1421.

Hodgkin, D. C., Kamper, J., Lindsey, L., Mackay, M., Pickworth, J., Robertson, J. H., Shoemaker, C. B., White, J. G., Prosen, R. J., and Trueblood, K. N., 1957, The structure of vitamin B_{12}. I. An outline of the crystallographic investigation of vitamin B_{12}, *Proc. R. Soc. London Ser. A* **242:**228.

Holt, L. E., Jr., and Snyderman, S. E., 1955, Protein and amino acid requirements of infants and children, *Nutr. Abstr. Rev.* **35:**1.

Holt, L. E., Jr., Hallac, E., Jr., and Kajdi, C. N., 1962, The concept of protein stores and its implication in the diet, *J. Am. Med. Assoc.* **181:**699.

Hopkins, L. L., Jr., and Mohr, H. E., 1971, The biological essentiality of vanadium, in: *Newer Trace Elements in Nutrition* (W. Mertz and W. E. Cornatzer, eds.), pp. 195–213, Dekker, New York.

Hopkins, L. L., Jr., Ransome-Kuti, O., and Majaj, A. S., 1968, Improvement of impaired carbohydrate metabolism by chromium(III) in malnourished infants, *Am. J. Clin. Nutr.* **21:**203.

Horowitz, H. S., 1973, A review of systematic and topical fluorides for the prevention of dental caries, *Comm. Dent. Oral Epidemiol.* **1:**104.

Horwitt, M. K., 1962, Interrelations between vitamin E and polyunsaturated fatty acids in adult men, *Vitam. Horm. (N. Y.)* **20**:541.

Horwitt, M. K., Harvey, C. C., Hills, O. W., and Liebert, E., 1950, Correlation of urinary excretion of riboflavin with dietary intake and symptoms of ariboflavinosis, *J. Nutr.* **41**:247.

Horwitt, M. K., Harvey, C. C., Rothwell, W. S., Cutler, J. L., and Haffron, D., 1956, Tryptophan–niacin relationship in man, *J. Nutr.* **60**(Suppl. 1):1.

Hume, E. M., and Krebs, H. E. (compilers), 1949, *Vitamin A Requirements of Human Adults,* Report of the Vitamin A Subcommittee of the Accessory Food Factors Committee, Medical Research Council (Great Britain), Special Report Series No. 264, H. M. Stationery Office, London.

ICNND (Interdepartmental Committee on Nutrition for National Defense), 1963, *Manual for Nutrition Surveys,* 2nd ed., National Institutes of Health, Bethesda, Md.

IUNS/CN (International Union of Nutritional Sciences, Committee on Nomenclature), 1970, Tentative rules for generic descriptors and trivial names for vitamins and related compounds, *Nutr. Abstr. Rev.* **40**:395.

J. Allergy, 1971, A statement on the question of allergy to fluoride as used in the fluoridation of community water supplies, **47**:347.

Jelliffe, D. B., 1969, Field anthropometry independent of precise age, *J. Pediatr.* **75**:334.

Jenkins, G. N., 1971, Mechanism of action of fluoride in reducing dental caries, in: *Fluoride in Medicine* (T. L. Vischer, ed.), pp. 88–94, Hans Huber, Bern.

Johnson, N. E., Alcantara, E. N., and Linkswiler, H. M., 1970, Effect of level of protein intake on urinary and fecal calcium and calcium retention of young adult males, *J. Nutr.* **100**:1425.

Keys, A., and Brozek, J., 1953, Body fat in adult man, *Physiol. Rev.* **33**:245.

Kodicek, E., 1966, Antivitamins of nicotinic acid and of biotin, in: *Antivitamins* (J. C. Somogyi, ed.), Bibliotheca Nutritio et Dieta, p. 8, S. Karger, Basel.

Kogl, F., and Tonnis, B., 1936, Uber das Bios-Problem Dartstellung von Krystallisiertem Biotin aus Eigelb, *Z. Physiol. Chem.* **242**:43.

Kovalsky, V. V., 1970, The geochemical ecology of organisms under conditions of varying contents of trace elements in the environment, in: *Trace Element Metabolism in Animals* (C. F. Mills, ed.), pp. 385–397, S. Livingstone, Edinburgh.

Krehl, W. A., 1964, Development and evaluation of nutritional deficiencies, *Med. Clin. North Am.* **48**:1129.

Krehl, W. A., Tepley, L. J., Sarma, S. A., and Elvejhem, C. A., 1945, Growth retarding effect of corn in nicotinic acid low rations and its counteraction by tryptophan, *Science* **101**:489.

Laitinen, H. A., 1972, Analytical methods for trace minerals: An overview, *Ann. N.Y. Acad. Sci.* **199**:173.

Lan, T. H., and Sealock, R. R., 1944, The metabolism *in vitro* of tyrosine by liver and kidney tissues of normal and vitamin C deficient guinea pigs, *J. Biol. Chem.* **155**:483.

Lane, M., and Alfrey, C. P., Jr., 1970, The anemia of human riboflavin deficiency, *Blood* **25**:432.

Larson, D. L., 1974, Oral zinc sulfate in the management of severely burned patients, in: *Clinical Applications of Zinc Metabolism* (W. J. Pories, W. H. Strain, J. M. Hsu, and R. L. Woosley, eds.), pp. 229–236, Charles C Thomas, Springfield, Ill.

Leach, R. M., 1971, Role of manganese in mucopolysaccharide metabolism, *Fed. Proc. Fed. Am. Soc. Exp. Biol.* **30**:991.

Leach, R. M., 1976, Metabolism and function of manganese, in: *Trace Elements in Human Health and Disease,* Vol. 2 (C. A. Prasad, ed.), pp. 235–247, Academic Press, New York.

Lee, Y. C., Gholson, R. K., and Raica, N., 1969, Isolation and identification of two new nicotinamide metabolites, *J. Biol. Chem.* **244**:3277.

Leone, N. C., 1970, Fluorides and general health. 2. Areas of the USA with a high natural content of water fluoride, in: *Fluorides and Human Health,* Chap. 8 pp. 274–284, World Health Organization, Geneva.

Levander, O. A., 1975, Selenium and chromium in human nutrition, *J. Am. Diet. Assoc.* **66**:338.

Leveille, G. A., 1972, Modified thiochrome procedure for the determination of urinary thiamin, *Am. J. Clin. Nutr.* **25**:273.

Levine, R. A., Streeter, D. H. P., and Doisy, R. J., 1968, Effect of oral chromium supplementation on the glucose tolerance of elderly human subjects, *Metabolism* **17**:114.

Levine, R. J., and Olson, R. E., 1970, Blood selenium in Thai children with protein–calorie malnutrition, *Proc. Soc. Exp. Biol. Med.* **134:**1030.

Liang, C. C., 1963, Tissue breakdown and glyoxylic acid formation, *Biochem. J.* **83:**101.

Lipmann, F., Kaplan, N. O., Novelli, G. C., Tuttle, L. C., and Guirard, B. M., 1947, Coenzyme for acetylation, a pantothenic acid derivative, *J. Biol. Chem.* **167:**869.

Lohmann, K., and Schuster, p., 1937, Untersuchungen uber die Cocarboxylase, *Biochem. Z.* **294:**188.

Loomis, W. F., 1967, Skin-pigmentation regulation of vitamin-D biosynthesis in man, *Science* **157:**501.

Lotz, M., Zisman, E., and Bartter, F. C., 1968, Evidence for a phosphorus depletion in man, *N. Engl. J. Med.* **278:**409.

Lowry, O. H., 1952, Biochemical evidence of nutritional status, *Physiol. Rev.* **37:**431.

Lowry, O. H., and Bessey, A. O., 1945, Microbiochemical methods for nutritional studies, *Fed. Proc. Fed. Am. Soc. Exp. Biol.* **4:**268.

Macara, I. G., Hoy, T. G., and Harrison, P. M., 1972, Formation of ferritin from apoferritin. Kinetics and mechanism of iron uptake, *Biochem. J.* **126:**151.

Margen, S., Chu, J. Y., Kaufman, N. A., and Calloway, D. H., 1974, Studies in calcium metabolism, I. The calciuretic effect of dietary protein, *Am. J. Clin. Nutr.* **27:**584.

Martin, H. E., Mehl, J., and Wertman, M., 1952, Clinical studies of magnesium metabolism, *Med. Clin. North Am.* **36:**1157.

Maugh, T. H., II, 1973, Vitamin B$_{12}$: After twenty-five years the first synthesis, *Science* **179:**266.

McCollum, E., 1957, *A History of Nutrition,* Houghton Mifflin, Boston.

McCormick, D. B., and Wright, L. D. (eds.), 1970, *Methods in Enzymology,* Vol. XVIII, Part A, Sect. II, Thiamine, Phosphates and Analogs, pp. 73–266, Academic Press, New York.

McCormick, D. B., and Wright, L. D., (eds.), 1971, *Methods in Enzymology,* Vol. XVIII, Part B, Sect. IX, Pteridines, Analogs and Pterin Coenzymes, pp. 599–816, Academic Press, New York.

Melhorn, D. K., and Gross, S., 1969, Relationship between iron-dextran and vitamin E in iron deficiency in children, *J. Lab. Clin. Med.* **74:**789.

Meneghello, J., and Neimeyer, H., 1950, Liver steatosis in undernourished Chilean children, III. Evaluation of choline treatment with repeated liver biopsies, *Am. J. Dis. Child.* **80:**905.

Merrill, A. L., and Watt, B. K., 1955, *Energy Value of Foods, Bases, and Derivation,* Hum. Nutr. Res. Branch, U.S. Dep. Agric., Agric. Res. Serv. Handb. No. 74.

Merrill, A. L., Adams, C. F., and Fincher, L. J., 1966, *Procedures for Calculating Nutritive Values of Home-Prepared Foods,* U.S. Dep. Agric. Agric. Res. Serv. Rep. No. ARS 62–13.

Mertz, W., 1969, Chromium occurrence and function in biological systems, *Physiol. Rev.* **49:**163.

Mertz, E., and Roginski, E. E., 1971, Chromium metabolism: The glucose tolerance factor, in: *Newer Trace Elements in Nutrition* (W. Mertz and W. E. Cornatzer, eds.), pp. 123–153, Dekker, New York.

Mertz, W., Toepfer, E. W., Roginski, E. E., and Polansky, M. M., 1974, Present knowledge of the role of chromium, *Fed. Proc. Fed. Am. Soc. Exp. Biol.* **33:**2275.

Messer, H. H., and Singer, L., 1976, Fluoride, in: *Present Knowledge in Nutrition,* 4th ed,, pp. 325–337, Nutrition Foundation, Washington, D.C.

Messer, H. H., Armstrong, W. D., and Singer, L., 1973, Fluoride, parathyroid hormone and calcitonin: Interrelationships in bone calcium metabolism, *Calcif. Tissue Res.* **13:**217.

Metropolitan Insurance Company, 1960, *How to Control Your Weight* (Supplement based on 1959 Body Build and Blood Pressure Study), New York.

Miller, R. F., and Engel, R. W., 1960, Interrelationships of copper, molybdenum and sulfate sulfur in nutrition, *Fed. Proc. Fed. Am. Soc. Exp. Biol.* **19:**666.

Minot, G. R., and Murphy, W. P., 1926, Treatment of pernicious anemia by a special diet, *J. Am. Med. Assoc.* **87:**470.

Mistry, S. P., and Dakshinamurti, K., 1964, Biochemistry of biotin, *Vitam. Horm. (N. Y.)* **22:**1.

Mitchell, H. K., Snell, E. E., and Williams, R. J., 1941, Concentration of "folic acid," *J. Am. Chem. Soc.* **63:**2284.

Mizrahi, A., London, R. D., and Gribetz, D., 1968, Neonatal hypocalcemia—its causes and treatment, *N. Engl. J. Med.* **278:**1163.

Moore, F. D., and Boyden, C. M., 1963, Body cell mass and limits of hydration of the fat-free body: Their relation to estimated skeletal weight, *Ann. N.Y. Acad. Sci.* **110:**62.

Moore, F. D., Olesen, K. H., McMurrey, J. D., Parker, H. V., Ball, M. R., and Boyden, C. M., 1963, *The Body Cell Mass and Its Supporting Environment,* W. B. Saunders, Philadelphia.

Morris, V. C., and Levander, O. A., 1970, Selenium content of foods, *J. Nutr.* **100:**1383.

Murphy, G. K., Rhea, U., and Peeler, J. T., 1971, Levels of antimony, cadmium, chromium, cobalt, magnesium and zinc in institutional diets, *Environ. Sci. Technol.* **5:**436.

Nielsen, F. H., 1971, Studies on the essentiality of nickel, in: *Newer Trace Elements in Nutrition* (W. Mertz and W. E., Cornatzer, eds.), pp. 215–253, Dekker, New York.

Nielsen, F. H., and Sandstead, H. H., 1974, Are nickel, vanadium, silicon, fluorine, and tin essential for man? A review, *Am. J. Clin. Nutr.* **27:**510.

Nordin, B. E. C., 1962, Calcium balance and calcium requirement in spinal osteoporosis, *Am. J. Clin. Nutr.* **10:**384.

NRC/CAN (National Research Council, Committee on Animal Nutrition), 1971, *Selenium in Nutrition,* National Academy of Sciences, Washington, D.C.

NRC/CMN (National Research Council, Committee on Maternal Nutrition), 1973, *Nutrition Supplementation and the Outcome of Pregnancy,* National Academy of Sciences, Washington, D.C.

Nutr. Rev., 1968, Epidemic cardiac failure in beer drinkers, **26:**173.

Nyhan, W. L., 1975, Prenatal treatment of methylmalonic acidemia, *N. Engl. J. Med.* **293:**353.

O'Dell, B. L., 1972, Dietary factors that affect biological availability of trace elements, *Ann. N.Y. Acad. Sci.* **199:**70.

O'Dell, B. L., 1976, Biochemistry on physiology of copper in vertebrates, in: *Trace Elements in Human Health and Disease,* Vol. 1 (A. S. Prasad, ed.), pp. 391–413, Academic Press, New York.

Oppel, T. W., 1942, Studies of biotin metabolism in man, *J. Med. Sci.* **204:**856.

Orten, J. M., and Neuhaus, O. W., 1975, *Human Biochemistry,* 9th ed., p. 551, C. V. Mosby, St. Louis, Mo.

Owen, C. A., Jr., 1971, Vitamin K group, X. Deficiency effects in animals and human beings, in: *The Vitamins,* Vol. III (W. H. Sebrell, Jr. and R. S. Harris, eds.), pp. 417–522, Academic Press, New York.

Page, L., and Phipard, E. F., 1957, *Essentials of an Adequate Diet,* U.S. Dep. Agric. Home Econ. Res. Rep. No. 3.

Parisi, A. F., and Vallee, B. L., 1969, Zinc metalloenzymes: Characteristics and significance in biology and medicine, *Am. J. Clin. Nutr.* **22:**1222.

Parsons, H. T., Lease, J. G., and Kelly, E., 1937, Interrelationship between dietary egg white and requirement for protective factor in cure of nutritional disorder due to egg white, *Biochem. J.* **31:**424.

Phipard, E. F., and Page, L., 1962, Meeting nutritional needs through food, *Borden Rev. Nutr. Res.* **23:**31.

Pike, R. L., and Smiciklas, H. A., 1972, A reappraisal of sodium restriction during pregnancy, *Int. J. Gynecol. Obstet.* **10:**1.

Pitkin, R. M., Kaminetsky, R. H., Newton, M., and Pritchard, J. A., 1972, Maternal nutrition, a selective review of clinical topics, *J. Obstet. Gynecol.* **40:**773.

Polansky, M. M., 1974, Properties of synthetic nicotinic acid–chromium complexes and purified concentrates from brewers yeast, *Fed. Proc. Fed. Am. Soc. Exp. Biol.* **33:**659.

Prasad, A. S., Miale, A., Jr., Farid, Z., Sandstead, H. H., and Schulert, A. R., 1963, Zinc metabolism in patients with the syndrome of iron deficiency anemia, hepatosplenomegaly, dwarfism and hypogonadism, *J. Lab. Clin. Med.* **61:**537.

Price, J. M., Thornton, M. J., and Mueller, L. M., 1967, Tryptophan metabolism in women using steroid hormones for ovulation control, *Am. J. Clin. Nutr.* **20:**452.

Prockop, D. J., 1971, Role of iron in the synthesis of collagen in connective tissue, *Fed. Proc. Fed. Am. Soc. Exp. Biol.* **30:**984.

Pugh, E. L., and Wakil, S. J., 1965, Studies on the mechanism of fatty acid synthesis, XIV. The prosthetic group of acyl carrier protein and the mode of its attachment to the protein, *J. Biol. Chem.* **240:**4727.

Raghuramulu, N., Srikantia, S. G., Narasinga Rao, B. S., and Gopalan, C., 1965, Nicotinamide nucleotides in the erythrocytes of patients suffering from pellagra, *Biochem. J.* **96:**837.

Rathbun, E. N., and Pace, N., 1945, Studies on body composition, I. The determination of total body fat by means of body specific gravity, *J. Biol. Chem.* **158:**667.

Recommended Dietary Allowances, 1968, 7th ed., National Research Council, National Academy of Sciences.

Recommended Dietary Allowances, 1974, 8th ed., National Research Council, National Academy of Sciences.

Recommended Dietary Allowances, 1979, 9th ed., National Research Council, National Academy of Sciences.

Reinhold, J. G., Faradji, B., Abadi, P., and Ismail-Beigi, F., 1976, Binding of zinc to fiber and other solids of wholemeal bread with a preliminary examination of the effects of cellulose consumption upon the metabolism of zinc, calcium and phosphorus in man, in: *Trace Elements in Human Health and Disease,* Vol. 1 (A. S. Prasad, ed.), pp. 163–180, Academic Press, New York.

Rheed, W. J., and Schrauzer, G. N., 1971, Risks of long-term ascorbic acid overdosage, *Nutr. Rev.* **29:**262.

Rickert, D. A., and Westerfeld, W. W., 1953, Isolation and identification of xanthine oxidase factor as molybdenum, *J. Biol. Chem.* **203:**915.

Rickles, E. L., Brink, N. G., Konwszy, F. R., Wood, T. R., and Folkers, K., 1948, Crystalline vitamin B_{12}, *Science* **107:**396.

Rindi, G., and Ventura, U., 1972, Thiamine intestinal transport, *Physiol. Rev.* **52:**821.

Roberts, L. J., 1958, Beginnings of the Recommended Dietary Allowances, *J. Am. Diet. Assoc.* **34:**903.

Robinson, F. A., 1966, *The Vitamin Co-Factors of Enzyme Systems,* pp. 638–666, Pergamon Press, New York.

Rogers, T. A., Setliff, J. A., and Klopping, J. C., 1963, The caloric cost and fluid and electrolyte balance in simulated subarctic survival situations, *Arct. Aeromed. Lab. U. S. Tech. Doc. Rep.* 13–16.

Roholm, K., 1937, *Fluorine Intoxication: A Clinical Hygienic Study,* Lewis, London.

Rose, W. C., 1957, The amino acid requirements of adult man, *Nutr. Abstr. Rev.* **27:**631.

Rotruck, J. T., Pope, A. L., Ganther, H. E., Swanson, A. B., Hateman, D. G., and Hoekstra, W. G., 1973, Selenium: Chemical role as a component of glutathione peroxidase, *Science* **179:**588.

Sandstead, H. H., 1973, Zinc nutrition in the United States, *Am. J. Clin. Nutr.* **26:**1251.

Sandstead, H. H., Prasad, A. S., Schulert, Z., Farid, Z., Maile, A., Jr., Bassilly, S., and Darby, W. J., 1967, Human zinc deficiency. Endocrine manifestations and the response to treatment, *Am. J. Clin. Nutr.* **20:**422.

Sarett, H. P., and Goldsmith, G. A., 1947, The effect of tryptophan on the excretion of nicotinic acid derivatives in humans, *J. Biol. Chem.* **167:**293.

Sass-Kortsak, A., 1965, Copper metabolism, *Adv. Clin. Chem.* **8:**1.

Sauberlich, H. E., 1967, Biochemical alterations in the thiamin deficiency—their interpretation, *Am. J. Clin. Nutr.* **20:**528.

Sauberlich, H. E., and Canham, J. E., 1973, Vitamin B_6, in: *Modern Nutrition in Health and Disease,* 5th ed. (R. S. Goodhart and M. E. Shils, eds.), pp. 210–220, Lea and Febiger, Philadelphia.

Sauberlich, H. E., Canham, J. E., Baker, E. M. Raica, N., Jr., and Herman, Y. F., 1972, Biochemical assessment of the nutritional status of vitamin B_6 in the human, *Am. J. Clin. Nutr.* **25:**629.

Sauberlich, H. E., Dowdy, R. P., and Skala, J. H., 1973, Laboratory tests for the assessment of nutritional status, *Crit. Rev. Clin. Lab. Sci.* **4:**215.

Scherp, H. W., 1971, Dental caries: Prospects for prevention, *Science* **173:**1199.

Schroeder, H. A., Balassa, J. J. and Tipton, I. H., 1962, Abnormal trace metals in man—chromium, *J. Chron. Dis.* **15:**941.

Schroeder, H. A., Balassa, J. J., and Tipton, I. H., 1966a, Essential trace metals in man: Manganese, a study of homeostasis, *J. Chron. Dis.* **19:**549.

Schroeder, H. A., Nason, A. P., Tipton, I. H., and Balassa, J. J., 1966b, Essential trace metals in man: Copper, *J. Chron. Dis.* **19**:1007.

Schroeder, H. A., Nason, A. P., and Tipton, I. H., 1967, Essential trace metals in man: Cobalt, *J. Chron. Dis.* **20**:869.

Schroeder, H. A., Balassa, J. J., and Tipton, I. H., 1970, Essential trace metals in man: Molybdenum, *J. Chron. Dis.* **23**:481.

Schwarz, K., 1971, Tin as an essential growth factor for rats, in: *Newer Trace Elements in Nutrition* (W. Mertz and W. E., Cornatzer, eds.), pp. 313–326, Dekker, New York.

Schwarz, K., and Foltz, C. M., 1957, Selenium as an integral part of Factor 3 against dietary necrotic liver degeneration, *J. Am. Chem. Soc.* **79**:3292.

Schwarz, K., and Mertz, W., 1957, A glucose tolerance factor and its differentiation from Factor 3, *Arch. Biochem. Biophys.* **72**:515.

Schwarz, K., and Mertz, W., 1959, Chromium(III) and the glucose tolerance factor, *Arch. Biochem. Biophys.* **85**:292.

Schwarz, K., and Milne, D. B., 1971, Growth effects of vanadium in the rat, *Science,* **174**:426.

Scott, D., 1958, Clinical biotin deficiency (egg white injury), *Acta Med. Scand.* **162**:69.

Scott, M. L., 1973, The selenium dilemma, *J. Nutr.* **103**:803.

Scrutton, M. C., Utter, M. F., and Mildvan, A. S., 1966, Pyruvate carboxylase, VI. The presence of tightly bound manganese, *J. Biol. Chem.* **241**:3480.

Sebrell, W. H., Jr., and Butler, R. E., 1938, Riboflavin deficiency in man: Preliminary note, *U.S. Public Health Serv. Rep.* **52**:2282.

Sebrell, W. H., Jr., and Harris, R. S., 1972, Riboflavin, in: *The Vitamins,* Vol. V (W. H. Sebrell, Jr. and R. S. Harris, eds.), Chap. 8, pp. 2–96, Academic Press, New York.

Sederholm, T., Kouvalainen, K., and Lamberg, B. A., 1968, Cobalt induced hypothyroidism and polycythemia in lipid nephrosis, *Acta Med. Scand.* **184**:301.

Seelig, M. S., 1971, Human requirements of magnesium: Factors that increase needs, Presented at 1st International Symposium on Magnesium Deficit in Human Pathology, Vittel, France.

Shapiro, J. R., 1972, Selenium and carcinogenesis: A review, *Ann. N.Y. Acad. Sci.* **192**:215.

Sherman, H. C., 1952, *Chemistry of Food and Nutrition,* 8th ed., Macmillan, New York.

Shils, M. E., 1964, Experimental human depletion, I. Clinical observations and blood chemistry alterations, *Am. J. Clin. Nutr.* **15**:133.

Shils, M. E., 1969, Experimental human magnesium depletion, *Medicine (Baltimore)* **48**:61.

Shils, M. E., 1973, Magnesium, in: *Modern Nutrition in Health and Disease,* 5th ed. (R. S. Goodhart and M. E. Shils, eds.), pp. 287–296, Lea and Febiger, Philadelphia.

Shimazono, N., and Katsura, E. (eds.), 1965, *Review of Japanese Literature on Beriberi and Thiamin,* Vitamin B Research Committee of Japan, Igaku, Shoin Ltd., Tokyo.

Siri, W. E., 1953, Fat, water and lean tissue studies, *Fed. Proc. Fed. Am. Soc. Exp. Biol.* **12**:133.

Smith, E. L., and Parker, L. F. J., 1948, Purification of antipernicious anaemia factor, *Biochem. J.* **43**:viii.

Snell, E. E., 1945, The vitamin B_6 group, *J. Biol. Chem.* **157**:491.

Stadtman, T. C., 1974, Selenium biochemistry, *Science* **183**:915.

State University of Iowa, 1943a, *Growth Curves for Boys from Birth to Six Years* (prepared from data compiled by Iowa Child Welfare Research Station), Iowa City, Iowa.

State University of Iowa, 1943b, *Growth Curves for Girls from Birth to Six Years* (prepared from data compiled by Iowa Child Welfare Research Station), Iowa City, Iowa.

Stiebling, H. K. 1933, Food Budget for Nutrition and Production Programs, US Dep. Agric. Misc. Publ. 183.

Stiller, E. T., Harris, S. A., Finkelstein, J., Keresztesy, J. C., and Folkers, K., 1940, Pantothenic acid, VIII. The total synthesis of pure pantothenic acid, *J. Amer. Chem. Soc.* **62**:1785.

Sutherland, J. M., Glueck, H. L., and Gleser, G., 1967, Hemorrhagic disease of the newborn. Breast feeding as a necessary factor in the pathogenesis, *Am. J. Dis. Child.* **113**:524.

Sydenstricker, V. P., Singal, S. A., Briggs, A. P., DeVaughn, N. M., and Isbell, H., 1942, Observations of the "egg white injury" in man and its cure with biotin concentrate, *J. Am. Med. Assoc.* **118**:1199.

Takaki, K., 1887, Health of the Japanese navy, *Lancet* **2**:86.

Tanaka, Y., Frank, H., and DeLuca, H. F., 1973, Role of 1,25-dihydroxycholecalciferol in calcification of bone and maintenance of serum calcium concentration in the rat, *J. Nutr.* **102:**1569.

Tank, G., and Storvick, C. A., 1960, Effect of naturally occurring selenium and vanadium on dental caries, *J. Dent. Res.* **39:**473.

Tarli, P., Benocci, S., and Neri, P., 1971, Gas chromatographic determination of pantothenates and panthenol in pharmaceutical preparations by pantoyl lactone, *Anal. Biochem.* **42:**8.

Taylor, C. M., and Pye, O. F., 1966, *Foundations of Nutrition,* Macmillan, New York.

Taylor, C. M., and Riddle, K. P., 1971, *An Annotated International Bibliography of Nutrition Education,* Teachers College Press, Columbia University, New York.

Thomson, A. D., Baker, H., and Leevy, C. M., 1971, Folate-induced malabsorption of thiamin, *Gastroenterology,* **60:**756.

Thomson, C. D., 1974, Recovery of large doses of selenium given as sodium selenite with and without vitamin E, *N. Z. Med. J.* **80:**163.

Tipton, I. H., Stewart, P. L., and Martin, P. G., 1966, Trace elements in diet and excreta, *Health Phys.* **12:**1683.

Underwood, E. J., 1971, *Trace Elements in Human and Animal Nutrition,* 3rd ed., Academic Press, New York.

USDHEW (United States Department of Health, Education and Welfare), 1953, *Basic Body Measurements of School Age Children,* Office of Education, Washington, D.C.

USDHEW (United States Department of Health, Education and Welfare), 1972, *Ten State Nutrition Survey 1968/1970. I. Historical Development; II. Demographic Data; III. Clinical, Anthropometry, Dental; IV. Biochemical; V. Dietary; and Highlights,* USDHEW Publ. Nos. (HMS) 72–8130, -8131, -8132, -8133, -8134.

Valberg, L. S., 1971, in: *Intestinal Absorption of Metal Ions, Trace Elements and Radionuclides* (S. C. Skoryna and D. Waldron-Edward, eds.), pp. 257–263, Pergamon Press, Montreal.

Van Etten, C. H., and Wolff, I. A., 1973, *Toxicants Occurring Naturally in Foods,* Monograph, 2nd ed., p. 214, Committee on Food Protection, Food and Nutrition Board, National Research Council, National Academy of Sciences, Washington, D.C.

Van Sydow, G., 1946, A study of the development of rickets in premature infants, *Acta Paediatr. Scand. Suppl.* **33:**2.

Wacker, W. E. C., and Parisi, A. F., 1968, Magnesium metabolism, *N. Engl. J. Med.* **278:**712.

Wakil, S., and Gibson, D. M., 1960, Studies on the mechanism of fatty acid synthesis, VIII. The participation of protein bound biotin in the biosynthesis of fatty acids, *Biochim. Biophys. Acta* **41:**122.

Wald, G., 1973, The biochemistry of vision, *Annu. Rev. Biochem.* **23:**497.

Waller, C. W., Hutchings, B. L., Mowat, J. H., Stokstad, E. L. R., Boothe, J. H., Angier, R. B., Semb, J., Subbarow, Y., Cosulich, D. B., Fabrenbach, M. J., Hultquist, M. E., Kuh, E., Northcy, E. H., Seeger, D. R., Sickels, J. P., and Smith, J. M., Jr., 1948, Synthesis of pteroylglutamic acid (liver *L. casei* factor) and pteroic acid, *J. Am. Chem. Soc.* **70:**19.

Walser, M., 1967, Magnesium metabolism, *Ergeb. Physiol.* **59:**185.

Watt, B. L., and Merrill, A. L., 1963, *Composition of Foods—Raw, Processed, Prepared,* U.S. Dep. Agric. Agric. Handb. No. 8.

Wetzel, N., 1940, *The Wetzel Grid for Evaluating Physical Fitness,* National Education Association, Cleveland Heights, Ohio.

Wilde, W. S., 1962, Potassium, in: *Mineral Metabolism,* Vol. 2, Part B (C. L. Comar and F. Bronner, eds.), pp. 73–107, Academic Press, New York.

Williams, R. H., 1943, Clinical biotin deficiency, *N. Eng. J. Med.* **288:**247.

Williams, R. J., Truesdail, J. H., Weinstock, H. H., Rohrmann, E., Lyman, C. M., and McBurney, C. H., 1938, Pantothenic acid, II. Its concentration and purification from liver, *J. Am. Chem. Soc.* **60:**2719.

Willis, M. R., 1973, Intestinal absorption of calcium, *Lancet* **1:**820.

Wills, L., 1933, The nature of the hemopoietic factor in marmite, *Lancet* **224:**1283.

Wooley, D. W., 1945, Production of nicotinic acid deficiency with 3-acetylpyridine in the ketone analogue of nicotinic acid, *J. Biol. Chem.* **157:**445.

Wooley, D. W., and Krampitz, L. O., 1943, Production of scurvy conditions by feeding of a compound structurally related to ascorbic acid, *J. Exp. Med.* **78:**333.

WHO/ECMANS (World Health Organization, Expert Committee on Medical Assessment of Nutritional Status), 1963, *W.H.O. Tech. Rep. Ser. No. 258.*

Yamagata, N., Muratan, S., and Morii, T., 1962, The cobalt content of the human body, *J. Radiat. Res. (Tokyo)* **3:**4.

Young, C. M., and Trulson, M. F., 1960, Methodology for dietary studies in epidemiological surveys, II. Strengths and weaknesses of existing methods, *Am. J. Public Health* **50:**803.

Young, C. M., Tensuan, R. S., Sault, F., and Holmes F., 1963, Estimating body fat of normal young women, *J. Am. Diet. Assoc.* **42:**409.

2

Energy: Caloric Requirements

Elsworth R. Buskirk and Jose Mendez

1. Brief History of Bioenergetics

The science of bioenergetics, as we know it today, was perhaps initiated with
the studies of Lavoisier (1743–1794), who discovered the principles of *in vivo*
and *in vitro* oxidation and combustion. He ascertained that the intensity of
metabolism was dependent upon physical work, environmental temperature,
and food intake. He devised an ice calorimeter to study the body heat ema-
nating from guinea pigs and determined that oxygen was utilized by the metabolizing body and that a gas was given off (carbon dioxide) that was the
same as that produced when acid was added to limestone. Lavoisier appreciated the chemical identity of the oxidative process of carbon and the burning
of carbon in a candle flame. At a time when the "phlogiston" theory was
widely adhered to, Lavoisier and Sequin wrote the following, as quoted by
Swift and French (1954):

> Respiration is only a slow combustion of carbon and hydrogen which is entirely
> similar to that which obtains in a lamp or lighted candle and from this point of view,
> animals which respire are truly combustible bodies which burn and consume them-
> selves. In respiration as in combustion it is the air which furnishes the oxygen—but
> in respiration it is the body substance which furnishes the heat—if animals do not
> repair constantly the losses of respiration, the lamp soon lacks oil, and the animal
> dies, as a lamp goes out when it lacks food.

The studies of von Liebig are of almost equal significance to those of
Lavoisier. Liebig showed that carbohydrates, fat, and protein were oxidized
in the body, and not hydrogen and carbon as thought by Lavoisier. He mis-
interpreted utilization of protein as an energy source and also that the breakdown of protein was caused exclusively by muscular work. It was not until

---

Elsworth R. Buskirk and Jose Mendez • Laboratory for Human Performance Research, Intercollege
Research Programs, The Pennsylvania State University, University Park, Pennsylvania 16802.

about 50 years later, about 1900, that von Voit showed that protein metabolism was not exclusively related to muscular work. During this same period, Rubner conclusively demonstrated the validity of the laws of conservation of energy in intact animals.

The original experiments on metabolic balance studies were conducted by Boussingault on dairy cows. He ascertained elemental levels in the feed and made a quantitative comparison with values in the milk, urine, and feces. The carbon, oxygen, nitrogen, and hydrogen that were unaccounted for were thought to be associated with respiratory processes.

The first indirect calorimeter or respiratory chamber was built by Pettenkofer. A man could live reasonably normally in this chamber—he could sleep, eat, and work for 24 or more hours. Voit and Pettenkofer performed several important experiments in their respiratory chamber and contributed much to the study of total body energetics. The first closed-circuit respiratory apparatus, for use in experiments of indirect calorimetry, was designed by Regnault and Reiset in 1849. The apparatus consisted of a glass bell jar into which the animal was placed. The animal breathed via a demand oxygen system in which the oxygen partial pressure was kept constant as indicated by a manometer, and potassium hydroxide absorbers were used to measure the elimination of carbon dioxide. Regnault and Reiset showed that the respiratory quotient (RQ) was a useful concept and that the RQ was dependent upon the nature of the foodstuffs oxidized and independent of the animal species. A chronology of the early history of bioenergetics appears in Table I.

Large direct and indirect calorimeters were constructed around the turn of the century in several locations. Atwater constructed the first human respiration calorimeter at Middletown, Connecticut. Armsby built a similar but larger apparatus for farm animals at The Pennsylvania State College in State College, Pennsylvania, and Benedict built a calorimeter at the Nutrition Laboratory of the Carnegie Institute of Washington at Boston, Massachusetts. The calorimeter at The Pennsylvania State University (formerly College) is still intact and usable, but all the others have been dismantled. Swift *et al.* (1958) used the Penn State calorimeter for research on human subjects as recently as 1958. Interestingly, two men, presumably the metabolic equivalent of one steer, had to be studied together to achieve the required metabolic input for study with the calorimeter. The Benedict calorimeter is currently in residence at the Smithsonian Institute in packing crates. The names of Armsby, Atwater, Benedict, DuBois, Lusk, and Murlin were those associated with the dynamic phase of calorimetric work from 1890 to 1935. Their studies, along with those of many others, form the basis of much of our understanding of energy metabolism, including behavior of the basal and resting metabolism in numerous diseases, particularly diabetes, disorders of the thyroid, and fevers of various origins. Following these early important experimental efforts, research on energy metabolism waned with certain exceptions, such as the extensive studies of energy metabolism and insensible water loss conducted by Newburgh and his colleagues (1945). The insensible-water-loss method could be used to estimate metabolism because insensible water loss tends to

Table I. Some Significant Early Dates in the Investigation of Energy as Related to Biological Systems[a]

180	Galen worked on circulation and speculated on value of respiration.
1592	Galileo's thermometer invoked pneuma and vital spirit.
1665	Boyle conceived of gas laws.
1670	Hooke and Boyle worked with first respiratory chambers.
1670	Mayow understood biological meaning of "nitro-aereal spirit," or oxygen.
1680	Borelli recognized that air dissolved in liquids can pass through membranes and that air can enter blood.
1693	Leibnitz identified the law of conservation of kinetic and potential energy.
1727	Hales developed pneumatic trough for collecting gases.
1764	Black investigated properties of "fixed air" (carbon dioxide) and proved that it was produced during respiration.
1772	Rutherford described what we now call nitrogen (Scheele also).
1774–1777	Priestley and Scheele, respectively, described "pure air" and "fire air" (oxygen). Oxygen was recognized as an element by Lavoisier. Scheele also prepared "vitiated-air," or nitrogen.
1777	Lavoisier compared animal respiration with a slow oxidation of carbon and measured the amount of carbon dioxide given off.
1785	Cavendish discovered hydrogen.
1789–1790	Lavoisier and Sequin studied metabolism using Sequin's methods for gas analysis.
1802–1805	Henry and Dalton developed laws of gas behavior in relation to partial pressures of gases.
1803	Spallazoni provided early evidence that oxidation occurs in tissues.
1817	Nysten separated inspired and expired air and collected expired air in a bladder to study chemistry of respiration.
1835	Becquerel and Breschet demonstrated temperature rise in the arm during exercise.
1837	Magnus set up improved methods for analyzing the gaseous content of blood.
1839	Boussingault initiated *in vivo* metabolic balance studies.
1842	Joule supplied experimental data which established the mechanical equivalent of heat.
1842	Liebig conceived of the processes of nutrition and oxidation of protein, carbohydrates, and fats in body and founded organic analysis.
1842	Mayer developed in general form the concept of conservation of energy and applied concept to biological systems in 1845.
1843	Liebig provided evidence that respiration occurs in tissues.
1845	Mayer laid down law of conservation of energy.
1847	Helmholtz included heat in law of conservation of energy as applied to biological systems and established the first law of thermodynamics.
1849	Regnault and Reiset perfected the closed-circuit type of metabolism apparatus.
1850	Regnault and Reiset reported respiration experiments with small animals.
1850	Clausius enunciated the first and second laws of thermodynamics.
1851	Kelvin formulated the second law of thermodynamics.
1852	Bidder and Schmidt calculated metabolic fate of various foodstuffs in the body.

(Continued)

Table I. (Continued)

1856	Bernard, from thermoelectric measurements, concluded that oxidations occur in tissues.
1856	Matteuci found increased respiration and heat during contraction of isolated muscle.
1857	Voit studied nitrogenous equilibrium and demonstrated that food nitrogen was not excreted as gaseous nitrogen.
1860	Bischoff and Voit showed metabolism of protein and fat in starvation.
1866	Pettenkofer and Voit successfully used a respiration chamber for experiments on humans and extended metabolic balance work.
1877	Pflüger studied the relationship of metabolism to respiration and calculated the respiratory quotient (RQ) in dogs.
1878	Bert established that physiological effects of gases depend on their partial pressures.
1879–1884	Stefan–Boltzmann law was discovered empirically by Stefan and deduced theoretically 5 years later by Boltzmann.
1883	Rubner demonstrated that fat and carbohydrate are interchangeable in terms of energy equivalence (isodynamic law).
1885	Stohmann published his research on the caloric values for foods, urea, etc.
1885	Rubner undertook valuable calorimetric determinations on equivalence of oxidizable substrates *in vivo* and heat value of protein and discovered "specific dynamic action."
1892	Haldane described first version of gas analyzer.
1892	Atwater and Rosa began development of their calorimeter.
1892	Magnus-Levy used Zuntz's apparatus and made pioneer metabolic studies of patients.
1901	Rubner and Voit discovered that rate of metabolism is related to body surface area and that law of conservation of energy applies to the living body.

a From Kleiber (1961), Lusk (1928), Perkins (1964), and Smith and French (1954).

be roughly proportional to heat loss in the absence of overt sweating. Significant studies were also accomplished in the area of body temperature regulation by Bazett, Burton, Hardy, Belding, Winslow, Gagge, and many others.

Within the past 20 years, a resurgence of interest in energy metabolism has taken place (Whedon, 1959). In terms of instrumentation, these efforts have involved direct calorimetry with a gradient-layer (heat-flow) calorimeter (Benzinger and Kitzinger, 1949), portable systems for indirect calorimetry (Kofranyi and Michaelis, 1941; Liddell, 1963; Müller and Franz, 1952; Passmore *et al.*, 1952), and the introduction of physical gas analyzers capable of continuously measuring a specific gas concentration in a gas stream. The gradient calorimeter employed by Benzinger and co-workers (Benzinger and Kitzinger, 1949, Benzinger *et al.*, 1958) has the advantage of being able to follow with high precision rapid changes in body heat loss, which was impossible with earlier direct calorimeters. The portable systems have made possible a wide range of energy-expenditure surveys under actual and simulated work conditions (Edholm *et al.*, 1955). The continuous-stream gas analyzers have made possible continuous measurement of gas exchange and energy expenditure under laboratory conditions. Rapid changes in rates of removal of oxygen and production of carbon dioxide can be assessed with considerable accuracy. In addition, much of the drudgery associated with discrete-sample gas analysis with chemical gas-absorption techniques has been eliminated; thus, the study of energy metabolism has become a more flexible and versatile procedure with applications in both laboratory and field situations.

A history of bioenergetics can be obtained by consulting the listed references (Chambers, 1942; DuBois, 1936; Kleiber, 1961; Lusk, 1922, 1928; Murlin, 1922; Perkins, 1964; Swift and French, 1954).

2. Definitions*

cal (calorie)—The amount of heat required to raise the temperature of 1 g of water from 15 to 16°C, or 1°C. Redefined in 1948 to equal 4.1840 abs J.

kcal (kilocalorie)—The amount of heat required to raise the temperature of 1 kg of water from 15 to 16°C, or 1°C. The common heat unit in nutrition. Redefined in 1948 to equal 4184 abs J. The joule is the International Unit of heat and may be adopted by nutritional organizations in the future (1 kcal = 4.184 kJ).

Specific heat—The amount of heat required to raise the temperature of 1 g of any substance 1°C. A measure of thermal capacity. The specific heat of pure water is 1.0. Because body tissues vary in specific heat, the average specific heat is usually taken as 0.83.

Heat—Randomized energy in molecules of matter; temperature is a measure of the extent of randomization in specific matter.

First law of thermodynamics, or law of conservation of energy—Energy is conserved in any transformation of energy. Energy can neither be created nor destroyed in any observable process. The overall total energy remains constant in any process.

Second law of thermodynamics—States, in effect, that heat will never flow of itself from a colder to a hotter body or from a region of lower to higher temperature. The net entropy change in heat transfer never becomes negative and may become positive. While application of the second law to biological systems is debatable, living organisms do increase the flux of degradation of energy in their environments.

Newton's law of cooling—Over moderate temperature ranges, the rate of cooling is proportional to the difference in temperature between the cooling body and the surrounding medium. May be expressed as a differential equation, the solution of which for the temperature T as a function of time t is

$$T = T_a + (T_0 - T_a) e^{-At}$$

---

* From Shephard (1967), Burton and Edholm (1955), Carlson (1954), Kleiber (1961), and Ubbelohde (1955).

where T_a is the air temperature and T_0 is the value of T at the beginning of the time interval t. A is a constant depending on the size, shape, material, and surface of the body. The law is only approximate but useful for calorimeter corrections.

Stefan–Boltzmann law—Total radiation (E) from a black body is proportional to the fourth power of its absolute temperature (T)–

$$E = \sigma T^4$$

where σ is the proportionality, or Stefan–Boltzmann constant, which has been determined experimentally as

$$5.672 \times 10^{-5} \text{ erg} \cdot \text{sec}^{-1} \cdot \text{cm}^{-2} \cdot \text{deg}^{-4}$$

Calorimeter—Any apparatus used to measure the quantity of heat.

Thermal conductivity—The flux of heat through a layer of any substance is proportional to the temperature gradient across the substance and to a factor called the "thermal conductivity" of the substance expressed in $\text{cal} \cdot \text{cm}^{-2} \cdot \text{cm}^{-1} \cdot \text{sec}^{-1} \cdot {}^\circ\text{C}^{-1}$. Representative conductivities of *in vitro* fat, skin, and muscle are in the ratio of $1:2:3$.

Thermal insulation—The reciprocal of thermal conductivity.

Convection—Refers to the exchange of heat between warmer to cooler objects by physical passage of a liquid or gas over the surface of the objects. Can be divided into natural and forced convection.

Metabolism—The interchange of materials between living organisms and the environment by which the body is built up and the energy for its vital processes secured.

Basal metabolic rate (BMR)—The metabolic rate expressed in $\text{kcal} \cdot \text{m}^{-2} \cdot \text{hr}^{-1}$ measured in a fasting individual at rest 12 to 15 hr after the last meal. A standard for comparison of metabolism.

Resting metabolic rate (RMR)—The metabolic rate during quiet supine, prone, or sitting rest. Averages approximately $1.1 \times$ BMR.

Met—Relative unit of metabolism or the ratio of actual metabolic rate to the resting metabolic rate.

kcal allowance—A recommended caloric allowance based on estimates from group requirements.

kcal requirement—The quantity of calories required to meet the energy demands of a specific person at a particular time in a given environment.

Work—Physical activity is usually expressed as a rate of working or power and involves the units of force, distance, and time or the equivalent heat or electrical energy. Acceptable units are $\text{kgm} \cdot \text{min}^{-1}$, $\text{kpm} \cdot \text{min}^{-1}$, $\text{kcal} \cdot \text{min}^{-1}$, or W. In a $1 g$ gravitational field, $1 \text{ kgm} \cdot \text{min}^{-1} = 1 \text{ kpm} \cdot \text{min}^{-1}$. $1 \text{ kcal} \cdot \text{min}^{-1} = 427 \text{ kgm} \cdot \text{min}^{-1} = 427 \text{ kpm} \cdot \text{min}^{-1} = 71.5 \text{ W}$.

3. Calorimetry

Calorimetry involves the determination of body heat loss either directly or indirectly. Direct calorimeters are traditionally expensive, permanent installations, although there have been recent attempts to construct calorimeter suits in which a person may work and live in a controlled laboratory environment (Webb, 1971).

3.1. Direct Calorimetry

The various general types of direct calorimeters as classified by Kleiber (1950) are listed in Table II. Kleiber's paper is also an excellent review of calorimetric measurements. In direct calorimetry, it is desirable to partition heat loss into conductive, convective, radiative, and evaporative components as well as to separate heat loss via the lungs and skin. The direct calorimeters constructed by Benedict and Carpenter (1918), Winslow *et al.* (1936), Murlin (1922), Murlin and Burton (1935), and Benzinger and Kitzinger (1949) were essentially partitional calorimeters. In general, these calorimeters, with the exception of Winslow *et al.*'s nine-paneled booth of copper, were thermal gradient calorimeters. The method is based on the fact that when heat is conducted across a layer of thermally conducting material, a difference in temperature exists between the two surfaces of the layer. By interlayering thermocouples above an insulating layer, the calorimeter provided a rapid thermal response, and continuous measurements of heat loss were possible. Incorporation of an air ventilation circuit provided for separation of evaporative thermal loss. A separate breathing circuit made possible separation of pulmonary heat loss.

3.2. Indirect Calorimetry

Numerous methods for indirect calorimetry are available. Some of them are listed in Table III. These methods are based primarily on the caloric values determined for the metabolism of various foodstuffs and the caloric equivalents of oxygen, carbon dioxide, and nitrogen when substrate mixtures are oxidized in the body. Measurements of oxygen consumption ($\dot{V}_{O_2}$), carbon dioxide production ($\dot{V}_{CO_2}$), and nitrogen elimination in the urine are required for precise measurements of energy expenditure. Tabulated values for caloric equivalents of carbon dioxide and oxygen for a given RQ are available (Consolazio, 1963). In practice, urinary nitrogen excretion is either corrected for or neglected and only $\dot{V}_{O_2}$ and $\dot{V}_{CO_2}$ measured. A shorter method has been proposed by Weir (1949), who suggested that only the expired minute volume and the fraction of oxygen in expired air need be measured. When indirect calorimetry is employed, gas volumes are usually expressed in STPD units, that is, standard temperature and pressure, dry.

Table II. Classification of Calorimeters[a,b]

Category of classification	No heat transfer to outside, adiabatic		Heat transfer to outside			
			Temperature in calorimeter nearly constant			
Temperature of calorimeter	Constant isothermal	Changing temperature	Heat flow through walls		Heat flow in circulating medium	
Major form of heat transferred	Latent heat	Sensible heat				
Quantity measured	Amount of substance melted or evaporated	Difference in temperature of calorimeter	Difference in temperature inside and outside of calorimeter as index for heat flow	Heat produced in compensating chamber so as to match thermal head of measuring chamber	Difference in temperature at entrance and exit of circulating medium and rate of flow of this medium	
					Cooling medium	
Calculation of results	km	$C\Delta T_1$	$L\Delta T_2$	i^2r	$c(\Delta m/\Delta t)\Delta T_3$	
Examples	Ice calorimeter	Bomb calorimeter; Hill's microcalorimeter	Richet's and Rubner's respiration calorimeter	Compensation calorimeter	Circulating water Respiration calorimeter of Atwater and Rosa; Armsby	Circulating air Respiration calorimeter of Murlin and Burton; Auquet; Lefvre

[a] Kleiber (1950).

[b] Abbreviations: k, latent heat per unit mass of melted or evaporated substance; m, mass of substance melted or evaporated; C, heat capacity of calorimeter (without jacket); ΔT_1, increase in temperature of calorimeter (without jacket); L, heat-transfer coefficient of calorimeter wall; ΔT_2, difference in temperature inside and outside of calorimeter wall; i, intensity of electric current in heater of compensating chamber; r, electrical resistance of heater in compensating chamber; c, specific heat of circulating medium; ΔT_3, increase in temperature by circulating through chamber.

Table III. Indirect Calorimetry: Methods for Measuring Pulmonary Ventilation ($\dot{V}$) and the Fraction of Oxygen (F_{O_2}), Carbon Dioxide (F_{CO_2}), or Nitrogen in Expired Air (F_{N_2})

	$\dot{V}$	F_{O_2}	F_{CO_2}	F_{N_2}
Methods of measurement or analysis	Douglas bags; meteorological balloons; spirometers; wet gas meter; dry gas meter; portable dry gas meter; flowmeter; pneumotachograph	Chemical[a]; paramagnetic; thermal conductivity; mass spectrometer; gas chromatograph	Chemical[a]; infrared; thermal conductivity; mass spectrometer; gas chromatograph	Chemical[a]; ionization; thermal conductivity; mass spectrometer; gas chromatograph
Systems	Open; closed; flowthrough or hood	Discrete samples; flowthrough cells	Discrete samples; flowthrough cells	Discrete samples; flowthrough cells

[a] The chemical methods of gas analysis include the traditional ones of Van Slyke, Haldane, and Scholander (Consolazio *et al.*, 1963).

Table IV. Representative Physiological Equivalents of Metabolic Exchange

	Body carbohydrate	Body fat	Body protein
kcal·g^{-1}	3.7–4.3	9.5	4.0–4.3
CO_2 (liter·g^{-1})	0.75–0.83	1.43	0.78
O_2 (liter·g^{-1})	0.75–0.83	2.03	0.97
RQ	1.00	0.70–0.71	0.80–0.82
O_2 (kcal·liter^{-1})	5.0	4.7	4.5
CO_2 (kcal·liter^{-1})	5.0	6.6	5.6

3.3. Principles of Indirect Calorimetry

The results of indirect calorimetric measurements can be expressed in calories, because appropriate caloric values for carbohydrate, protein, and fat have been determined experimentally. These values are reported in Table IV. Oxygen reacts with these metabolic fuels in certain definite proportions to form carbon dioxide and water. For example, in the oxidation of glucose,

$$C_6H_{12}O_6 + 6\ O_2 \longrightarrow 6\ CO_2 + 6\ H_2O$$

$$180\ g + (6 \times 32\ g,\ or\ 192\ g) \longrightarrow (6 \times 44\ g,\ or\ 264\ g) + (6 \times 18\ g,\ or\ 108\ g)$$

the respective gram-molecular and total weights are involved as indicated. Each gram-molecular weight of a gas occupies 22.4 liters; hence, 6 O_2 and 6 CO_2 each occupy a volume of 6 × 22.4, or 134.4 liters. Dividing 134.4 by 180 g of glucose gives 0.75 liters O_2 or CO_2·g^{-1} glucose. The reaction liberates 3.74 kcal·g^{-1}. Therefore, 3.74 kcal divided by 0.75 liter·g^{-1} yields 5 kcal·liter^{-1} O_2 or CO_2. Starch and glycogen yield more kcal·g^{-1} than glucose, but more oxygen is required, and the kcal·g^{-1} remains about the same.

Similarly, oxidation of a typical fat:

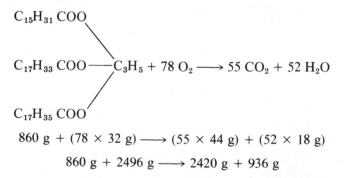

$$860\ g + (78 \times 32\ g) \longrightarrow (55 \times 44\ g) + (52 \times 18\ g)$$

$$860\ g + 2496\ g \longrightarrow 2420\ g + 936\ g$$

A gram-molecular weight of the fat (860 g) combines with 22.4 × 78, or 1747.2, liters O_2 and forms 22.4 × 55, or 1232, liters CO_2. Thus, 1 mol of fat uses 1747.2 liters O_2/860 g, or 2.03 liters O_2·g^{-1}, and yields 1232 liters CO_2/860 g, or 1.43 liters CO_2·g^{-1}. In the process, 9.5 kcal·g^{-1} is liberated or 4.7 kcal·liter^{-1} O_2 or 6.6 kcal·liter^{-1} CO_2.

Similar calculations are possible for the oxidation of protein, but they are more complex because of differences in chemical structure of protein and the elimination from the body of protein metabolic remnants. Appropriate caloric values for protein are listed in Table IV.

In the application of the principles of indirect calorimetry, grams of nitrogen excretion are multiplied by 6.25 to give the grams of protein oxidized. The grams of protein can be converted to O_2 and CO_2 equivalents in liters by utilizing the respective factors of 0.97 and 0.78. From the total respiratory exchange, the volumes of O_2 and CO_2 used in protein metabolism are subtracted, leaving the volumes of O_2 and CO_2 associated with the utilization of carbohydrate and fat or the nonprotein utilization. Nonprotein $\dot{V}_{CO_2}/\dot{V}_{O_2}$ yields the nonprotein RQ. Tables are available for converting nonprotein RQ into the proportion of carbohydrate and fat oxidized and appropriate caloric equivalents for O_2 and CO_2. Application of this process *in toto* is known as indirect calorimetry and is treated elsewhere in more detail (Consolazio *et al.,* 1963; Kleiber, 1961; Lusk, 1928). Representative physiological equivalents of metabolic exchange are provided in Table IV.

3.4. Use of Respiratory Quotient (RQ)

Use of the respiratory quotient on $\dot{V}_{CO_2}/\dot{V}_{O_2}$ based exclusively on respiratory exchange of gases may be influenced by factors other than those mentioned. For example, hyper- or hypoventilation or alterations in CO_2 transport in blood, such as occurs in metabolic acidosis or alkalosis, can influence the measured RQ. Short-term ventilatory changes should really be referred to as the respiratory exchange ratio (*R*) and not as RQ, because true metabolic effects are not measured (Farhi and Rahn, 1955). In addition, interconversions of foodstuffs and oxidation of foodstuffs in specialized tissues such as brain have direct effects on RQ. Nevertheless, if experimental conditions are carefully controlled and the measurement time is long enough, the RQ can provide useful metabolic information.

3.5. Interrelationships

A scheme for the examination of the relationships among metabolic fuel mixtures and the attendant heat equivalents adapted from the schematic presented by MacHattie (1960) is presented in Fig. 1. The substrate fraction is that portion contributed by one average fuel to the total metabolic energy. C, P, and F, respectively, represent carbohydrate, protein, and fat. A negative value indicates synthesis. Unsteady states, such as body cooling or heat production from anerobic processes, cause the apparent position of the triangle CPF to shift to the left. Detailed plotting of metabolic events is presented and discussed by MacHattie (1960).

Daily energy turnover expressed in kcal·day^{-1} or in relation to daily oxygen consumption as determined for various diets is given in Fig. 2. At equal caloric value, a high-fat diet requires utilization of more oxygen than does a carbohydrate or mixed diet.

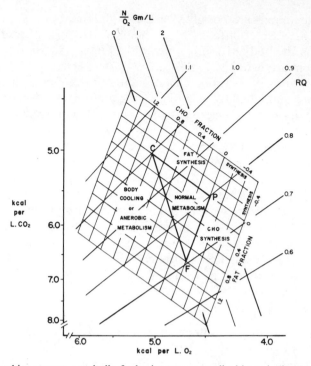

Fig. 1. Relationships among metabolic fuel mixtures as applicable to indirect calorimetry. From MacHattie (1960). Reproduced by permission of the author and the *Journal of Applied Physiology*.

4. Assessing Energy Balance

4.1. Methods of Assessment

In the study of energy balance, a variety of procedures can be used, depending on the precision of measurements required. Since energy intake must equal energy expenditure plus losses and storage, as shown in Table V,

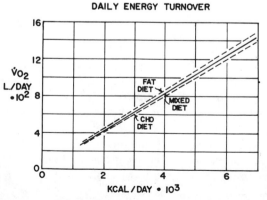

Fig. 2. Energy turnover per day. The relationship between oxygen consumption and caloric utilization when consuming different diets. Adapted from Webb (1964).

Table V. Caloric or Energy Balance

Intake	= Expenditure $\pm E$
Food intake	= Heat Work $\pm$ Tissue storage or
	depletion
	Lipid
	Protein
	Carbohydrate
	Other substrate
	Heat storage or
	loss
	Other losses

either side of the equation can be evaluated to obtain an estimate of energy requirements. The various general methods for measuring caloric intake and expenditure are given in Table VI. Classification of the methods of assessment of caloric intake can be broken down further (FAO, 1964). An attempt at finer classification of useful methodology for measuring caloric intake is provided in Table VII.

Similarly, the various methods for measuring energy expenditure are presented in Table VIII. A number of procedures are available which vary considerably in accuracy and ease of application. A ranking of several of the procedures for measuring energy expenditure, with respect to accuracy and practicality, is provided in Table IX. Presentation of a variety of methods for the indirect assessment of energy expenditure appeared in a special issue of the *American Journal of Clinical Nutrition* (1971).

Table VI. Methods of Measurement of Energy
Intake and Expenditure

Energy intake
  Dietary survey
  Survey with weighed portions
  Chemical analysis of food samples
  Liquid formulas and powdered meals
  Bomb calorimetry of replicate meals or aliquots
Energy expenditure
  Activity diary
  Activity survey
  Indirect heat loss
  Direct heat loss
  Heat production, indirect
  Activity related, i.e., heart rate, pulmonary ventilation,
    etc.
  Body-fat and lean-tissue loss

Table VII. Methods of Assessment of
Caloric Intake

Dietary survey
  Self-administered recall questionnaires
  Self-prepared food-intake record
  Recall questionnaire with interview
  Observation during meal eating
  Group logistics and ration analysis
Survey with weighed portions[a]
  Individual data
  Group data
    Procurement
    Distribution
    Waste
Chemical analysis of weighed samples
  Individual meals
  Daily composites
  Multiday composites
  Analysis of food purchases and food inventory
Liquid formulas and prepared meals
  Diets from recipes
  Commercial preparations
    Liquid
    Powder
    Meals such as TV dinners
Bomb calorimetry
  Sample foods
  Composite meals
  Multiday composites

[a] Recipe method of food inventory; records of food preparation and consumption.

4.2. Energy Losses

The energy losses from the body are subtle and many. Rarely are all energy losses measured in conventional studies of energy balance. The usual losses considered are those in feces and urine. The following energy losses are usually disregarded: sweat, male and female discharges, tears, nail clippings, epidermal sloughing, hair loss, sputum, flatus, and blood loss. These energy losses may, under certain circumstances, be appreciable but are usually considered of minor significance. Thus, in comparative studies of energy balance in which both intake and expenditure are controlled and measured, energy expenditure (as conventionally measured) should tend to be less than energy intake because of the disregarded energy losses.

5. Caloric Allowances

The Food and Nutrition Board of the National Academy of Sciences/ National Research Council in the United States has developed formulations

Table VIII. Methods for Measurement of Caloric or Energy Expenditure and Heat Production

Activity diary
  Recall at intervals
  Personal time–motion study
  15 min check sheet
Activity survey
  Self-administered recall questionnaire
  Questionnaire with interview
  Direct-observation time–motion study
  Photographic time–motion study
  Job classification
Activity related
  Pedometers
  Cumulative heartbeats
  Cumulative joint rotation
  Cumulative extremity acceleration
  Ventilation ($\dot{V}$)
Indirect heat loss
  Body temperature changes
    Radiometers
    Thermocouples
    Thermistors
  Environmental heat changes
    Measurement of environmental change
    Measurement of input to control change
Direct heat loss
  Calorimeters: Partial or total body
    Radiation
    Convection and conduction
    Evaporation

Heat production, indirect
  Oxygen consumption ($\dot{V}_{O_2}$) and/or carbon
    dioxide production ($\dot{V}_{CO_2}$)
$\dot{V}$ and F_{O_2}
$\dot{V}_{O_2}$, $\dot{V}_{CO_2}$, and nitrogen excretion
^{18}O turnover ($^{2}H_2{}^{18}O$ or $^{3}H_2{}^{18}O$)
Insensible water loss
Body fat loss (change in)
  Relative body weight
  Anthropometric measurements
  Somatotype
  Subcutaneous fatness
    Skinfolds
    Soft-tissue X ray
    Ultrasonic pattern
    Tissue impedance
  Fat-soluble gas dilution
  Densitometry, air and water
  Total-body water and other fluids
Lean-tissue loss (change in)
  Representative biopsies
  Nitrogen excretion
  Creatinine excretion
  ^{40}K and exchangeable ^{42}K
  Potassium balance

Table IX. Daily Measurement of Energy Expenditure Using Methods for Activity Assessment Ranking[a]

Activity	I	II	Energy expenditure	I	II
Free diary	5	2	Dietary assessment	5	4
Questionnaires	4	1	Heart-rate accumulation	4	2
Prepared diary	3	3	Joint rotation	3	3
Time–motion, observer	2	5	$\dot{V}$ monitor	2	1
Time–motion, movie	1	4	Continuous $\dot{V}_{O_2}$, $\dot{V}_{CO_2}$	1	5

[a] I, accuracy of measurement; superior, 1; inferior, 5; II, practicality of measurement; relatively easy, 1; difficult, 5.

for Recommended Dietary Allowances since 1940 (FNB, 1964). The published allowances have been based on the best evidence available at the time and the considered judgment of a committee of nutritional scientists. The published allowances are only intended to serve as guides in planning food supplies and for the interpretation of food intake records of groups of people. The last edition of the Recommended Dietary Allowances was published in 1974 (FNB, 1974). The following information is based largely on the authors interpretation of the proposed revised section on calories.

Food energy is required by the body to support various processes and activities, including resting metabolism, growth, tissue maintenance and restoration, pregnancy, lactation, and physical activity. In the formulation of the United States caloric allowances, the concept of the "reference" man and woman has been utilized. These reference individuals live in a thermally neutral environment and are employed in an office or work in light industry. They engage in only modest amounts of recreational activity. Allowances for others who live, work, and play under different conditions can easily be compared to the reference caloric allowances. The objective of the recommended allowances is to provide energy in amounts sufficient to support the processes and activities mentioned as well as to promote and maintain physical and mental well-being and health. A summary of the factors affecting caloric requirements is presented in Table X.

5.1. Reference Individuals

The reference average man is 23 years old, weighs 70 kg, and is 172 cm tall; the reference average woman is also 23 years old but weighs 58 kg and is 162 cm tall. These weights are regarded as the weights for mature individuals. Thermal neutrality of their microenvironment is maintained by controlling heat and air-conditioning systems and by adjusting the amount and kind of clothing worn. The reference individual's job involves light activity in an office, industrial, or laboratory setting and he/she engages in recreational activities no more than 4 hr·wk^{-1} at a caloric level not exceeding 300 kcal·hr^{-1}. Thus, the average

Table X. Factors Modifying Caloric Requirement or Level of Energy Balance[a]

Physical activity
Body size
Growth and development
Aging
Sex
Climate or environment
Disability
Pregnancy and lactation

[a] Not all factors listed are independent.

daily caloric allowance for the reference average man is 2740 kcal·day^{-1} and for the reference woman 2030 kcal·day^{-1} (FNB, 1974).

5.2. Adjustments in Caloric Allowance

Appropriate caloric adjustments must be made for groups of individuals and populations who differ from the reference average man and woman with respect to body size, age, and physical activity. In addition, caloric allowance adjustments must be made for bed rest, pregnancy, and lactation. Procedures for making these adjustments are discussed in turn.

5.2.1. Adjustment for Body Size

Larger caloric allowances must be provided for those of greater body size and smaller allowances for those of lesser body size than the reference man or woman. Fat stored in obesity tissue should not be considered when making caloric allowance adjustments for body size, but only the fat-free body mass together with an average body fatness. Thus, an objective assessment of body fatness would be a valuable procedure to incorporate for more precise estimates of caloric allowances. An average body fatness has been arbitrarily taken as 14% for men and 20% for women.

In men or women who are overweight because of fatness, the increased metabolic cost of transporting obesity tissue may be compensated for by reduced physical activity (see Section 9). Physical growth is essentially complete in men by age 21 and in women by age 17, although body weight, on the average, increases until about age 60 (National Center for Health Statistics, 1965). The latter weight gain serves no useful purpose, and considerable evidence suggests that it is mostly body fat and therefore undesirable. Appropriate caloric adjustments for body size appear in Table XI. Data for the reference average man and woman are given in boldface, and the values in Table XI differ slightly from the recommended allowances published in 1974 (FNB, 1974) but essentially match those published in 1968 (FNB, 1968). Interpolation between body weights can be performed if necessary.

5.2.2. Adjustment for Age in Adults

Caloric adjustments for age in adults are also provided in Table XI. The adjustments at a specific age are given, and interpolation to intermediate ages can be accomplished if desired. The allowance for age 22 really covers the age range 18–25. Following age 25, or the postcollege years, energy requirements decline progressively. This decline is associated with both a decrease in resting metabolism as well as reduced physical activity. The factors for the percentage adjustment with age are also provided in Table XI. The recommended kcal allowance published in 1974 for ages greater than 50 years has been reduced to 2400 for men and 1800 for women (FNB, 1974). The value for men at age 55 is somewhat less than that in Table XI, but the value for women is the same.

Table XI. Adjustment of Calorie (kcal) Allowances[a] for Adult Individuals of Various Body Weights and Ages[b,c]

Body weight		RMR[d] at age 22	Age (yr)[e]						85 and over
kg	lb		22	35	45	55	65	75	
Men									
50	110	1540	2200	2100	2050	2000	1950	1900	1850
55	121	1617	2350	2250	2200	2150	2100	2050	2000
60	132	1716	2500	2400	2350	2300	2250	2200	2150
65	143	1815	2650	2500	2450	2400	2350	2300	2250
70	**154**	**1881**	**2800**	**2650**	**2600**	**2550**	**2500**	**2450**	**2400**
75	165	1969	2950	2800	2750	2700	2650	2550	2500
80	176	2016	3100	2950	2900	2800	2750	2700	2650
85	187	2112	3250	3100	3000	2950	2900	2850	2750
90	198	2211	3400	3250	3150	3100	3050	2950	2900
95	209	2288	3550	3400	3300	3250	3150	3100	3000
100	220	2376	3700	3500	3450	3350	3300	3200	3150
Women									
40	88	1276	1550	1500	1450	1400	1400	1350	1300
45	99	1375	1700	1600	1600	1550	1500	1500	1450
50	110	1463	1800	1700	1650	1650	1600	1550	1550
55	121	1562	1950	1850	1800	1750	1750	1700	1650
58	**128**	**1617**	**2000**	**1900**	**1850**	**1800**	**1800**	**1750**	**1700**
60	132	1639	2050	1950	1900	1850	1800	1800	1750
65	143	1738	2200	2100	2050	2000	1950	1900	1850
70	154	1826	2300	2200	2150	2100	2050	2000	1950

[a] kcal allowance (men) = (RMR + 13W) × (% adjustment for age); kcal allowance (women) = (RMR + 7W) × (% adjustment for age). W = weight in kilograms. Values rounded to nearest 50 kcal.

[b] At a mean environmental temperature of 20–25°C (68–77°F), assuming appropriate clothing and light physical activity.

[c] Basis for preparation of table: The formulation described for estimating caloric allowance in this table is based on the assumption that a resting metabolic rate (RMR) can be validly estimated. This RMR is equivalent to the basal metabolic rate (BMR) plus 10% and corresponds to the BMR plus an activity increment that is only partially weight dependent. The activity increment in metabolic rate is weight dependent (for fixed tasks where speed is constant), and a coefficient of 13 is empirically proposed for men and 5 for women. Thus, the concept is that there is a minimal daily metabolic rate associated with living that is only partially weight dependent (proportional to lean body mass, active tissue mass, or surface area—body weight 3/4—depending on one's choice of reference) plus an increment in metabolic rate associated with activity that is weight dependent because it involves body movement and the performance of work in the upright position. The adjustment for age applies both to RMR and to the total daily metabolic rate because there is a downward trend in BMR with age and a trend for older individuals to be less active. It is noteworthy that the 1974 recommended allowances (FNB, 1974) assume no decrement in BMR or RMR over the age range 23–50 yr.

[d] RMR, resting metabolic rate; approximately equivalent to the metabolic rate measured under basal conditions plus 10%.

[e] Age adjustments: Age—22, 35, 45, 55, 65, 75, 85 and over; % adjustment—100, 95, 93, 91, 89, 87, 85. In general, the age adjustment amounts to approximately 50 kcal·decade^{-1} after age 35, except for the heavier individuals.

5.2.3. Adjustment for Age in Children

A schematic presentation of the caloric allowances (CA) for groups of boys and girls is provided in Fig. 3. The resting metabolic rate (RMR) is

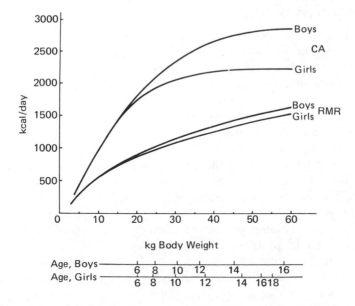

Fig. 3. Suggested average caloric allowances (CA) for children in the United States. Constructed from data in the following references: Passmore and Durnin (1955), Hathaway (1957), and Macy (1942).

provided for reference. The smoothed curves were fitted from data reported by Hathaway (1957), Passmore and Durnin (1955), and Macy (1942). One can estimate the average allowance from either age in years or body weight. Average body fatness was assumed. It should be recognized, however, that more appropriate allowances for individual children can be made through observation of growth pattern and development of body fatness. Children exhibit marked variations in energy intake and expenditure, and the average values provided do not adequately cope with the range in physical activity that children exhibit. It is recognized that certain inactive children could develop obesity on the caloric allowances provided. A separation in the caloric allowance curves for boys and girls is evident at about age 6. Major differences in growth rate begin to appear by age 6 and are more apparent by age 9 or 10. It has been estimated that the caloric allowance should provide about 100–120 kcal·kg^{-1} at birth and 90–100 kcal·kg^{-1} by the end of the first year. Thereafter, the caloric allowance stays between 80 to 100 kcal·kg^{-1} up to 10 years.

The asymptote for the caloric allowance data for mature boys and girls slightly exceeds the caloric allowance for adults of equivalent weight. Perhaps this increment is not unreasonable if one assumes that mature boys and girls tend to be more physically active than adults. The 1974 recommended allowances (FNB, 1974) give values for boys and girls for three different age groups: 11–14, 15–18, and 19–22 years. These values average, respectively, for boys and girls 63.6 and 54.5 kcal·kg^{-1} for the younger group to 44.7 to 36.2 kcal·kg^{-1} for the oldest groups.

5.2.4. Adjustment for Physical Activity

Physical activity is the major variable affecting caloric expenditure and intake. Normally, the responsiveness of the appetite mechanism is sufficiently precise to compensate for changes in daily physical activity, so that body weight and composition remain relatively constant. An important exception is the inactive person who may consume more than the required calories to remain in energy balance and who therefore becomes obese (Jacobs *et al.*, 1963). Arbitrary examples of energy expenditure by the reference average man and woman are provided in Table XII. Persons engaged in heavier activity must have higher allowances, although this allowance need seldom exceed 1500 extra kcal. A plot of caloric surveys of expenditure and/or intake is presented in Fig. 4 to provide a frame of reference for the variety of caloric requirements that exist in our culture. In practice, the caloric requirements of a moderately active person might be increased by about 300 kcal, but for very active persons it could go to 600–900 kcal·day^{-1}.

An interesting scheme for evaluating the caloric requirement of adults who work in various occupations appears in Fig. 5. This scheme was prepared by DuBois (1960). The figure caption explains how the data in the figure can be used.

It is common in the United States, particularly in urban areas, to encounter people who are even more sedentary than the reference average man or woman. Caloric allowances should be adjusted downward for these individuals.

The problem of assessing how much work people do is important. Competent evaluations of civilian working groups are rare. Economic, social, and

Table XII. Arbitrary Examples of Energy Expenditure by Reference Man and Woman

Activity	Time (hr)	Man Rate (kcal·min^{-1})	Total	Woman Rate (kcal·min^{-1})	Total
Sleeping[a] and lying	8	1.1	528	1.0	480
Sitting[b]	7	1.5	630	1.1	462
Standing[c]	5	2.5	750	1.5	450
Walking[d]	2	3.0	360	2.5	300
Other[e]	2	4.5	540	3.0	360
			2808		2052

[a] Essentially BMR plus some allowance for turning over or getting up or down.
[b] Includes normal activity carried on while sitting, e.g., reading, driving automobile, eating, playing cards, and desk or bench work.
[c] Includes normal indoor activities such as standing and occasional walking in a limited area, e.g., personal toilet and moving from one room to another.
[d] Includes purposeful walking, largely outdoors, e.g., from home to commuting station and to work site and shopping.
[e] Includes occasional activities such as limited stair climbing, occupational activities involving light physical work, and recreational exercise. This category may also include weekend swimming, golf, tennis, or picnics.

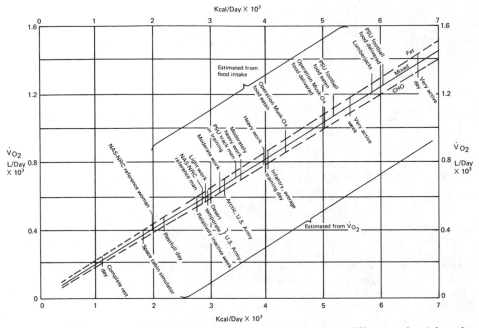

Fig. 4. Daily energy turnover in various groups of men who performed different tasks. Adapted from *Bioastronautics Data Book,* Webb (1964).

psychological factors plus recreational facilities and climate affect energy expenditure patterns, but we do not know precisely how. Neither do we understand in a comprehensive way who does how much or how or why they do it (Bullen *et al.,* 1964; Buskirk, 1960; Durnin and Passmore, 1967; Edholm, 1966; Passmore and Durnin, 1955).

5.2.5. Adjustment for Climate

The ambient temperature range to which the majority of people in the United States are exposed probably falls within 20–25°C (68–77°F). Most people are protected against environmental extremes by central heating and air-conditioning systems, variable amounts of protective clothing, and even-conditioned vehicles for transportation. While not everyone can avoid environmental extremes, few groups would require an adjustment in caloric allowance because of climatic exposure. On the basis of current evidence (Burton and Edholm, 1955; Buskirk and Boyer, 1965; Carlson, 1954; FNB, 1974), it is recommended that no more than a 2–5% increase in caloric allowance need be provided for those living in very cold environments. The increase in caloric demands for those living in cold areas, compared to those living in more thermally neutral regions, is associated with the hobbling effect of bulky clothing, the necessity to wear heavy boots on the feet where load carriage is inefficient, periodic chilling and shivering, and uncertain or difficult footing in

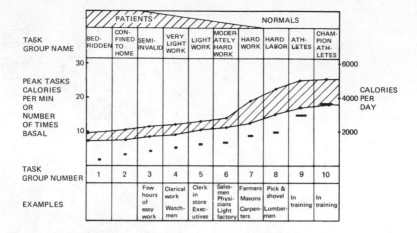

Fig. 5. Classification of physical effort and caloric requirements. At the top of the figure, there is a box giving the division of experimental subjects into patients and normals. The line between these is made diagonal to show that there is a overlapping of patients and normals when it comes to light work and even hard work, since some who are classified as patients can do harder work. The second box at the top of the chart shows ten grades of activity, ranging from bedridden patients to athletes, and at the bottom of the chart, the various occupations are graded arbitrarily in the same manner. In the middle of the graph, there are two curves representing roughly the estimated 24-hr caloric expenditure for the people in the different groups. This estimate of caloric expenditure has been made from the survey of the food consumption of people in different occupations, a method that is not very accurate but the best available at the present time. Below these two curves comes a series of bars representing roughly the peak tasks of the different occupations in terms of calories per minute or number of times the basal. These could be represented in terms of oxygen consumption per minute, but the calories are used because they are better understood by most people. The height of the bar above the baseline represents the calories of the peak tasks, and the length of the bar is intended to represent roughly the duration of the task. Much more information is needed regarding the peak tasks in various American occupations, and there are great differences in various institutions and individuals. From DuBois (1960).

traversing ice or snow (Buskirk *et al.*, 1963b; Welch *et al.*, 1958). In very hot environments, the energy requirements are slightly increased in men performing physical work (Buskirk and Boyer, 1965). This extra energy expenditure for standardized work is probably associated with maintenance of high blood flow through skin, the energy for sweat elaboration and for extra movements associated with mopping one's brow, etc., plus a temperature-induced Q-10 effect. However, the desire to exercise in the heat is less, and the extra energy required to perform work may be compensated for by extended rest. If an adjustment in caloric allowance is made for hot climates, it has been estimated that a $0.5\%\cdot°C^{-1}$ increase is a sufficient adjustment for men or women working in environments above 30°C. Thus, the temperature of the environment alone has little effect on caloric allowance, because people can easily secure adequate environmental protection and use volitional means to insure thermal balance. It is fortunate that a major correction for climate is unnecessary, for climate is difficult to describe precisely. An adequate description involves consideration of temperature, air movement, humidity, and effective radiation.

Figure 6 grossly summarizes much of the literature with respect to the effects of climate and physical activity (Buskirk and Boyer, 1965). It is readily apparent that physical activity is the dominant factor of the two.

5.2.6. Adjustment for Bed Rest

With bed rest, an appropriate reduction in caloric allowance must be made. It is suggested that the caloric allowance be reduced to the level of the RMR. Febrile reactions, burns and sepsis, and various disabilities may increase caloric needs over the RMR (Kinney, 1959).

5.2.7. Adjustment for Pregnancy and Lactation

Pregnancy increases the caloric requirement because of an increased RMR during the second and third trimesters, increased work done by the mother in moving an increased body weight (although the mother may voluntarily decrease her physical activity during the third trimester), and the increased energy required for growth and development of the placenta and fetus. The 1974 Recommended Dietary Allowances (FNB, 1974) suggest that

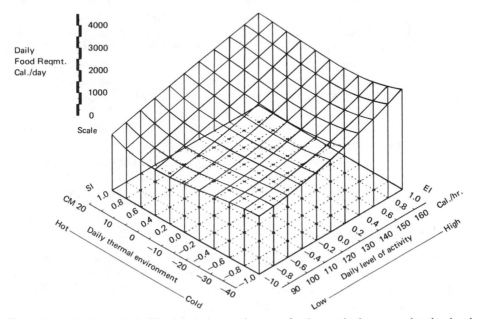

Fig. 6. Isometric illustration of food (caloric) requirements for the standard man as related to level of activity and thermal environment. From Buskirk and Boyer (1965).

there is a net caloric cost to support the pregnancy, over and above former maintenance requirements, of approximately 80,000 kcal for 40 wk, or about 300 kcal·day^{-1}. A value of 80,000 kcal appears excessive to us, and, indeed, Emerson *et al.* (1972) found that the total cumulative extra energy cost solely attributable to pregnancy was 27,120 ± 2175 kcal. This value was quite constant regardless of dietary intake or body build.

The caloric requirements for lactation and breast-feeding are proportional to the quantity of milk produced. An appropriate caloric allowance is 120 kcal for each 100 ml of milk. If the mother produces 850 ml of milk per day, then approximately 1000 kcal more per day is required by the mother. Assessment of the mother's body weight and composition change during the breast-feeding period is a helpful guide as to the adequacy of the caloric supplementation.

5.2.8. Special Adjustments

There are many conditions for which special adjustments of caloric allowances are necessary (Orr and Leitch, 1938; FAO, 1947). For the most part, appetite and satiety provide sufficient control of caloric intake. Nevertheless, regular assessment of body weight and body composition (probably skinfolds) helps the physician and nutritionist to detect caloric imbalance and to prepare their recommendations.

Resting metabolism may be increased as much as 50% in hyperthyroidism and reduced 50% in hypothyroidism (D. DuBois and DuBois, 1916; E. F. DuBois, 1948, 1954). Hyperthyroid patients may fail to eat sufficient calories, and the hypothyroid patient may take in an excess. The associated weight changes are readily apparent. When the normal hunger, appetite, and satiety mechanisms are inaccurate, the patient requires assistance for control of his/ her disease, including appropriate guidance concerning caloric intake.

As a result of injury or disease to bone, nerve, or muscle, normal movements and movement patterns may be altered. Studies of walking and other means of body progression in disabled patients indicate an increased energy expenditure for these movements (Durnin and Passmore, 1967). Whether an increased caloric requirement is necessary depends on the work–rest patterns of activity throughout the day. Again, careful assessment of body weight and body composition is the best way to determine appropriate caloric allowance. Hemiplegic and paraplegic subjects are special cases of disability. Each must be evaluated carefully if problems arise with respect to their caloric balance.

5.3. Comparative Allowances

Comparative caloric standards or allowances for the various countries that have prepared them are given in Table XIII. It can be seen that there is rather wide variation in the standards. No common methods for preparing the standard tables have been adopted. The standards are based primarily on recommendations of committees of local nutritional scientists, who have presumably scaled their standards to their local population and environment.

Table XIII. Comparative Caloric Standards for Adults in Selected Countries and as Recommended by FAO (1958) and FNB (1974)

Country	Sex	Age (yr)	Weight (kg)	kcal·day^{-1}	Physical activity and Environment
United States (FNB)	M	23–50	70	2740	Persons normally active in a temperate climate
	F	23–50	58	2030	
FAO	M	25	65	3200	Men—employed 8 hr/day in an occupation requiring no more than occasional hard physical labor. Off work he may walk 1½ hr and spends 1½ hr on recreation or household work.
	F	25	55	2300	Women—engaged in general household duties or light industry. Spends 1 hr walking and 1 hr of recreation on nonstrenuous sport. Mean annual temperature, 10°C.
Australia	M	25	65	2700	Similar to FAO
	F	25	55	2300	Mean annual temperature, 18°C
Canada	M	25	72	2850	Men—office, laboratory, shop and mill work, technical trades or crafts
	F	25	57	2400	Women—most household chores
Central America and Panama	M	25	55	2700	Moderate work. Mean annual temperature, 20°C
	F	25	50	2000	
India	M	25.4	55	2800	Light industrial occupation
	F	21.5	45	2300	
Japan	M	Not specified	56	3000	Moderate work
	F		48.5	2400	
The Netherlands	M	20–29	70	3000	Light work, average requirements
	F	20–29	60	2400	
Norway	M	25	70	3400	Not specified
	F	25	60	2500	
The Philippines	M	Not specified	53	2600	Moderate work
	F		45	2300	
South Africa	M	Not specified	73	3000	Moderate work
	F		60	2300	
United Kingdom	M	20 and up	65	3000	Medium work of 8 hr at 100 kcal·hr^{-1} and 1 hr traveling at 130 kcal·hr^{-1}
	F	20 and up	56	2500	

5.4. Classification of Physical Effort

The diversity of tasks performed by men and women is enormous, yet we are faced with the problem of classifying physical effort in order to provide useful guidance to those responsible for health and welfare throughout the world. Many tasks are not easily classified because intricate movements difficult to analyze, are involved. In many cases, a human engineering study must be made of the human–machine–task interrelationships before further progress can be made. On the other hand, many tasks can be grouped according to the way the task affects common physiological variables such as oxygen consumption and heart rate.

Numerous arbitrary schemes have been devised for the classification of physical effort. For the most part, each scheme has only served the immediate needs of those responsible for it. A "classification" vocabulary acceptable to most potential users has not been adopted. To various investigators, hard work may mean a level of work varying between 2 and 15 kcal·min^{-1}. Nor is it likely that an acceptable set of words and definitions will be adopted unless the lead of the respiratory physiologists is followed: they defined lung volumes and pulmonary function tests and standardized their commonly used terminology. The classifications and definitions should be simple, convenient to use, and acceptable to physiologists, nutritionists, clinicians, and industrialists alike.

At least two useful attempts to grade physical work have appeared in the literature. Christensen's classification (1953) is based on the effort expended in various industrial jobs in the Swedish steel industry, which could easily be applied to professional, recreational, and athletic activities as well. The physiological variables in Christensen's classification are metabolic rate and heart rate. Wells *et al.* (1957), in an extension of the classification presented by Christensen, added criteria for the assessment of work level: ventilation rate and volume, RQ, and blood lactic acid content. Wells *et al.* changed the terminology used by Christensen, dropped the 2.5 kcal·min^{-1} level, and added a 15 kcal·min^{-1} level.

Because of the low energy expenditure in many occupations, it is our feeling that an intermediate classification between 1 and 5 kcal·min^{-1} is necessary. Ventilation volume should prove to be a useful addition to an adopted scheme because of the possibility of telemetering ventilation volume as well as heart rate (Wolff, 1956). There is a direct relationship between heart rate, ventilation volume, and metabolic rate under a wide variety of work conditions; therefore, these simple measurements should prove valuable in assessing "on-the-job" energy expenditure. It is felt that the measurement of blood lactic acid content would not add appreciably to a classification scheme because of the difficulties of drawing blood at the proper times for evaluation of the anaerobic component for a given job. Then, too, the anaerobic component for any given job is likely to be negligible, because people only infrequently work that hard.

On the basis of the ideas presented by Christensen and by Wells *et al.*, review of the literature, and firsthand experience, the combined scheme presented as Table XIV may satisfy many requirements.

Table XIV. Classification of Physical Effort[a,b,c]

Classification	$\dot{V}_E$	$\dot{V}_{O_2}$	MR	HR
Very light	<10	<0.5	<2.5	<80
Light	10–20	0.5–1.0	2.5–5.0	80–100
Moderate	20–35	1.0–1.5	5.0–7.5	100–120
Heavy	35–50	1.5–2.0	7.5–10.0	120–140
Very heavy	50–65	2.0–2.5	10.0–12.5	140–160
Unduly heavy	60–85	2.5–3.0	12.5–15.0	160–180
Exhausting	≥85	≥3.0	≥15.0	≥180

[a] Buskirk (1960).
[b] Abbreviations: $\dot{V}_E$, ventilation volume (liter·min^{-1}); $\dot{V}_{O_2}$, oxygen consumption (liter·min^{-1}); MR, caloric expenditure (kcal·min^{-1}); HR, heart rate (beats·min^{-1}).
[c] The values listed apply to steady-state work and also to peak effort.

The physiological variables related to metabolic rate that are easily measured, ventilation volume and heart rate, are included in Table XIV. Since one purpose of such a table would be the estimation of the energy expenditure for a given task, it follows that the measurement of choice in this instance would be oxygen consumption. Measurement of oxygen consumption should take into account the oxygen debt, unless the work can be done in a "steady-state" and measurements are made during the "steady-state" period.

It should be emphasized that under certain conditions, one or more of the variables listed in Table XIV may fail to yield a valid estimate of the severity of the work, but rather emphasize an emotional component, such as fear, or fail to emphasize a "fitness" component, such as nutritional status during semistarvation. If heart rate is the variable in question, fear or apprehension may yield abnormally high working heart rates, whereas semistarvation produces abnormally low heart rates.

The deficiency in Table XIV is that body mass or size and physical fitness have not been taken into account (Buskirk and Taylor, 1957). Perhaps the simplest method to correct for body size would be to express oxygen consumption and metabolic rate as the number of times RMR, that is, two times, four times, etc. This type of expression has appeared in the literature on temperature regulation, and the term "Met" for 1.1 times the BMR has been adopted. This approach assumes, however, that one either knows or is willing to estimate BMR from surface area. For the time being, it is felt that a correction for body size, while desirable, might only hinder wide adoption of a simple scheme for the classification of physical effort. Physical fitness for work is more difficult to cope with. In general, classification problems associated with age, disease, etc., could be considered as those basically arising because of different levels of physical fitness. Perhaps the easiest way to express the values in a classification table is in relative terms, that is, relative to the maximal aerobic capacity. Tests are available for assessing maximal aerobic capacity; however, one would have to either measure maximal aerobic capacity or estimate it from submaximal values. Again, this may be an unnecessary complication and would thwart wide acceptance of a simple classification of physical effort.

6. Metabolic Size

The metabolic rate of animals of various sizes, including humans, although highly correlated with body weight, is more closely related to a fractional power of the body weight and is often predicted on the basis of body surface area. One prediction of surface area that has gained wide acceptance for humans is the DuBois equation (DuBois and DuBois, 1916):

$$m^2 = 0.007184 + L^{0.725} + W^{0.425}$$

where m^2 = square meters of body surface; L = body length (height) in cm; and, W = body weight in kg.

Body heat is lost through the body surface, which probably explains the relationship to heat production when the body is at rest and under the carefully controlled conditions specified for determination of basal metabolism. It has been shown that heat production per unit surface area is not a constant for humans, but varies with age, sex, maturation, etc.

In addition, a man's radiating body surface and his surface for convective heat loss is altered by posture, and clothing alters his insulation. The problems inherent in application of the surface law were avoided by Harris and Benedict (1919) when they developed their empirical multiple-regression equations for predicting BMR. They were for men and women, respectively:

$$BMR = 66.4730 + 13.751W + 5.0033L - 6.7550A$$

$$BMR = 655.0955 + 9.463W + 1.8496L - 4.6756A$$

where W = body weight in kg; L = body length (height) in cm; A = age in years; and BMR = kcal·24 hr^{-1}. These empirical equations are useful, but nevertheless leave something to be desired in terms of physiological significance. Brody (1945) and Kleiber (1947, 1965) have been the principal supporters of use of the 0.75 power of the body weight as an index of a metabolically active mass. For interspecies comparison, the following equation has been proposed:

$$BMR = 70W^{0.75}$$

where BMR is expressed in kcal·24^{-1} and W = body weight in kg. Kleiber (1975) has indicated that weight to the 0.75 power can be used as well for physically active animals, including humans. The Canadians (Canadian Bulletin on Nutrition, 1964) have based their caloric allowance tables on this concept of metabolic body size. While this concept is useful, assuming equal densities in large and small individuals, it is well known that body fat is added with age and that there are rather wide interindividual differences in body composition at any age. As lipid is stored in adipose tissue cells, it has not been demonstrated that adipose cellular material is added, and it is unknown to what extent muscle is developed to support the stored lipid. Thus, when body lipid is stored or utilized, there is not necessarily a change in metabolically active tissue, yet body weight to the 0.75 power would change, but no change in BMR need occur.

Kleiber (1947, 1965) has postulated that a combined theoretical and empirical approach can be employed within a species such as man and summarized the data of Harris and Benedict (1919) for men and women, respectively, as follows:

$$\text{BMR} = 71.2\,W^{0.75}\left[1 + 0.004(30 - A) + 0.010\left(\frac{L}{W^{0.33}} - 43.4\right)\right]$$

$$\text{BMR} = 65.8\,W^{0.75}\left[1 + 0.004(30 - A) + 0.018\left(\frac{L}{W^{0.33}} - 42.1\right)\right]$$

where BMR is expressed in kcal·24 hr^{-1}; W = body weight in kg; L = body length (height) in cm; and A = age in years.

In recent years, an effort has been made to actually measure the metabolically active mass. This has been done in several ways using the procedures for determining body composition. For example ^{40}K, a naturally occurring isotope, is primarily located within cells and randomly mixed with other potassium-containing compounds. The presence of ^{40}K, a gamma-emitting isotope, can be detected in a whole-body counter. A determination of the quantity of ^{40}K in the body should, therefore, be proportional to the body cell mass. Unfortunately, the ^{40}K content per unit cell mass varies from tissue to tissue, and those tissues most active metabolically do not necessarily have the highest concentrations of ^{40}K. The relationship of the amount of ^{40}K, as an indication of total-body potassium, to metabolic rate is not one to one. Similarly, fat-free body weight can be used as a reference. Body fatness can be assessed with skinfold procedures or body density measurements, and body water can be determined with deuterium or tritium dilution procedures. "Lean body mass" is the term given to tissues exclusive of stored lipid, and "active tissue mass" excludes stored lipid, bone minerals, and extracellular fluid (Grande, 1961). Miller (1954) has summarized these concepts for conditions involving maximal metabolism. The body composition reference standards suffer from equating all tissues in terms of their metabolism. Each tissue tends to have a different metabolic rate. Yet use of a concept such as the active tissue mass has some validity, particularly when conditions of energy balance are different in members of a single species such as man (Buskirk and Taylor, 1957). Practical considerations have precluded widespread utilization of body composition procedures for metabolic reference purposes.

7. Caloric Content of Foods

Food contains energy in a chemical form bound in molecules of its chief organic components. These energy-yielding organic components are carbohydrates, proteins, and fats.

If food is ignited under high oxygen pressure in a bomb calorimeter, all the organic matter is burned, and heat is released. This is the *heat of combustion* of foods and is a measure of the *gross energy value*.

7.1. Energy Partition of Ingested Foods

The heat of combustion, or gross energy value, of foods does not represent the net energy that is available to the body as fuel energy. Part of the food ingested remains in the gastrointestinal tract and is lost in the feces ("fecal energy"). The "digestible or absorbable energy" is transformed in the body through different metabolic pathways to "metabolizable energy," but part of the absorbable energy is lost in the urine and other excretory processes, as the body does not retain it.

Within the metabolic framework of energy transformation, some energy is lost as heat and is unavailable for metabolic processes. This heat loss is the heat increment of foods, or the "calorigenic effect." The energy trapped, which appears as the savings of body substances, is the "net energy." Figure 7 presents Kleiber's scheme (1961) for the partition of food energy.

7.1.1. Food Energy (Heat of Combustion)

Average heats of combustion determined by direct calorimetry are usually employed in energy calculations of foodstuffs. Although these values serve most purposes, calculations derived from proximate composition of a particular food can differ significantly from the heat of combustion determined by direct calorimetry. The differences arise mainly from the variety of individual organic substances within each broad division of food components. The caloric value of fat depends on the fatty acid composition of glycerides and the associated amount of non-energy-yielding lipid materials. The heats of combustion of individual carbohydrates also differ significantly. In foods derived from their nitrogen components, the heat of combustion depends on the kind of protein and the proportion of nonprotein nitrogen material.

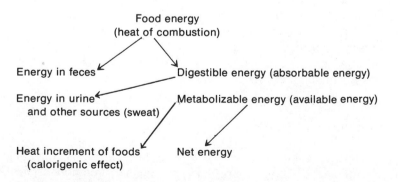

Fig. 7. Scheme for classifying energy in food. From Kleiber (1961).

Occurrence in foods of organic acids such as acetic, citric, lactic, and malic may be an important source of heat in special circumstances. These compounds are usually included with carbohydrates, but their caloric value is lower. Other organic acids, such as tartaric and oxalic, are not utilized by the body, although their heat of combustion is usually included under carbohydrates.

7.1.2. Digestible Energy

During the process of digestion, not all the energy-yielding substances are absorbed, and a portion is excreted in the feces. This unavailable energy must be accounted for in calculating the caloric value of foods. The feces contain, in addition to the undigested food residues, bacteria and their products, residual digestive secretions, and detached epithelium from the lining of the digestive tract. The total heat of combustion of feces includes both metabolic and undigested energy food components. In addition, the digestibility of single food components varies significantly with the makeup of the diet. Complex carbohydrates such as cellulose, pentosans, and other polysaccharides are undigestible and give bulk to the feces.

It has been a customary practice to express digestibility of foods in terms of "apparent digestibility," or a "digestibility coefficient," which takes into account total food intake and feces elimination as follows:

$$\text{Percentage apparent digestibility} = \frac{\text{Intake} - \text{Feces}}{\text{Intake}} \times 100$$

Correction of total energy intake for the digestibility of food gives the digestible or absorbable energy. In calculating digestibility, feces collections are usually made with the aid of markers. One of many undigestible dyes (e.g., carmine) may be used as a marker. A tracer such as chromium oxide or nonabsorbable salts of the radioisotopes ^{51}Cr and ^{46}Sc may also be used.

7.1.3. Metabolizable Energy

Carbohydrates and fats are almost completely oxidized in the body. Carbon dioxide and water are produced in a manner similar to direct combustion in the bomb calorimeter. Urinary losses of organic molecules such as ketone bodies, citrate, lactate, and pyruvate can make a substantial contribution to overall energy loss. The oxidation of proteins in the body is incomplete, and nitrogen components derived from the protein metabolism (i.e., urea) are excreted in the urine. These nitrogen metabolites yield energy that is not utilized by the body. Nitrogen metabolites may also escape oxidation by being excreted in sweat, tears, and other body discharges.

Correction for the energy-yielding components excreted or lost from the body in different ways yields the metabolizable or available energy from food that the body is able to transform into useful work and heat.

7.1.4. Heat Increment of Foods and Net Energy

Heat production increases above the resting level when a food is ingested (Buskirk *et al.,* 1957). This increased heat production has been known for many years as "specific dynamic action," or the "heat increment," of foods. The onset of this heat effect starts within minutes after the food has been eaten, reaching a maximum in about the third hour, and may be maintained at low levels for several hours (Buskirk, 1960; Buskirk *et al.,* 1960). There are marked differences in the heat increment of individual food components. It has been shown that protein exerts the highest heat increment of all food components, which under certain conditions, may reach 30% above the resting metabolic level. Carbohydrate and fat, on the other hand, cause an average rise for the postprandial period of about 6 and 4%, respectively.

The heat increment of foods is extra heat that is not utilized by the body for the production of mechanical work or other forms of energy. It may be considered as waste heat that is only added to other heat produced by the body, but it may be important to thermoregulation. It is interesting that the heat increment of proteins is not usually seen when the organism is in positive nitrogen balance, as when new tissue is being formed.

The heat increment of a mixed diet cannot be accounted for from the sum of the effects of the dietary components. The total heat increment is lower than the sum of the individual values. It has been customary when calculating energy yield from mixed well-balanced diets to utilize an average heat increment of about 6% of the total available energy. Nutritionally deficient diets may produce a much higher average heat increment.

The mechanisms that explain the heat increment of foods remain obscure, although several theories have been advanced. Energy utilized for digestion of foods can only be considered a small part of the heat increment. Included in the energy used for digestion are the energy costs of secretion, increased muscle tone, and absorption. The main contribution apparently comes from intermediate metabolism, including the synthesis of compounds of lower energy content from energy-rich compounds, that is, glucose into glycogen or fatty acids into fat. In addition, the deamination of amino acids is involved. Additional energy is required for ATP formation when protein rather than carbohydrate and fat is used in oxidative processes (Krebs, 1964; Pike and Brown, 1975).

Because the heat increment involves energy that is unavailable for metabolic purposes, it is substracted from the available energy of food in order to arrive at an energy value which represents the energy that the organism utilizes in its metabolic processes. This latter energy is net energy (Kleiber, 1961).

7.2. Energy Values of Food Components

The energy values of food components, as generally used in most calculations, are values representing the metabolizable or available energy. This is obtained by deducting from the heat of combustion of foods the losses in the feces and other excretions. The systems proposed by Rubner (1855, 1901),

Atwater and Bryant (1899), and Atwater and Snell (1903) have been the basis for today's caloric conversion factors.

Rubner's factors of 9.3 and 4.1 kcal·g^{-1} for fat and carbohydrate, respectively, are usually considered as expressing the physiological fuel values of these components in the diet. The value for protein of 4.1 kcal·g^{-1} in Rubner's calculations included nitrogen excretion in feces, which was assumed to be entirely of metabolic origin. Rubner's conversion factor for protein represented the heat of combustion of the protein eaten minus the heat of combustion of the materials excreted in both urine and feces. Rubner made a distinction between protein of animal and vegetable origin and obtained values for both of them. We also calculated the caloric value of protein in a mixed diet, taking into account the makeup of the diet and caloric values for proteins of animal and vegetable origin. The mixed-dietary value for protein was 4.12 kcal·g^{-1}. The heat of combustion of starch of 4.12 kcal·g^{-1} was selected as the best representative estimate for energy-yielding carbohydrate in the diet. The values of heat of combustion for animal fat, butterfat, and olive oil were taken as representative of fat in a mixed diet.

From digestibility studies of humans on mixed diets, Atwater and Bryant (1899) proposed caloric coefficients to use in dietary calculations. Available energy was determined from average values of heat of combustion and apparent digestibility. Atwater and Bryant found values of 4.0 and 8.9 kcal·g^{-1} for carbohydrate and fat, respectively. The average heat of combustion for protein was corrected for apparent digestibility and further corrected for urinary caloric loss. Thus, the value of 4.0 kcal·g^{-1} for protein in a mixed diet was proposed for the calculation of available energy.

When feces and urine are directly analyzed for caloric losses associated with protein, fat, and carbohydrate, these losses can be deducted from the caloric values given corresponding nutrients as eaten. Atwater and Bryant calculated the caloric value of the available nutrients in a mixed diet. The caloric factors for carbohydrate, fat, and protein were 4.15, 9.4, and 4.4 kcal·g^{-1}, respectively.

The general caloric values of 4, 9, and 4 kcal·g^{-1} calculated by Atwater and Bryant for protein, fat, and carbohydrate, respectively, are rounded values and were designed to facilitate the estimation of the metabolizable or available energy in average diets. These values have been critically tested and analyzed and have been found to be quite suitable factors in general dietary and energy calculations (FAO, 1947).

There have been several fine reviews covering the energy values of foods and the bases of their derivation (FAO, 1947; Merrill and Watt, 1955; Morey, 1936). Modified factors for energy-yielding components of individual foods have also been suggested (Merrill and Watt, 1955; Watt and Merrill, 1963), and food composition tables based on these factors have been prepared.

In addition to the nutrients already discussed, alcohol is commonly found in most dietary records, and the energy derived from alcohol must be taken into account in calculating total energy intake. Energy from alcohol is made available via the normal oxidative pathways after initial oxidation to acetaldehyde. There is a caloric loss in urine and expired air of about 5% of the caloric value of ingested alcohol. The caloric value of ethyl alcohol is 7.0 $kcal \cdot g^{-1}$ (5.6 $kcal \cdot ml^{-1}$).

When food caloric values are determined from proximate composition, as contrasted to the caloric values of individual components, the calculated values may differ from the direct heat of combustion determined in the bomb calorimeter. This caloric difference is due, in part, to the difference in analytical methods.

Fat in foods is given as "total fat" obtained from solvent extraction, acid hydrolysis and solvent extraction, or from saponification and solvent extraction. Total fat by solvent extraction includes not only glycerides but also other materials with the same solubility, such as sterols, chlorophyll, and other pigments. In addition, the usual solvents do not completely extract all food fat. In methods involving acid hydrolysis or saponification with additional solvent extraction, it is assumed that all fatty acids are combined as triglyceride. Protein determination by total-nitrogen analysis does not differentiate between protein nitrogen and nonprotein nitrogen. The nitrogen content of many commonly occurring proteins in food is approximately 16%, and the factor often used to convert nitrogen content to protein is 6.25. It has been shown, however, that the nitrogen content of different kinds of proteins may deviate significantly from this value (Jones, 1931). Total carbohydrate in foods is commonly determined as the difference between total food weight and the quantity of water, protein, fat, and ash in the food. This method is called the "carbohydrate-by-difference" method. With this method, it is difficult to differentiate between various kinds of carbohydrates. Starch and mono- and disaccharides are well utilized by the body, but cellulose, pentosans, and other complex carbohydrates are not. Determination of crude fiber allows calculations of "carbohydrate by difference minus fiber," or "nitrogen-free extract." This fraction is a closer estimate of energy-yielding carbohydrate in the body. To avoid the uncertainty in using the carbohydrate-by-difference method, the quantity of total reducing sugars in foods has been measured. The quantity of sugars, starches, and dextrins is measured, but pentoses and hemicelluloses are excluded. There is a possibility, however, that the presence of noncarbohydrate reducing substances leads to falsely high results. In addition, the reducing capacity of different kinds of carbohydrates is not uniform. In the future, other specific methods to determine the quantity of carbohydrate in foods should become available. Van Soest and McQueen (1973) discussed the problems in devising new methods for dietary fiber determination.

8. Methods of Determining Caloric Content of Foods

The potential energy of foods can be estimated by direct or indirect methods. The direct method measures the heat release during combustion and is called direct calorimetry. The indirect methods involve (1) oxygen consumption during combustion (oxycalorimetry), (2) oxidation by strong chemical oxidants, (3) chemical analyses of foodstuffs for energy-yielding components, or (4) estimation from food composition tables and calculation of energy from caloric values of individual food components or from average caloric values of mixed diets.

8.1. Direct Calorimetry

The determination of the energy in food is obtained directly by burning a weighed sample under high oxygen pressure in a calorimeter and measuring the heat evolved during combustion. The most common type of direct calorimeter used in the determination of heats of combustion of foods has been the oxygen-bomb calorimeter.

The bomb calorimeter consists of a small, strong, steel vessel in which a known weight of the dried foodstuff is placed. The steel container is filled with oxygen under pressure. Electrical ignition is employed to suddenly combust the sample. The heat evolved during combustion is usually measured by recording the temperature rise in a weighed amount of water in which the bomb is immersed. Adiabatic conditions can be employed in special adiabatic calorimeters that have a second controlled water bath. The adiabatic calorimeter eliminates the need for radiation and other heat-loss corrections that must be made when using a plain calorimeter. The temperature rise in the calorimeter is usually measured with thermometers, thermistors, or thermocouples which are capable of determining temperature to the nearest 0.001°C. The heat of combustion is calculated from the temperature rise, the weight of the water, the "water equivalent" of the apparatus, and the weight of the sample burned. A correction is made for the moisture content of the sample. The water equivalent of the bomb calorimeter is determined by burning a sample of pure organic substance of known calorie content. Purified benzoic acid, thermochemical grade, has been recommended as a reference substance for this purpose. The heat evolved during combustion of a sample can be also measured as the maximum temperature rise occurring in a ballistic bomb calorimeter. Details on the use of both types of calorimeters can be found elsewhere (Gallenkamp and Co. Ltd., 1961; Parr Instruments Co., 1965).

In a comparison of the accuracy of the conventional and ballistic calorimeters, it was found that the ballistic type has slightly less accuracy, but the samples could be analyzed faster. With the ballistic unit, 12 determinations per hour can be made, as compared to two per hour with a conventional bomb calorimeter (Robinson *et al.*, 1964).

8.2. Indirect Calorimetry (Oxycalorimetry)

The heat production per liter of oxygen consumed during combustion can be estimated from thermal equivalents of the foodstuff. Thermal equivalents represent the heat release in terms of $kcal \cdot liter^{-1}$ of oxygen consumed or of carbon dioxide produced during combustion. The thermal equivalents vary from food component to food component, but the variation in diverse components within single nutrient groups (e.g., carbohydrates) is relatively small. Thermal equivalents derived from carbon dioxide production have greater variation than those derived from oxygen consumption. Therefore, measurement of oxygen consumption during combustion is preferred over measurement of carbon dioxide production if accurate results are desired. The thermal equivalent of oxygen when carbohydrate (mainly starch) is being burned is generally given as 5.04 kcal·liter^{-1} and for fat about 4.68. For protein, an average value of 4.5 kcal·liter^{-1} of oxygen consumed has been used.

The determination of the energy value of foods by oxygen consumption, as given above, has been made in an oxycalorimeter. The sample is ignited electrically in a combustion chamber connected to a gas-spirometer-type apparatus containing oxygen in a closed system. The carbon dioxide produced during combustion is taken up in an appropriate absorbent. The heat of combustion is then calculated from the oxygen consumed and the thermal equivalent of oxygen. In the determination of the calorie content of a mixed diet, an average thermal equivalent figure of 4.8 kcal·liter^{-1} has been generally used. This value has been suggested since the thermal equivalents of single food components only vary from 4.5 to 5.0.

8.3. Oxidation by Chemical Oxidants

Oxidation of foods with chromic acid has been accomplished (Chermnykh *et al.*, 1966; Rozental, 1957), and the obtained caloric values for fats and carbohydrates are similar to the heats of combustion determined in the bomb calorimeter (Rozental, 1957). The oxidation of proteins is incomplete, and the caloric values are closer to the actual combustion in the organism. In this method, the excess dichromate after oxidation is determined iodimetrically, and the amount of dichromate required for the oxidation is calculated. The caloric value of specific foods is calculated from the caloric equivalent of dichromate obtained from the oxidation of foodstuff of known caloric content. The determination of excess dichromate may be obtained by titration with ferrous ammonium sulfate solution and *o*-phenylanthranilic acid as indicator (Chermnykh *et al.*, 1966).

Another oxidation method (Ghimicescu and Mustesta-Ghimicescu, 1961) is based on the ability of potassium iodate in the presence of concentrated sulfuric acid at 200°C to oxidize the organic substances from foods. The liberated iodine is steam-distilled in an alkaline solution of potassium iodide and titrated with sodium thiosulfate in the presence of a starch indicator.

One of the criticisms of the chemical oxidation procedures is based on the lack of specificity for physiological energy-yielding components. The same kind of criticism can be made, however, for the other procedures already discussed.

8.4. Chemical Analyses and Food Composition Tables

The caloric content of a diet can be estimated from the energy-yielding food components in the diet by applying caloric values to the individual components. A good estimate is obtained if direct chemical analyses are performed on the dietary components, feces, and urine. Total calories are computed by applying appropriate heat-of-combustion factors. The latter is preferred over the former method to achieve accurate assessment of the available energy in the diet.

There are numerous food composition tables (Chatfield, 1949, 1954; Gilpin *et al.*, 1965; Kiernat *et al.*, 1964; Leichsenring and Wilson, 1951; Leung and Flores, 1961; Leung *et al.*, 1952; Leverton and Odell, 1958; McCance and Widdowson, 1947; Miller and Branthoover, 1957; Platt, 1962; USDA, 1959, 1964; Watt and Merrill, 1963; Widdowson and McCance, 1960) which include a large variety of foods. These tables have been prepared for several countries and different geographical areas of the world. Food composition data can also be handled by computer technology, and programs have been written to speed up calculation of dietary information.

Despite new knowledge of food composition and the derivation of appropriate calorie conversion factors, there are several uncertainties in the estimation of contained energy from food composition. The reasons for these uncertainties are as follows: (1) the selection of a representative sample is difficult, (2) the accuracy of the chemical analyses and the specificity of the methods leave something to be desired, (3) there is variation in moisture content of different samples, (4) a choice must be made in the selection of conversion factors for nitrogen to protein and for available energy, and (5) there is variation in food due to species variety, growing conditions, etc. Nevertheless, the results obtained with the compositional method are of practical value and are routinely employed in the calculation of dietary caloric content in nutrition, food science, and agricultural economics.

There are other sources of error when computing caloric intake from food composition tables. These errors arise mainly from the method of obtaining the dietary information and may or may not be important depending upon the final use of the computed values.

9. Calories and Obesity

Animal experimentation in particular has emphasized the many genetic and environmental factors that are important in the processes for deposition and utilization of body fat. In general, these studies have shown that the basic tenets of energy exchange and thermodynamics are not violated in obese individuals, but that many energy-transfer steps remain to be elucidated (Glennon, 1966). Humans are vulnerable to many neurological, metabolic, and psychological disturbances which alter the mechanisms for appetite, satiety, and physical activity (Brobeck, 1948). If energy intake exceeds heat loss and energy expenditure, energy is retained by the body, usually in the form of excess body fat. Nevertheless, the multiple etiology of obesity should be stressed so that a proper search is conducted for the underlying causes of obesity. Inborn and environmentally induced errors of metabolism are currently under study. In addition, identification of environmental and stress factors, as they affect overt energy-regulatory mechanisms, is also progressing, but perhaps somewhat slowly (Bray, 1975).

In recent years, simple obesity has been classified by (1) large total-body adipocyte number, or "hyperplastic obesity" and (2) large mean size of adipose tissue cells, or "hypertrophic obesity." An exaggerated increase in total-body adipocyte number in early childhood predisposes the child to development of obesity. This type has been called "early-onset" or "childhood-onset" obesity. In contrast, an increase in adipocyte size by lipid accumulation in adipose tissue with no change in total cell number commonly occurs in the adult and has been called "adult-onset" obesity. This distinction provides a new vista of the obesity problem and has stimulated research in both treatment and prevention. Whereas these two types depict environmentally induced obesity, a study of monozygotic and dizygotic twins has shown that genetic factors apparently play a decisive role in the origin of obesity, for the monozygotic twins showed much greater sameness in their body fatness than did the dizygotic twins (Börjeson, 1976).

9.1. Prevalence of Obesity

Overweight and obesity are prevalent in the United States in both children and adults. Obese children form a potential pool of those highly eligible for obesity in later life. While "ideal" or "best" weight tables are not precise, from 25 to 34% of men and 25 to 45% of women in the age range 30–69 were labeled more than 20% overweight by a Metropolitan Life Insurance Co. survey (USDHEW, 1960). In general, excess fat deposition increased with advancing years, at least in the age range 20–60.

Obesity is difficult to define precisely, but we have frequently selected the arbitrary percentage-body-fat values of 20% for men and 30% for women as indicating obesity or excess body fatness (Christian *et al.*, 1964). Whenever body weight is gained and body fat stored, energy balance is positive. If this process continues, obesity will result.

9.2. Energy Balance

It is, of course, very difficult to prove that the living organism always obeys the laws of thermodynamics and conservation of energy (Miller and Mumford, 1966). The labor involved in constructing an appropriate balance sheet of energy turnover is immense and may never be achieved except in superficial ways. There is no evidence to suggest that the classical calorimetric experiments are fallacious nor that gross energy balance in humans cannot be followed to within 1 or 2% with careful experimentation, appropriate instruments and equipment, and well-defined regimens. The obese do not have a built-in source of energy of unknown origin, extremely efficient metabolic turnovers, or special energy-conservation mechanisms. While these conclusions are continually contested, no evidence has been forthcoming to alter them (Berlin *et al.*, 1962; Bloom and Eidex, 1967a,b; Dabney, 1964; Thomson *et al.*, 1961; Widdowson *et al.*, 1954; Wilson *et al.*, 1967).

9.3. Caloric Equivalent of Weight Loss

When body weight is lost on diet-restriction regimens that promote weight loss, the caloric equivalent of the weight loss is not constant but depends on the following factors: amount of body fatness; severity of the dietary restriction; amount of physical activity; magnitude of the negative caloric balance; composition of the diet; salt intake, particularly for sodium chloride; resident environment; adequacy of fluid intake; drug therapy; and status of hormonal and neurogenic control mechanisms, including pituitary–thyroid–adrenal function. When body weight is lost by obese individuals, the major constituents of the weight loss are water and fat. The caloric equivalent of the weight lost varies with the experimental or therapeutic conditions, but it is always less than either the caloric equivalent of stored body fat (9.5 kcal·g^{-1}) or even the commonly used average nutritional value of 9.0 kcal·g^{-1}. The highest reported values for the kcal equivalent in well-conducted energy balance and body composition studies have not exceeded 8.5 kcal·g^{-1}. Representative data from the literature on the caloric equivalents of weight loss are presented in Table XV. It should be noted that the caloric equivalent of the weight loss changes during the course of weight reduction. Early weight loss involves a higher proportion of water and a lower proportion of fat than is the case later. Individual differences in composition of the weight loss are apparent (Grande, 1964). Dole *et al.* (1955) studied the caloric value of obesity tissue lost in very short periods of weight reduction and gain. Their experimental design may only slightly exaggerate the body weight fluctuations experienced by many people. Weight gains and losses were produced in their subjects by changing their caloric intake 485 kcal·day^{-1} every four days. Under these conditions, the caloric value of the weight gain or loss varied in five subjects from 2.16 to 3.61 kcal·g^{-1} (see Table XV), a value vastly different from the commonly

Table XV. Mean Caloric Equivalents of Body Weight Loss in Humans[a]

Study	n	Days	kcal·kg^{-1}	Reference
Undernutrition	5	4	2730	Dole *et al.* (1955)
Starvation	1	4	2840	Benedict (1907)
Starvation with work	6	5	2840	Taylor *et al.* (1954)
Semistarvation with work	6	12	4300	Brozek *et al.* (1957)
Semistarvation with work	13	24	5320	Brozek *et al.* (1957)
Reducing	12	63	6170	Grande (1964)
Prolonged semistarvation	32	168	7510	Keys *et al.* (1956)
Obesity, reducing		14 or more	6000–8200	Berlin *et al.* (1962) Buskirk *et al.* (1963a), Gilder *et al.* (1967), and several others

[a] n, number of subjects; days, days on experimental regimen; kcal·kg^{-1}, kcal deficit/kg weight loss.

assumed 9–9.5 kcal$\cdot^{-1}$. The other extreme is provided by the obese person who has been subjected to caloric undernutrition and negative caloric balance for two or more weeks. Several studies in the literature indicate a caloric value for the weight lost by these obese subjects to range between 6.0 and 8.2 kcal$\cdot$g^{-1}.

9.4. Energy Expenditure during Weight Loss

As body weight is lost, and as body cellular constituents are lost in the process, metabolic requirements are reduced during both rest and exercise. A lighter body enables the individual to perform a given task involving gross body movement with less expenditure of energy (Dempsey *et al.*, 1966). The complexity of human behavior when on a weight-loss regimen can be illustrated by the findings in a study of caloric balance of obese subjects. On certain days, a period of exercise was included in the daily regimen that required expenditure of approximately 800 kcal. It was found that daily caloric expenditure increased no more than 400 kcal. Thus, the obese subjects compensated for the added exercise by being less active during the remainder of the day. In the same study, the energy required to perform a standard exercise was 860 kcal before weight reduction and 590 kcal several weeks later (Buskirk *et al.*, 1963a). For these reasons, among others, it is difficult to precisely predict the extent of weight loss for a given individual even if he/she is participating in a specific weight-loss regimen. The prediction becomes more difficult if the subject lives at home, away from the controlled conditions of the laboratory or clinic (Bullen *et al.*, 1964).

10. Undernutrition

When consumed for a prolonged time, diets which provide energy below the caloric requirements will cause undernutrition. These diets produce not only negative caloric balance with loss of body weight, but also nutritional deficiencies associated with a low intake of other nutrients.

The loss of weight in chronic calorie undernutrition is due mainly (but not exclusively) to the loss of body fat as the body meets the caloric demands from its depot fat (Elkinton and Huth, 1959). Also, as the caloric deficit increases, the body proteins are partially utilized as fuel, and negative nitrogen balance results. In undernutrition, therefore, marked changes in body composition are observed, involving losses of weight, body fat, and fat-free body mass (Grande, 1961; Rath and Mazek, 1966; Taylor *et al.*, 1954).

In addition to the changes in body composition, severe undernutrition produces a reduction in BMR as body protein components are lost. The metabolic rate per unit body mass is diminished. There is a decline in cardiac work, tone of skeletal muscle, and body temperature. The skin vessels are constricted in an attempt to conserve heat by reducing the dissipation of heat through radiation and convection. The subjects become abnormally sensitive to cold. Basal pulse rate and hemoglobin decrease, and, in some cases, serum proteins diminish and edema develops. An increased susceptibility to infection is possible.

Chronic undernutrition diminishes muscular strength and work capacity, and the subjects are readily fatigued. There are also marked psychological changes as evidenced by mental apathy, depression, introversion and other changes in personality, and lower intellectual capacity.

Calorie undernutrition produces retardation of growth and development in children. When a diet very low in calories and proteins is consumed for a prolonged time, detention of growth is observed with loss of weight, fat, and fat-free body tissue, producing emaciated bodies with most of the changes described for undernutrition in adults. This type of severe undernutrition in children has been identified as marasmus. On the other hand, a diet either low or adequate in calories, but very low in protein or containing protein of poor biological value, produces detention of growth. In severe chronic cases, low serum proteins, edema, skin lesions, depigmentation and changes in the texture of hair, and fatty liver are observed. This is known clinically as kwashiorkor. Different degrees of severe protein–calorie undernutrition give rise to mixed symptomatology of kwashiorkormarasmus. Severe undernutrition in children also has been classified according to the weight deficit as first-, second-, and third-degree malnutrition if the weight deficit is more than 10, 25, and 40%, respectively (Gomez *et al.*, 1956).

Mild calorie deprivation in children produces a slower rate of growth, as evidenced by body height and weight, and small skinfold thickness is also observed. Stunting in many population groups is believed to be a consequence of deficiency of calories and other vital nutrients as well, although the genetic influence cannot be ruled out completely. The slower rate of physical growth is accompanied by a retardation of physiological development, as shown by a delayed maturation in children.

11. Source of Calories during Exercise

Carbohydrates and fat constitute the major fuels in the body. During work, the relative contribution of carbohydrates and fat to the total energy output varies according to the duration and intensity of work. Glucose and carbohydrate reserves are the immediate sources of energy to support increased metabolism. Fatty acids from body fat stores, however, provide a significant proportion of the fuel required for the metabolism of the resting muscle and for physical activity. Carbohydrates have been referred to as the preferred source of energy, but when the muscle glycogen reserves are depleted, free fatty acids are used as fuel to a greater extent, and a drop in RQ is observed.

Carbohydrates as fuel are theoretically more efficient than fats on the basis that they provide more calories per liter of oxygen consumed despite their lower caloric density.

In short periods of work, carbohydrates are used as the principal fuel, and in endurance exercise, energy is derived primarily from fatty acids, although glycogen reserves are readily depleted by prolonged work at intensities greater than 60% of aerobic capacity. The trained individual may use more fat for muscular work than the untrained. Diets rich in carbohydrates have been recommended to maintain higher work efficiency. It has been shown that in prolonged exercise, high-carbohydrate diets are superior to high-fat diets and that endurance capabilities are closely related to the concentration of muscle glycogen (Hultman, 1967).

Protein, on the other hand, contributes very little to the total energy expenditure during exercise, and, therefore, high-protein diets do not improve work performance or endurance. The protein requirement during heavy work is not increased, and protein nitrogen is only necessary to meet the general demands of growth, development, and maintenance. There is no reason, therefore, to justify a high-protein diet for heavy work or sports unless body tissues are damaged and excessive tissue repair and rebuilding is required. For general reviews, see Hultman (1967) and Buskirk (1971).

12. References

American Journal of Clinical Nutrition, 1971, *Symposium: Assessment of Typical Daily Energy Expenditure* (R. B. Bradfield, ed.), Part I. **25**:1109; Part II. **25**:1403.

Atwater, W. O., and Bryant, A. P., 1899, *The Availability and Fuel Value of Food Materials,* Agric. Exp. Stn. Annu. Rep. No. 73.

Atwater, W. O., and Snell, J. F., 1903, Description of a bomb-calorimeter and method of its use, *J. Am. Chem. Soc.* **25**:659.

Benedict, F. G., 1907, *The Influence of Inanition on Metabolism,* Carnegie Inst. Washington Publ. No. 77.

Benedict, F. G., and Carpenter, T. M., 1918, *Respiration Calorimeters for Studying the Respiratory Exchange and Energy Transformations of Man,* Carnegie Inst. Washington Publ. No. 123.

Benzinger, T. H., and Kitzinger, C., 1949, Direct calorimetry by means of the gradient principle, *Rev. Sci. Instr.* **20**:849.

Benzinger, T. H., Huebscher, R. G., Minard, D., and Kitzinger, C., 1958, Human calorimetry by means of the gradient principle, *J. Appl. Physiol.* **12**:S1–4.

Berlin, N. I., Watkin, D. W., and Gevirtz, N. R., 1962, Measurement of changes in gross body composition during controlled weight reduction in obesity by metabolic balance and body density–body water technics, *Metabolism* **11**:302.

Bloom, W. L., 1965, A mechanical device for measuring human energy expenditure, *Metabolism* **14**:955.

Bloom, W. L., and Eidex, M. F., 1967a, Inactivity as a major factor in adult obesity, *Metabolism* **16**:679.

Bloom, W. L., and Eidex, M. F., 1967b, The comparison of energy expenditure in the obese and lean, *Metabolism* **16**:685.

Börjeson, M., 1976, The aetiology of obesity in children: A study of 101 twin pairs, *Acta Paediatr. Scand.* **65**:279.

Bray, G. (ed.), 1975, *Obesity in Perspective,* USDHEW Publ. No. (NIH) 75–78.

Brobeck, J. R., 1948, Regulation of energy exchange, *Annu. Rev. Physiol.* **10**:315.

Brody, S., 1945, *Bioenergetics and Growth,* Reinhold, New York.

Brozek, J. F., Grande, F., Taylor, H. L., Anderson, J. T., Buskirk, E. R., and Keys, A., 1957,

Changes in body weight and body dimensions in men performing work on a low caloric carbohydrate diet, *J. Appl. Physiol.* **10**:412.

Bullen, B. A., Reed, R. B. and Mayer, J., 1964, Physical activity of obese and nonobese adolescent girls appraised by motion picture sampling, *Am. J. Clin. Nutr.* **14**:211.

Burton, A. C., and Edholm, O. G., 1955, *Man in a Cold Environment*, Edward Arnold, London.

Buskirk, E. R., 1960, Problems related to the caloric cost of living, *Bull. N.Y. Acad. Med.* **36**:365.

Buskirk, E. R., 1971, Nutrition and college athletics, in: *Administration of Athletics in Colleges and Universities* (E. S. Steitz, ed.), pp. 186–205, American Association for Health, Physical Education, and Recreation, National Education Association, Washington, D.C.

Buskirk, E. R., and Boyer, L. L., 1965, in: *Proceedings of the Symposium on Arctic Biology and Medicine. V. Nutritional Requirements for Survival in the Cold and at Altitude* (L. Vaughn, ed.), pp. 49–84, Arctic Aeromedical Laboratory, Fort Wainwright, AK.

Buskirk, E. R., and Taylor, H. L., 1957, Maximal oxygen intake and its relation to body composition with special reference to chronic physical activity and obesity, *J. Appl. Physiol.* **11**:72.

Buskirk, E. R., Iampietro, P. F., and Welch, B. E., 1957, Variations in resting metabolism with changes in food, exercise, and climate, *Metabolism* **6**:144.

Buskirk, E. R., Thompson, R. H., Moore, R., and Whedon, G. D., 1960, Human energy expenditure studies in the National Institute of Arthritis and Metabolic Diseases metabolic chamber. I. Interaction of cold environment and specific dynamic effect. II. Sleep. *Am. J. Clin. Nutr.* **8**:602.

Buskirk, E. R., Thompson, R. H., Lutwak, L., and Whedon, G. D., 1963a, Energy balance of obese patients during weight reduction: Influence of diet restriction and exercise, *Ann. N.Y. Acad. Sci.* **110**:918.

Buskirk, E. R., Thompson, R. H., and Whedon, G. D., 1963b, Metabolic response to cooling in the human: Role of body composition and particularly of body fat, in: *Temperature—Its Measurement and Control in Science and Industry*, Vol. 3, American Institute of Physics, Reinhold, pp. 429–442, New York.

Canadian Bulletin on Nutrition, 1964, *Dietary Standard for Canada*, Vol. 6, No. 1, Department of Public Printing and Stationery, Ottawa.

Carlson, L. D., 1954, *Man in a Cold Environment*, United States Department of Commerce, Office of Technical Services, Washington, D.C.

Chambers, W. H., 1942, Energy metabolism, *Annu. Rev. Physiol.* **4**:139.

Chatfield, C., 1949, 1954, *Food Composition Tables for International Use*, FAO Nutr. Stud. Nos. 3 and 11.

Chermnykh, N. I., Popovtseva, A. A., and Roshchevskii, M. P., 1966, Determination of the caloric value of fodders by using chromic acid as an oxidant, *Khim. Selsk. Khoz. Bashk.* **4**:307.

Christensen, E. H., 1953, Physiological valuation of work in Nykroppa iron works, in: *Symposium on Fatigue* (W. F. Floyd and A. T. Wolford, eds.), pp. 93–108, H. K. Lewis, London.

Christian, J. E., Combs, L. W., and Kessler, W. V., 1964, The body composition of obese subjects: Studies of the effect of weight loss on the fat and lean body mass, *Am. J. Clin. Nutr.* **15**:20.

Consolazio, C. F., Johnson, R. B., and Pecora, L. J., 1963, *Physiological Measurements of Metabolic Functions in Man*, McGraw-Hill, New York.

Dabney, J. N., 1964, Energy balance and obesity, *Ann. Intern. Med.* **60**:689.

Dempsey, J. A., Reddan, W., Balke, B., and Rankin, J., 1966, Work capacity determinants and physiologic cost of weight support work in obesity, *J. Appl. Physiol.* **21**:1815.

Dole, V. P., Schwartz, I. L., Thorn, N. A., and Silver, L., 1955, The caloric value of labile body tissue in obese subjects, *J. Clin. Invest.* **4**:590.

DuBois, D., and DuBois, E. F., 1916, Clinical calorimetry: A formula to estimate the approximate surface area if height and weight be known, *Arch. Intern. Med.* **17**:863.

DuBois, E. F., 1936, *Basal Metabolism in Health and Disease*, Lea and Febiger, Philadelphia.

DuBois, E. F., 1948, *Fever and Regulation of Body Temperature*, Charles C. Thomas, Springfield, Ill.

DuBois, E. F., 1954, Energy metabolism, *Annu. Rev. Physiol.* **16**:125.

DuBois, E. F., 1960, An attempt to classify occupations in ten task groups according to physical exertion or according to the amount of physical exertion demanded, *Proc. Am. Philos. Soc.* **104**:111.

Durnin, J. V. G. A., and Passmore, R., 1967, *Energy, Work and Leisure,* Heinemann Educational Books, London.

Edholm, O. G., 1966, The assessment of habitual activity, in: *Physical Activity in Health and Disease,* (K. Evans and K. L. Andersen, eds.,) Oslo University Press, Oslo.

Edholm, O. G., Fletcher, J. G., Widdowson, E. M., and McCance, R. A., 1955, The energy expenditure and food intake of individual men, *Br. J. Nutr.* **9**:286.

Elkinton, J. R., and Huth, H. J., 1959, Body fluid abnormalities in anorexia nervosa and under-nutrition, *Metabolism* **8**:376.

Emerson, K., Jr., Saxena, B. N., and Poindexter, E. L., 1972, Caloric cost of normal pregnancy, *Obstet. Gynecol.* **40**:786.

FAO (Food and Agriculture Organization of the United Nations), 1947, *Energy-Yielding Components of Food and Computation of Caloric Values,* Washington, D.C.

FAO (Food and Agriculture Organization of the United Nations), 1957, *Caloric Requirements, Report of the Second Committee on Caloric Requirements,* FAO Nutr. Stud. No. 15.

FAO (Food and Agriculture Organization of the United Nations), 1958, *Caloric Requirements,* FAO Nutr. Stud. No. 15.

FAO (Food and Agriculture Organization of the United Nations), 1964, *Program of Food Consumption Surveys,* Food and Agriculture Organization, Rome.

Farhi, L. E., and Rahn, H., 1955, Gas stores of the body and the unsteady state, *J. Appl. Physiol.* **7**:472.

FNB (Food and Nutrition Board), 1964, *Recommended Dietary Allowances,* 6th ed., National Academy of Sciences, National Research Council, Washington, D.C.

FNB (Food and Nutrition Board), 1968, *Recommended Dietary Allowances,* 7th ed., National Academy of Sciences, National Research Council, Washington, D.C.

FNB (Food and Nutrition Board), 1974, *Recommended Dietary Allowances,* 8th ed., National Academy of Sciences, National Research Council, Washington, D.C.

Gallenkamp, A., and Co. Ltd., 1961, *Ballistic Bomb Calorimeter Instructions CB-370,* London.

Ghimicescu, G. H., and Mustesta-Ghimicescu, C., 1961, A new micromethod for determination of the caloric value of foods and food rations, *Acad. Repub. Pop. Rom. Stud. Cercet. Chim.* **9**:513.

Gilder, H., Cornell, G. N., Graff, W. R., MacFarlane, J. R., Asaph, J. W., Stubenbord, W. T., Watkins, G. M., Rees, J. R., and Throbjarnson, B., 1967, Components of weight loss in obese patients subjected to prolonged starvation, *J. Appl. Physiol.* **23**:304.

Gilpin, G. L., Murphy, H. W., Marsh, A. C., Dawson, H. H., Bowman, F., Kerr, R. G., and Snyder, D. G., 1965, *Meat, Fish, Poultry and Cheese: Home Preparation Time, Yield and Composition of Various Market Forms,* U.S. Dep. Agric. Home Econ. Res. Rep. No. 30.

Glennon, J. A., 1966, Weight reduction: An enigma, *Arch. Intern. Med.* **118**:1.

Gomez, F., Ramos-Galvan, R., Frank, S., Cravioto, J., Chavez, R., and Vasquez, J., 1956, Mortality in second and third degree malnutrition, *J. Trop. Pediatr.* **2**:77.

Grande, F., 1961, Nutrition and energy balance in body composition studies, in: *Techniques for Measuring Body Composition* (J. Brozek and A. Henschel, eds.), pp. 168–188, National Academy of Sciences, National Research Council, Washington, D.C.

Grande, F., 1964, Man under caloric deficiency, in: *Handbook of Physiology,* Section 4, *Adaptation to the Environment* (D. B. Dill, ed.), pp. 911–937, American Physiological Society, Washington, D.C.

Harris, J. A., and Benedict, F. G., 1919, *A Biometric Study of Basal Metabolism in Man,* Carnegie Inst. Washington Publ. No. 279.

Hathaway, M. L., 1957, *Heights and Weights of Children and Youth in the United States,* U.S. Dep. Agric. Home Econ. Res. Rep. No. 2.

Hultman, E., 1967, Studies on muscle metabolism of glycogen and active phosphate in man with special reference to exercise and diet, *Scand. J. Clin. Lab. Invest.* **19** (Suppl. 94):1.

Jacobs, D., Heald, F., White, P. L., and McGanity, W. J., 1963, The prevention of obesity, *J. Am. Med. Assoc.* **186**:28.

Jones, D. B., 1931, *Factors for Converting Percentages of Nitrogen in Foods and Feeds into Percentages of Protein,* U.S. Dep. Agric. Circ. No. 183.

Keys, A., Brozek, J., Henschel, A., Mickelsen, O., and Taylor, H. L., 1956, *The Biology of Human Starvation,* 2 vols., University of Minnesota Press, Minneapolis.

Kiernat, B. H., Johnson, J. A., and Siedler, A. J., 1964, *A Summary of the Nutrient Content of Meat,* Am. Meat. Inst. Found. Bull. No. 57.

Kinney, J. M., 1959, Influence of intermediary metabolism on nitrogen balance and weight loss: Some considerations basic to understanding of injury, *Metabolism* **8**:809.

Kleiber, M., 1947, Body size and metabolic rate, *Physiol. Rev.* **27**:511.

Kleiber, M., 1950, Calorimetric measurements, in: *Biophysical Research Methods* (F. Uber, ed.), Interscience, New York.

Kleiber, M., 1961, *The Fire of Life,* John Wiley and Sons, New York.

Kleiber, M., 1965, Respiratory exchange and metabolic rate, in: *Handbook of Physiology,* Section 3, *Respiration,* Vol. 2 (W. O. Fenn and H. Rahn, eds.), pp. 927–938, American Physiological Society, Washington, D.C.

Kleiber, M., 1975, Metabolic turnover rate: A physiological meaning of the metabolic rate per unit body weight, *J. Theor. Biol.* **53**:199.

Kofranyi, E., and Michaelis, H. F., 1941, Ein trägbarer Apparat zur bestimmung des Gasstoffwechsels, *Arbeitsphysiologie* **11**:148.

Krebs, H. A., 1964, The metabolic fate of amino acids, in: *Mammalian Protein Metabolism* Vol. 1 (H. N. Munro and J. B. Allison, eds.), pp. 125–176, Academic Press, New York.

Leichsenring, J. M., and Wilson, E. D., 1951, Food composition table for short method of dietary analysis (2nd revision), *J. Am. Diet. Assoc.* **27**:386.

Leung, W. T. W., and Flores, M., 1961, *Food Composition Table for Use in Latin America,* Institute of Nutrition of Central America and Panama–ICNND, National Institutes of Health, Bethesda, Md.

Leung, W. T. W., Pecot, R. K., and Watt, B. K., 1952, *Composition of Foods Used in Far Eastern Countries,* U.S. Dep. Agric. Agric. Handb. No. 34.

Leverton, R. M., and Odell, G. V., 1958, *The Nutritive Value of Cooked Meat,* Okla. Agric. Exp. Stn. Misc. Publ. No. MP-49.

Liddell, F. D. K., 1963, Estimation of energy expenditure from expired air, *J. Appl. Physiol.* **18**:25.

Lusk, G., 1922, A history of metabolism, in: *Endocrinology and Metabolism,* Vol. 3, D. Appleton, New York.

Lusk, G., 1928, *The Elements of the Science of Nutrition,* W. B. Saunders, Philadelphia.

MacHattie, L. A., 1960, Graphic visualization of the relations of metabolic fuels:heat: O_2:CO_2:H_2O:urine N, *J. Appl. Physiol.* **15**:677.

Macy, I. G., 1942, Energy metabolism in childhood, in: *Nutrition and Chemical Growth in Childhood,* Vol. 1, Charles C. Thomas, Springfield, Ill.

McCance, P. A., and Widdowson, R. M., 1947, *The Chemical Composition of Foods,* Chemical Publishing, London.

Merrill, A. L., and Watt, B. K., 1955, *Energy Value of Foods: Basis and Derivation,* U.S. Dep. Agric. Agric. Handb. No. 74.

Miller, A. T., 1954, Energy metabolism and metabolic reference standards, *Methods Med. Res.* **6**:74.

Miller, C. D., and Branthoover, B., 1957, *Nutritive Value of Some Hawaiian Foods in Household Units and Common Measures,* Hawaii Agric. Exp. Stn. Circ. No. 52.

Miller, D. S., and Mumford, P., 1966, Obesity: Physical Activity and nutrition, *Proc. Nutr. Soc.* **25**:100.

Morey, N. B., 1936, An analysis and comparison of different methods of calculating the energy values of diets, *Nutr. Abstr. Rev.* **6**:1.

Müller, E. A., and Franz, H., 1952, The measurement of energy consumption in occupational work with an improved respiratory-gasmeter, *Arbeitsphysiologie* **14**:499.

Murlin, J. R., 1922, Normal processes of energy metabolism, in *Endocrinology and Metabolism,* Vol. 3, D. Appleton, New York.

Murlin, J. R., and Burton, A. C., 1935, Human calorimetry. I. A semiautomatic respiration calorimeter, *J. Nutr.* **9:**233.

National Center for Health Statistics, 1965, *Weight, Height, and Selected Body Dimensions of Adults, United States, 1960-1962, Series 11, No. 8,* U.S. Public Health Serv. Publ. No. 1000.

Newburgh, L. E., Johnston, M. W., and Newburgh, J. B., 1945, *Some Fundamental Principles of Metabolism,* Edwards Brothers, Ann Arbor, Mich.

Orr, J. B., and Leitch, I., 1938, The determination of the caloric requirements of man, *Nutr. Abstr. Rev.* **7:**509.

Parr Instrument Co., 1965, *Oxygen Bomb Calorimetry and Oxygen Bomb Combustion Methods,* Parr Manual No. 1200, Moline, Ill.

Passmore, R., and Durnin, J. V. G. A., 1955, Human energy expenditure, *Physiol. Rev.* **35:**801.

Passmore, R., Thompson, J. G., and Warnock, G. M., 1952, A balance sheet of the estimation of energy intake and energy expenditure as measured by indirect calorimetry, using the Kofranyi-Michaelis calorimeter, *Br. J. Nutr.* **6:**253.

Perkins, J. F., 1964, Historical development of respiratory physiology, in: *Handbook of Physiology,* Section 3, *Respiration,* Vol. 1 (W. O. Fenn and H. Rahn, eds.), pp. 1–62, American Physiological Society, Washington, D.C.

Pike, R. L., and Brown, M. L., 1975, *Nutrition: An Integrated Approach,* 2nd ed., John Wiley and Sons, New York.

Platt, B. S., 1962, *Tables of Representative Values of Food Commonly Used in Tropical Countries,* Med. Res. Counc. (G. B.) Spec. Rep. Ser. No. 302.

Rath, R., and Masek, J., 1966, Changes in the nitrogen metabolism in obese women after fasting and refeeding, *Metabolism* **15:**1.

Robinson, D. W., Cole, D., Clarke, M. H., and Bayley, H. S., 1964, Estimation of gross energy values in nutritional studies using a conventional and ballistic bomb calorimeter, *Proc. Nutr. Soc.* **23:**57.

Rosen, G., 1955, Metabolism: Evolution of a concept, *J. Am. Diet Assoc.* **31:**861.

Rozental, L., 1957, A new and quick method of estimating the energy value of foods, *Rocz. Panstw. Zakl. Hig.* **8:**27.

Rubner, M., 1885, Calorimetrische Utersuchungen II, *Z. Biol.* **21:**337.

Rubner, M., 1901, Der Enerwert der Kost des Menschen, *Z. Biol.* **42:**261.

Shephard, R. J., 1967, Glossary of specialized terms and units, *Can. Med. Assoc. J.* **96:**912.

Swift, R. W., and French, C. E., 1954, *Energy Metabolism and Nutrition,* Scarecrow Press, New Brunswick, N.J.

Swift, R. W., Barron, G. P., Fisher, K. H., French, C. E., Hartsook, E. W., Hershberger, T. V., Keck, E., Long, T. A., and Magruder, N. D., 1958, The effect of high versus low protein equicaloric diets on the heat production of human subjects, *J. Nutr.* **65:**89.

Taylor, H. L., Henschel, A., Mickelson, O., and Keys, A., 1954, Some effects of acute starvation with hard work on body weight, body fluids, and metabolism, *J. Appl. Physiol.* **6:**613.

Thomson, A. M., Billewica, W. Z., and Passmore, R., 1961, The relation between calorie intake and body weight in man, *Lancet* **1:**1027.

Ubbelohde, A. R., 1955, *Man and Energy,* George Brozilleo, New York.

USDA (United States Department of Agriculture), 1959, *Food,* U.S. Dep. Agric. Yearb. Agric.

USDA (United States Department of Agriculture), 1964, *Nutritive Values of Foods,* U.S. Dep. Agric. Home Gard. Bull. No. 72.

USDHEW (United States Department of Health, Education, and Welfare), 1960, *Obesity and Health,* United States Public Health Service, United States Government Printing Office, Washington, D.C.

Van Soest, P. J., and McQueen, R. W., 1973, The chemistry and estimation of fibre, *Proc. Nutr. Soc.* **32:**123.

Watt, B. K., and Merrill, A. L., 1963, *Composition of Foods—Raw, Processed, Prepared,* U.S. Dep. Agric. Agric. Handb. No. 8.

Webb, P. (ed.), 1964, *Bioastronautics Data Book,* NASA Scientific and Technical Information Division, Washington, D.C.

Webb, P., 1971, Metabolic heat balance for 24 hour periods, *Int. J. Biometeorol.* **15**:151.

Weir, J. B., deV., 1949, New methods for calculating metabolic rate with special reference to protein metabolism, *J. Physiol.* **169**:1.

Welch, B. E., Buskirk, E. R., and Iampietro, P. F., 1958, Relation of climate and temperature to food and water intake in man, *Metabolism* **7**:141.

Wells, J. G., Balke, B., and Van Fossan, D. D., 1957, Lactic acid accumulation during work: A suggested standardization of work classification, *J. Appl. Physiol.* **10**:51.

Whedon, G. D., 1959, New research in human energy metabolism, *J. Am. Diet. Assoc.* **35**:682.

Widdowson, E. M., and McCance, R. A., 1960, *The Composition of Foods,* Med. Res. Counc. (G. B.) Spec. Rep. No. 297.

Widdowson, E. M., Edholm, O. G., and McCance, R. A., 1954, The food intake and energy expenditure of cadets in training, *Br. J. Nutr.* **8**:147.

Wilson, N. L., Farber, S. M., Kimbrough, L. S. and Wilson, R. H. L., 1967, The development and perpetuation of obesity, *Lancet* **87**:13.

Winslow, C. E. A., Herrington, L. P., and Gagge, A. P., 1936, A new method of partitional calorimetry, *Am. J. Physiol.* **116**:641.

Wolff, H. S., 1956, Modern techniques for measuring energy expenditure, *Proc. Nutr. Soc.* **15**:77.

3

Suppliers of Energy: Carbohydrates

Ian Macdonald

1. Introduction

Carbohydrates are the cheapest form of energy available to the human metabolism. They form the energy stores of most plants, and hence they are easily cultivated and harvested. In fact, the affluence of a community, and indeed of a family unit, could be judged by the proportion of its energy intake present as carbohydrate.

Unlike the plant kingdom, humans store very little carbohydrate (about 650 kcal in an average man) in contrast to fat, whose reserve of energy in a nonoverweight person may be 100,000 kcal (Cahill and Owen, 1968). As carbohydrates form a large proportion of the energy in the diet (45% of calories in Western man and much greater in inhabitants of tropical climes), and as there are virtually no stores, the carbohydrates we consume must be metabolized rapidly.

The role of carbohydrates in metabolism is soley as an energy provider, and they can be consumed in large or small quantities with little ill effect. Only in fairly recent times has it been learned that there are some apparently healthy people who are "sensitive" to the carbohydrate in their diet in that their fasting serum triglyceride level fluctuates directly with their carbohydrate intake. A diet that is carbohydrate free is theoretically compatible with survival, but it would be existence rather than living.

It is not possible to define what is meant by carbohydrate (it shares this in common with lipid) with any exactitude, but, despite this, there seems to be little controversy over meaning, even in the scientific community. The reason possibly lies in the rather limited number of common compounds in mammalian metabolism receiving the label "carbohydrate," and, furthermore, these compounds do bear a chemical relationship to one another.

Ian Macdonald • Department of Physiology, Guy's Hospital Medical School, London SE1 9RT, United Kingdom.

According to Pigman (1957), "an oversimplified definition of the carbo-hydrates is that they are composed of the polyhydroxyaldehydes, ketones, alcohols, acids, their simple derivatives, and the polymers having hemiacetal polymeric linkages. The nonpolymeric carbohydrates are the five-, six-, and high-carbon members of the several homologous series." The nutritionist and the physiologist may have limited understanding of this definition in the ab-sence of a basis in biochemistry, but in terms of practice, most of the carbo-hydrates commonly eaten and metabolized are glucose and its polymers (mal-tose through polysaccharides and dextrins to starch), fructose (and its polymer inulin), galactose, and the disaccharides sucrose (glucose and fructose) and lactose (glucose and galactose). There is a confusion of names for these com-pounds, and some of the more commonly used names are given in Table I.

The plant kingdom, in the main, stores its energy in the form of starch either in the seed, for a new separate existence, or in the root for times of shortage. Small amounts of starch are found in animals, as glycogen in liver and muscle, but are inconsequential as energy stores compared with fat. The plant starches are of two main biochemical types, depending on the linkage of the glucose molecules within the starch—the so-called 1:4 (amylose) and the 1:6 (amylopectin) linkages—and the importance of these to the nutritionist is that some amylases in the gut are more efficient at hydrolyzing one than the other. As an example of the specificity of gut enzymes, cellulose, the most abundant polysaccharide in the world. cannot be split in the gastrointestinal tract except by bacteria in the host gut.

In more recent times, with the advent of modern microbiological and enzyme techniques, some carbohydrate alcohols have become available for general consumption. As the monosaccharides (whether as hexoses or pen-toses) are aldehydes or ketones, they can be converted to their corresponding alcohols (Table II), and these compounds are being used in manufactured foods. Sorbitol, for example, is a sweetener used in "diabetic" foods. Other carbohydrate mixtures which are commercially available as a result of modern technology are glucose syrups, where the degree of hydrolysis of the starch (usually corn) can be stopped at any stage with a resultant mixture whose content of glucose polymers can be almost tailor made. It is not clear whether the metabolic effects of consuming these glucose syrups of varying composi-tion vary, but there is suggestive evidence that this might be so in rats (Birch and Etheridge, 1973). Also available are syrups with varying proportions of

Table I. Synonyms of Some Common Carbohydrates

Glucose	Dextrose, grape sugar, corn sugar
Fructose	Levulose, fruit sugar
Maltose	Malt sugar
Lactose	Milk sugar
Sucrose	Cane sugar, beet sugar, "sugar"
Glucose syrup (a partial hydrolysate of starch)	Corn syrup, liquid glucose

Table II. Some Common Monosaccharides
and Their Corresponding Alcohols

Monosaccharide	Alcohol
Glucose	Sorbitol
Fructose	Sorbitol
Galactose	Dulcitol (galactitol)
Mannose	Mannitol
Xylose (a pentose)	Xylitol

glucose and fructose, and the response of the body to these mixtures is, in the absence of evidence to the contrary, presumably similar to that seen after consuming either glucose or fructose and slightly different from that seen after sucrose (see below).

One physiological feature of soluble carbohydrates is their ability to stimulate the taste buds for sweetness, and it is this property, perhaps over all others, that makes the simple carbohydrates so attractive—perhaps overattractive—to the consumer. The ability to stimulate the taste buds varies, and approximate relative sweetness values are give in Table III. The theory of sweetness is discussed succinctly by Shallenberger (1971) and the psychology of sweetness by Moskowitz (1974).

2. Digestion

The hydrolysis of those carbohydrates that need to be reduced to the six-carbon chain length to make absorption possible is accelerated by enzymes secreted in the mouth, pancreas, and intestinal wall. However, in order that the enzymes secreted at these sites can function efficiently, it is necessary to have the carbohydrate available to the enzyme for hydrolysis. For example, when raw, many starches are covered by a layer which is impermeable to the enzymes, and hence their breakdown is limited. Large quantities of raw starch can lead to intestinal hurry; for example, potato starch is poorly utilized by humans unless cooked (Langworthy and Deuel, 1922). In the rat, raw starch

Table III. Relative Sweetness of
Some Common Carbohydrates

Sucrose	100
Maltose	33
Lactose	16
Glucose	67
Fructose	110
Sorbitol	54
Xylitol	120

from wheat, maize, rice, and cassava is readily digestible, whereas raw starch from potato, arrowroot, and sago is not, due to the nature of the cover of the starch granules (Booher *et al.,* 1951; Jelinek *et al.,* 1952).

2.1. Starch Hydrolysis

The enzyme present in saliva and pancreatic juice is amylase. It can hydrolyze starch to the disaccharides isomaltose and maltose but seems incapable of splitting these disaccharides to their constituent monosaccharide, glucose. Salivary amylase is relatively unimportant, and, in fact, dog saliva is devoid of this enzyme (Davenport, 1961). As the food in humans is only briefly in the mouth, most of the salivary amylase activity takes place in the stomach—in the center of the bolus—until such time as the low pH of gastric juice penetrates the bolus and inactivates the enzyme, whose optimal pH is 6.9.

The bulk of the breakdown of starch to disaccharide is carried out by pancreatic amylase, and, under normal circumstances in humans, this is a very swift and efficient procedure. The blood glucose levels following the ingestion of a given dose, either cooked starch, glucose syrup, or glucose per se, are virtually identical (Dodds *et al.,* 1959; Sun and Shay, 1961).

The rate of production of the salivary and pancreatic amylases is dependent on the amount of carbohydrate in the diet. This was demonstrated for saliva some time ago (Simon, 1907) and more recently for pancreatic juice (Reboud *et al.,* 1966).

2.2. Disaccharide Digestion

It was suggested by Starling in 1906 that the main site of action of the disaccharidases was not in the gut lumen but in the wall of the small intestine, and, despite confirmatory evidence (Cajori, 1933), this idea was not followed up until 1957 (Borgstrom *et al.,* 1957), when it was realized that the disaccharidase activity in the gut lumen was too small to account for the hydrolysis achieved. The site of the disaccharidases has been located in the brush border of the small intestine (Borgstrom *et al.,* 1957; Jos *et al.,* 1967; Miller and Crane, 1961), with the activity of the enzyme being greatest at the tip of the villus and least in the crypt (Dahlqvist and Nordstrom, 1966). The enzymes which are present in the brush border include maltase, sucrase, isomaltase, and lactase, and, apart from the latter, these enzymes can be induced (Rosensweig and Herman, 1968).

2.3. Monosaccharide Absorption

Humans have an enormous capacity to absorb glucose, and Crane (1975) has calculated from studies of human subjects (Holdsworth and Dawson, 1964) that over 20 lb, or more than 50,000 kcal, can be absorbed in 24 hr. In order to achieve such high levels of absorption, the process must be active, and, from the classical observations of Crane *et al.* (1965), it is now known that glucose crosses the intestinal cell membrane using a simple Na^+-dependent carrier system whereby the sodium moves, with the glucose, down a transmembrane gradient.

Glucose and galactose share and compete for the same transport system, with galactose having the lower affinity. The apparent competitiveness of glucose and galactose has led to findings which would suggest that there may be two modes of entry for glucose, one of which is coupled to the disaccharidases (Crane *et al.*, 1970). Xylose also shares the glucose pump, but, as it has a high K_m, it is transported relatively slowly (Alvarado, 1966) and hence was long considered to be absorbed by diffusion only. Fructose is not absorbed by simple diffusion because other monosaccharides with similar physical characteristics are much slower in their uptake, and *in vitro* studies have suggested that there is a carrier mechanism for fructose which is different from that of glucose but which, nevertheless, uses the transmembrane sodium gradient (Gracey *et al.*, 1972).

There does not seem to be any reason for considering that disaccharidase activity is rate limiting in carbohydrate absorption— for example, sucrose is absorbed as rapidly as glucose plus fructose (Gray and Ingelfinger, 1966)— and, in fact, the opposite may be true (Cook, 1973; Macdonald and Turner, 1968). No correlation is present between lactose, maltose, and sucrose absorption and the corresponding mucosal disaccharidase level, supporting the view that the latter is not normally rate limiting (McMichael *et al.*, 1967).

3. Carbohydrate Tolerance Tests

An often-used clinical test is the so-called tolerance test, in which the subject is given, by mouth, a standard amount of a substance, and its appearance or altered level in the serum is monitored for varying lengths of time after the introduction. The glucose tolerance test used in the diagnosis of inadequate insulin output is possibly the most commonly used such test among the dietary carbohydrates. As in all oral tolerance tests, the blood (or serum) level may depend on various factors, such as the rates of gastric emptying, absorption from the gut, removal or conversion by the liver, and removal from the blood or serum. Despite these variables, the oral glucose tolerance test is reasonably reproducible, and, within wide limits, the dose or concentration of glucose will not affect the result (Jourdan, 1972). In normal subjects, differences in glucose loads from 0.25 to 2.5 g/kg body weight and in concentrations from 6.25 to 50 g/100 ml cannot be regarded as important factors in producing variation in the oral glucose tolerance test.

Unlike the glucose levels, the insulin levels are related to the amount of glucose in the test, as are the fructose levels after both fructose and sucrose tolerance tests (Macdonald *et al.*, 1978).

The response to an oral glucose load can, however, be modified in humans by, for example, the amount of carbohydrate consumed in the days preceding the test—a high carbohydrate intake improves the glucose tolerance (Himsworth, 1933)—an effect that does not seem to be so marked in animals (Uram *et al.*, 1958). The type of carbohydrate may also modify the glucose tolerance test in animals (Cohen and Teitelbaum, 1964). A useful test of pancreatic function is to follow the blood glucose levels after ingesting cooked starch. In normal states, the curves of blood glucose after such a meal and after glucose itself are indistinguishable, but with pancreatic amylase deficiency, the rise and fall of the blood glucose after consuming cooked starch are much decreased (Sun and Shay, 1961). Uncooked starch (see above) tends to give a flat curve even in healthy subjects.

4. Carbohydrate Metabolism in the Liver

In the liver, glucose has no difficulty crossing the hepatocyte wall, and the main factors influencing the entry of glucose are the concentration of glucose in the portal blood and the ability of the enzymes within the cell to dispose of the glucose. Many of the enzymes are rate limiting for glucose (Weber *et al.*, 1966). The hepatocyte is also capable of delivering glucose to the circulation, by gluconeogenesis from glycogen, protein, glycerol, and the balance between uptake and release of glucose is probably determined by enzymes. Two of these key enzymes are glucokinase and glucose-6-phosphatase (Cahill *et al.*, 1959), both of which are subject to endocrine control (Ashmore and Weber, 1968). This implies that the control mechanism is relatively slow, needing at least a few hours to be fulfilled. The more acute regulation of glucose uptake or output by the liver cell is probably via phosphorylase activation through the adenyl cyclase system (Sutherland *et al.*, 1965) or via the free fatty acid levels (Randle *et al.*, 1963), which, in turn, are influenced by catecholamines, glucagon, and growth hormone.

It should be borne in mind that much of the glucose that is absorbed from the gut never reaches the peripheral circulation, as it is taken up by the liver. This process may be affected by the high concentration of insulin also present in portal blood which may, in some way, have actions different from that elsewhere in the body (Samols and Holdsworth, 1968).

4.1. Control of Carbohydrate Metabolism in the Liver

The control is entirely hormonal, acting via the enzymes, and can be considered in two temporal phases: the immediate alterations in metabolic response to acute fluctuations in glucose presented to the liver cells and the more long-term changes which take place in response to larger and more continuous changes in glucose levels. The latter changes are likely to be due to enzyme induction, whereas the acute changes in response cannot be due to induction but rather to some more complicated interplay between inhibitory and/or facilitative factors (Ashmore and Weber, 1968).

Glucagon. This hormone, which is secreted by the α cells of the pancreas, produces a rapid breakdown of liver glycogen. As might be expected, in conditions where the glycogen content of the liver is low, the effect of glucagon is less marked. Glucagon increases the production of AMP, which then stimulates liver phosphorylase activity, this enzyme catalyzing the first step in glycogen breakdown. Thus, glucagon also stimulates insulin production.

Adrenaline. Like glucagon, adrenaline stimulates the breakdown of glycogen, but, unlike glucagon, it is able to do this in muscles also. It does not stimulate insulin release.

Insulin. In contrast to glucagon in the liver, insulin lowers AMP, thus accelerating glycogen formation by keeping phosphorylase in a relatively inactive state. In the long term, a low level of insulin causes a fall in the enzymes of glycolysis, and the insulin-induced rise in these important enzymes can be blocked by inhibitors of protein synthesis, thus suggesting that enzyme synthesis is affected, in the long term, by insulin. In this respect, the effect of insulin on liver is different from its effect on other tissues, where it is concerned with the transfer of glucose into cells and not with synthesis of enzymes.

Glucocorticoids. These hormones maintain and aid gluconeogenesis. In the acute response, the effect is probably mediated via the compounds released as a result of the peripheral effects of these hormones. With chronic administration, there is an increase in several enzymes concerned with hepatic gluconeogenesis.

4.2. Fructose Metabolism in the Liver

Most of the fructose taken by mouth reaches the liver in humans as such (Cook, 1969), though the mucosa of the small intestine contains glucose-6-phosphatase (Ockerman, 1964) and can therefore convert fructose to glucose by a pathway similar to that in the liver (Ginsberg and Hers, 1960). Hepatic removal of fructose is probably responsible for the low plasma fructose levels after oral administration compared with intravenous administration, where fructose utilization seems to be directly proportional to the blood fructose concentration (Samols and Holdsworth, 1968).

One of the early references comparing fructose with glucose metabolism in the liver (Oppenheimer, 1912) referred to the greater ability of fructose to form lactate. This observation has since been reported and noted many times in humans, and it has been suggested that this is a good reason for not using intravenous fructose in parenteral therapy (Woods and Alberti, 1972). Unlike glucose, fructose administration leads to an increase in plasma uric acid levels and has been used as a screening test for gout (Schonthal *et al.*, 1972). The reason for the increase in urate production is that the rapid phosphorylation of fructose in the liver decreases the level of ATP, and this, in turn, causes the breakdown of AMP with the formation of adenosine and inosine and then uric acid.

In the liver, fructose is converted to triglyceride to a greater extent than is glucose. Both sugars stimulate the esterification of fatty acids, but fructose has a greater effect than glucose, possibly by the increased formation of the glyceride–glycerol moiety of the triglyceride molecule from fructose (Kupke and Lamprecht, 1967).

In patients who had a myocardial infarction, the formation of triglyceride–glycerol was greater after fructose than after glucose ingestion (Maruhama, 1970). Also, in animals in the fed state, fructose is converted to fatty acids by the liver at a greater rate than is glucose (Zakim, 1973).

It is well known that fructose, whether given orally (Brown *et al.,* 1972) or intravenously, speeds up the metabolism of ethanol, and it is used in the treatment of acute alcohol poisoning. The mechanism whereby this is achieved is not certain but probably involves the ethanol-induced changes in the hepatic redox state (Ylikahri *et al.,* 1972). Galactose also has a metabolic relationship with alcohol, but, unlike fructose, it is the alcohol which affects galactose metabolism rather than the other way around. Alcohol inhibits galactose metabolism, probably due to the inhibition, by ethanol, of the enzyme required for the conversion of UDP-galactose to UDP-glucose (Isselbacher and McCarthy, 1960).

5. Carbohydrates and Muscle

The muscle mass, because of its size, its responsiveness to hormones, and its sheer need, plays a principal role in carbohydrate metabolism. It is widely known that the uptake of glucose by muscle cells is insulin sensitive, and the biochemistry of glucose in the muscle cell is now standard textbook material. Less is known, however, about other carbohydrates which are likely to be presented to the muscle cell. It would seem that the absence of fructokinase in skeletal muscle (Villar-Palasi and Sols, 1957) would minimize any fructose metabolism within the cell. The value of fructose in support of muscle work is most likely to be due to the pyruvate and lactate formed from fructose in the liver (Butterfield *et al.,* 1964) and not by direct utilization.

6. Carbohydrates and Adipose Tissue

The pioneering studies of Lawes and Gilbert in 1852, showing that dietary carbohydrate can be converted to depot fat, are now a fact accepted by all nutritionists and, indeed, by most overweight people. Perhaps the pendulum has swung too far, as it is commonplace to know and read of carbohydrates having a predilection for adipose tissue with the impression that these are more adipogenic than other foods. Per unit weight, fat contains more than twice as much energy as carbohydrate, and the only claim that carbohydrates can have to being more "fattening" than other foods resides in their palatability and not in their potential energy. Historically, the respiratory quotient was used as an indicator that fat was being laid down following carbohydrate ingestion. Benedict and Lee (1937) found that geese that had been force-fed corn had a respiratory quotient of 1.4. They concluded from this that the carbohydrate was being converted to fat, though it was not apparent whether the conversion was taking place in liver or in adipose tissue or in both sites.

Glucose and fructose are the two carbohydrates usually presented to the adipocyte, and the extent of their uptake by the cell is closely controlled by hormones, the most influential of which is insulin. (For details of glucose metabolism in adipose tissue, see Jeanrenaud and Hepp, 1970, and Renold and Cahill, 1965). Part of the glucose taken up by adipose tissue is converted to glycogen (1–5%), which can be used locally for fatty acid or glycerol-3-phosphate formation. The main fate of glucose in the adipocyte is in the formation of fatty acids. Factors which reduce the rate of this synthesis are fasting (Jansen *et al.*, 1966) and a high-fat diet (Hausberger and Milstein, 1965), whereas an increase in fatty acid synthesis from glucose follows a high-carbohydrate diet (Jansen *et al.*, 1966) and an increase in insulin levels (Hausberger, 1958).

Fructose, when presented to the adipocyte, can pass into the cell, but its transport across the cell membrane is hindered by the carrier having a relatively high K_m, so that only at high levels of fructose do significant quantities enter the adipocyte (Froesch, 1972). In the intact animal, very little fructose enters the adipose tissue as such, most of the fructose being converted in the liver to triglyceride, which then enters the adipose tissue (Maruhama and Macdonald, 1973).

For a more detailed discussion of carbohydrate–fat relationships, see Chapter 5.

7. Carbohydrate Metabolism in the Brain

The brain seems to be entirely dependent on glucose and oxygen, and a reduction in either will rapidly lead to irreversible cerebral damage. The brain removes a fixed amount of glucose per unit time irrespective of the plasma concentration of glucose (Sokoloff, 1959), and the uptake of glucose is not dependent on insulin. However, in prolonged starvation, the brain's dependence on glucose diminishes, and it adapts to the utilization of ketone bodies as energy sources (Owen *et al.*, 1967).

An indirect effect of dietary carbohydrate on brain metabolism has been demonstrated in experimental animals, where it has been shown that the insulin release accompanying glucose ingestion increases the level of serotonin (5-HT) in brain tissue (Fernstrom, 1977). Serotonin is reputed to diminish the sensation of pain and produce a feeling of well-being.

8. Carbohydrate Metabolism in the Fetus and Neonate

As fat does not traverse the placenta to any extent (Goldwater and Stetten, 1947), the fat present in the fetus at birth must have been synthesized in the fetus from glucose or amino acid. Glucose probably passes across the placenta by facilitated rather than active transport. The maternal blood glucose is always higher than the corresponding fetal level (Davies, 1955), and changes in maternal glucose levels are quickly reflected in the fetal blood; diffusion cannot explain the observed kinetics (Davies, 1957). In contrast to glucose, fructose cannot cross the placenta. Many animal species, such as goat, horse, pig, and cow, have fructose as the principal blood sugar in the fetus, and this transformation of glucose to fructose is probably carried out in the placenta.

In the neonate, the sole source of carbohydrate is the lactose in milk. As an energy source, lactose is quantitatively small and represents, in the rat, for example, only 10% of the energy needed (Hahn *et al.*, 1961). No specific role for lactose is known, though it has been suggested that calcium absorption is enhanced by lactose (Duncan, 1955).

9. Carbohydrate and Lipid Metabolism*

Although it has been known for some time that dietary carbohydrate can be converted to triglyceride by the body and laid down as such in the adipose tissue, it is only comparatively recently that attention has been directed to the effects that dietary carbohydrate may have on blood lipids. Following the acute ingestion of carbohydrate in humans, the concentration of triglyceride in the serum falls (Havel, 1957). This was considered to be due to the effect of the increase in the level of insulin on lipoprotein lipase activity (Kessler, 1963), an explanation that now seems unlikely (Macdonald *et al.*, 1978).

In contrast, chronic or long-term ingestion of high-carbohydrate diets in humans raises the level of triglyceride in fasting serum (Ahrens *et al.*, 1961), a rise that, over the weeks, tends to return to pre-high-carbohydrate-diet levels (Antonis and Bersohn, 1961). The triglycerides present in fasting serum or in serum after ingestion of a fat-free meal are those produced in the body. They are generally considered to be of greater prognostic value in disease than the exogenous, or dietary, triglycerides found during the absorptive phase of a fat-containing meal, hence the tendency to study serum triglycerides in the fasting state, when they are endogenous. Moreover, after a 12–14 hr fast, the triglyceride concentration in the serum may be at its highest during a normal 24 hr period, as the intermittent intake of carbohydrate will tend to lower the serum triglyceride level (Schlierf *et al.*, 1971).

Not all dietary carbohydrates have the same effect on lipid metabolism, a fact first noted in 1913 in animals (Togel *et al.*, 1913) and in 1919 in humans (Higgins, 1919); fructose has a more marked effect on raising the level of fasting serum triglyceride than does glucose (Macdonald, 1973b). As the disaccharide sucrose is composed of a molecule of fructose as well as glucose, then the lipogenic properties of fructose are seen after consuming sucrose. The main difference in lipid metabolism between glucose and fructose is that fructose forms triglycerides in the liver to a greater extent than glucose, whereas the reverse occurs in adipose tissue.

10. Factors Affecting the Metabolic Response to Dietary Carbohydrate

There is a tendency to consider the nutritional component of the diet in isolation and to present it to the experimental animal or human in quantities

* For full coverage, see Macdonald (1973a).

and circumstances that are a distortion of the real world situation. There are, obviously, many variables that are uncontrolled and uncontrollable in any biological experiment, but there are some variables which can be altered deliberately and carefully, and when this is done the findings may have some meaning. In the context of dietary carbohydrates, there are several variables in the real world situation that do modify the metabolic response to the carbohydrate ingested.

10.1. Sex of the Consumer

A feature of coronary artery disease is the relatively low incidence in premenopausal women as compared to men. If dietary carbohydrate is playing a role in the etiology of this condition, then it might be reasonable to expect that there would be a sex difference in the metabolic response to carbohydrate, at least up to the menopause. This is, in fact, so. The increase in fasting serum triglyceride levels which is seen in men after a diet high in fructose is not seen in young women but is found in postmenopausal women (Macdonald, 1966). This difference may reside in the fact that, though both sexes increase hepatic lipogenesis when ingesting fructose, the premenopausal women are able to clear the serum triglycerides more rapidly (Kekki and Nikkila, 1971).

In the current state of knowledge, there seem to be some contradictory findings on the effect of female hormones on carbohydrate–lipid relationships. Though the reproductive age in a woman seems to protect her against coronary artery disease and elevated fasting serum triglyceride, the latter is increased during pregnancy, oral contraceptive therapy (Wynn and Doar, 1969), and estrogen therapy (Glueck *et al.*, 1972). In experimental primates, the combination of oral contraceptive and sucrose raises the level of fasting serum triglyceride to a greater extent than does either alone (Stovin and Macdonald, 1973).

10.2. Type of Fat Accompanying the Carbohydrate

As the usual diet in Western communities contains only slightly more energy from carbohydrate than from fat, and it is well known that the amount and type of fat consumed can affect serum cholesterol levels, then it is perhaps not surprising to learn that dietary fats can influence fasting serum triglyceride levels. A synergistic effect of sucrose and animal fat on fasting serum triglyceride has been shown in subjects with abnormally elevated serum lipids (Antar *et al.*, 1970), and similar effects have been observed in normolipemic men (Nestel *et al.*, 1970). The increase in fasting serum triglyceride levels seen after high-sucrose diets is abolished by a polyunsaturated fat (sunflower-seed oil) but not by a saturated fat (cream) (Macdonald, 1972).

10.3. ''Sensitivity'' of the Consumer

In any biological system, there is a wide variation in a standard response, and the metabolic response to dietary carbohydrate is no exception. At one end of the range are persons whose fasting serum triglyceride response to a high-carbohydrate diet is negligible, whereas at the other end are found those whose level of triglyceride in fasting serum is very "sensitive" to the amount and type of carbohydrate in the diet (Fredrickson's type IV; Fredrickson *et al.*, 1967) and whose therapy is the reduction of the carbohydrate intake (Kaufmann *et al.*, 1966).

10.4. Dietary Protein

Some *in vitro* work has shown that sucrose increases the absorption of leucine and lysine (Reiser *et al.*, 1975) and that the recovery of serum albumin after protein deficiency is slower with sucrose in the diet than with starch (Grimble, 1975). Earlier, in men, it was found that sucrose caused a greater fall in albumin and a greater rise in globulin than did starch (Coles and Macdonald, 1966). When amino acids were given to men on a high-sucrose diet as the sole nitrogen source, the resulting triglyceridemia was far greater than when the nitrogen intake was from casein, albumin, or gelatin (Coles and Macdonald, 1972). The interplay between dietary carbohydrate and protein becomes more interesting when it is appreciated that arginine and leucine are insulinogenic (Fajans *et al.*, 1967).

10.5. Species of Animal

There is a difference not only between species but also within species in the response to dietary carbohydrate. For example, about 70 g fructose (1 g/ kg body weight), ingested as such by a man, will probably induce an osmotic diarrhea. A comparable dose in rats has no such effect, probably due to different rates of absorption of fructose (Dahlqvist and Thompson, 1963). The response to dietary carbohydrate also varies with the strain of rat used (Duraud *et al.*, 1968).

11. Carbohydrates in the Etiology of Disease

11.1. Obesity

Reference to this has already been made, and, in the main, dietary carbohydrate has no particular role in the cause of obesity. It has been suggested that insulin enables storage of the carbohydrate to be rapid, whereas no such accelerative agent is present for dietary fat storage in the body. On the other hand, energy is required for carbohydrate breakdown in the gut and its conversion to triglyceride. Dietary fat can be placed in the fat depots with less expenditure of energy.

Although it is true that all dietary carbohydrates contain the same amount of energy, it may not be true that they are all handled with equal efficiency by the body. There is experimental evidence in animals that sucrose is metabolized more efficiently than other carbohydrates and hence contributes to greater weight gain (Allen and Leahy, 1966; Brook and Noel, 1969).

There is some tentative evidence in humans to suggest that the efficiency of the metabolic handling of the carbohydrate depends on the nature of that carbohydrate, in that isoenergetic but reduced intakes of sucrose and glucose do not produce the same effect on body weight loss (Macdonald and Taylor, 1973).

11.2. Diabetes

There is considerable disagreement as to whether a high-carbohydrate, in particular, sucrose, intake can give rise to or accelerate the onset of diabetes mellitus, and there is much speculation in this area (Cleave, 1974; Trowell, 1975). The fact that Yemenite Jews who had resided in Israel for 25 years had a higher incidence of diabetes than recent arrivals was put forward in support of the hypothesis that a high-sucrose diet was responsible for the increased incidence (Cohen, 1961). In this case, however, there was another variable in those who had resided in Israel, namely, increased weight, and this could have been responsible.

It has been known for some time that the response to the glucose tolerance test can be improved by a high intake of carbohydrate for a few days prior to the test (Himsworth, 1933). Also, it would be difficult to implicate sucrose in the etiology of diabetes in view of the fact that in a tolerance test, the insulin response to sucrose is less than that to an equivalent does of glucose (Swan *et al.*, 1966). Obesity is probably the main etiologic factor in maturity-onset diabetes (West and Kalbfleisch, 1971).

11.3. Hypertriglyceridemia

It has been postulated that a high intake of carbohydrate, notably sucrose, leads to hypertriglyceridemia. As mentioned earlier, this is true for men and postmenopausal women, but in the majority of persons the effect is transient. Approximately 10% of the adult male population, as shown in surveys in the United States (Wood *et al.*, 1972) and in the United Kingdom (Stone and Dick, 1973), have elevated fasting triglycerides in their serum. In these persons, a high carbohydrate intake is liable to lead to disease, and it is possible that fructose, as such or in sucrose, may accelerate this progress in this small group of persons. If an increased level of serum triglycerides predisposes to atherosclerosis (Fredrickson *et al.*, 1967), then those whose intake of carbo-hydrate is reflected in the serum triglyceride levels should eat little carbohy-drate. (For a full account of dietary carbohydrates and lipids disorders in humans see Albrink, 1973.)

There are many people throughout the world whose diets contain large quantities of carbohydrate, and yet they do not seem to develop hyperlipemia (Higginson and Pepler, 1954). There are several explanations for this, including the fact that the population is not sensitive to carbohydrate-induced hypertriglyceridemia, it does not live long enough to manifest the consequences of hyperlipemia, the high-carbohydrate diet contains little fructose, and the high-carbohydrate diet is accompanied by some vegetable (polyunsaturated) fats which prevent the rise in serum triglycerides. In other words, there are many factors which can modify the body's response to dietary carbohydrate.

Dietary carbohydrate also influences the serum cholesterol, in patricular, the cholesterol carried in the high-density lipoprotein fraction. It is considered that an increase in the concentration of this cholesterol fraction is associated with a decreased risk of coronary artery disease (Miller and Miller, 1975). Dietary carbohydrate lowers the concentration of this cholesterol fraction (Schonfeld *et al.*, 1976), and it is therefore perhaps not surprising that a rise in serum triglyceride concentration is very often accompanied by a fall in the high-density cholesterol level.

11.4. Dental Caries

It is widely accepted that dietary carbohydrates, along with tooth composition, shape, and position, can produce dental caries and that the organisms responsible for this disease prefer sucrose to all other dietary carbohydrates.

11.5. Skin Disorders

In older textbooks of medicine the treatment recommended for pustules or greasy skin was a reduction in the intake of carbohydrates. This advice, based on clinical observation, received some more support in the demonstration that the type of carbohydrate can affect the amount of lipid on the surface of the skin in humans (Llewellyn, 1966) and that *in vitro* metabolism of glucose and fructose in the skin from rats on diets high in glucose or fructose differs (Rebello and Macdonald, 1974).

11.6. Cataracts

When large quantities of galactose are given to rats, cataracts develop in the lens. The length of time required for this to arise depends not only on the amount of galactose in the diet but also on the age of the rat (Richter and Duke, 1970). Cataracts also occur with congenital galactosemia in humans. They can also arise if other monosaccharides gain access to the circulation in large quantities. This is unlikely in humans, however, as are galactose cataracts, because the limited levels of lactase in the intestine prevent large quantities (from lactose) from gaining access to the blood. The cataracts are due to the formation in the lens of dulcitol, the alcohol of galactose, and this formation being more rapid than its removal leads to water retention in the lens.

11.7. Kwashiorkor

Kwashiorkor is a kind of malnutrition seen in young children living in warm climates. It is characterized not only by reduced growth but also by relatively large amounts of fat in the liver and adipose tissue. The latter distinguishes it from marasmus. The fat in the depots and liver of a child with kwashiorkor must have had its origin in dietary carbohydrate, and such a child often has a history of a near adequate intake of energy (as carbohydrate) but an inadequate intake of protein. It is difficult to be convinced that the kwashiorkor seen in countires where sucrose is the staple carbohydrate (e.g., the West Indies) differs at all from the disease seen in those lands where maize is grown. There does, however, seem to be a lower incidence of kwashiorkor than might be expected in those countries where rice is the staple.

12. Other Carbohydrates Consumed by Humans

With modern food technology, previously expensive carbohydrates such as fructose and xylitol are now becoming economically competitive with sucrose and glucose.

Xylitol, a pentitol, follows the pentose-shunt pathway. As a compound, it is sweet, nontoxic by mouth (though leading to osmotic diarrhea if too much is ingested at one time), insulin independent, antiketogenic, and able to be given parenterally. (For review, see Horecker *et al.,* 1969.)

Sorbitol is the alcohol of both glucose and fructose and is converted in the body to fructose for futher metabolism. It is the source of fructose found in seminal fluid. It has been used clinically as a non-insulin-stimulating carbohydrate for diabetics and, for a similar reason, has been incorporated into parenteral nutrition mixtures. Sorbitol is not actively absorbed from the gut and therefore can produce osmotic diarrhea if given in sufficiently large amounts (about 50 g or more at one time in an adult).

13. Conclusion

The scientific neglect that has befallen dietary carbohydrates is being remedied, and no longer are medical students taught that all dietary carbohydrates are converted to glucose and handled by the body as such. The early days of diabetes did not help to make dietary carbohydrates respectable, and perhaps their comparative simplicity in structure and metabolism, as compared to protein, and their supposed failure to dramatically influence a disease as emotionally charged as coronary artery disease have not made them seem very exciting to the young research worker. The recent acknowledgment that dietary carbohydrates can and do play a role in disease processes in humans and that not all carbohydrates have the same metabolic effect should help to foster a greater interest in this major component of the diet.

14. References

Ahrens, E. H., Hirsch, S., Oettle, K., Farquhar, J. W., and Stein, Y., 1961, Carbohydrate-induced and fat-induced lipemia, *Trans. Assoc. Am. Physicians* **74:**134.

Albrink, M. J., 1973, Dietary carbohydrates in lipid disorders in man, *Prog. Biochem. Pharmacol.* **8:**242.

Allen R. J. L., and Leahy, J. S., 1966, Some effects of dietary dextrose, fructose, liquid glucose and sucrose in the adult male rate, *Br. J. Nutr.* **20:**339.

Alvarado, F., 1966, D-Xylose active transport in the hamster small intestine, *Biochim. Biophys. Acta* **112:**292.

Antar, M. A., Little, J. A., Lucus, P., Buckley, G. C., and Csima, A., 1970. Interrelationships between the kinds of dietary carbohydrate and fat in hyperlipoproteinemic patients, *Atherosclerosis* **11:**191.

Antonis, A., and Bersohn, I., 1961. The influence of diet on serum triglycerides in South African white and Bantu prisoners, *Lancet* **1:**3.

Ashmore, J., and Weber, G., 1968, Hormonal control of carbohydrate metabolism in the liver, in: *Carbohydrate Metabolism and Its Disorders,* Vol. 1 (F. Dickens, P. J. Randle, and W. J. Whelan, eds.), pp. 336–374, Academic Press, London.

Benedict, F. G., and Lee, R. C., 1937, *Lipogenesis in the Animal Body,* Carnegie Inst. Washington Publ. No. 489.

Birch, G. G., and Etheridge, I. J., 1973, Short-term effects of feeding rats with glucose syrup fractions and dextrose, *Br. J. Nutr.* **29:**87.

Booher, L. E., Behan, I., and McMeans, E., 1951, Biological utilizations of unmodified and modified food starches, *J. Nutr.* **45:**75.

Borgstrom, B., Dahlqvist, A., Lundh, G., and Sjovall, J., 1957, Studies of intestinal digestion and absorption in the human, *J. Clin. Invest.* **36:**1521.

Brook, M., and Noel, P., 1969, Influence of dietary liquid glucose, sucrose and fructose on body fat formation, *Nature (London)* **222:**562.

Brown, S. S., Forrest, J. A. N., and Roscoe, P., 1972, A controlled trial of fructose in the treatment of acute alcoholic intoxication, *Lancet* **2:**989.

Butterfield, W. J. H., Sargeant, B. M., and Whichelow, M. J., 1964, The metabolism of human forearm tissues after ingestion of glucose, fructose, sucrose of liquid glucose, *Lancet* **1:**574.

Cahill, G. F., and Owen, O. E., 1968, Some observations on carbohydrate metabolism in man, in: *Carbohydrate Metabolism and Its Disorders,* Vol. I (F. Dickens, P. J. Randle, and W. J. Whelan, eds.), pp. 497–522, Academic Press, London.

Cahill, G. F., Ashmore, J., Renold, A. E., and Hastings, A. B., 1959, Blood glucose and the liver, *Am. J. Med.* **26:**264.

Cajori, F. J., 1933, The enzyme activity of dogs' intestinal juice and its relation to intestinal digestion, *Am. J. Physiol.* **104:**659.

Cleave, T. L., 1974, *The Saccharine Disease: The Master Disease of Our Time,* John Wright, Bristol, England.

Cohen, A. M., 1961, Prevalence of diabetes among different ethnic Jewish groups in Israel, *Metab. Clin. Exp.* **10:**50.

Cohen, A. M., and Teitelbaum, A., 1964, Effect of dietary sucrose and starch on oral glucose tolerance and insulin-like activity, *Am. J. Physiol.* **206:**1.

Coles, B. L., and Macdonald, I., 1966, The effect of high carbohydrate diets on the serum proteins of healthy men, *Clin. Sci.* **30:**37.

Coles, B. L., and Macdonald, I., 1972, The influence of dietary protein on dietary carbohydrate: lipid interrelationships, *Nutr. Metab.* **14:**238.

Cook, G. C., 1969, Absorption products of D(−) fructose in man, *Clin. Sci.* **37:**675.

Cook, G. C., 1973, Comparison of absorption rates of glucose and maltose in man *in vivo, Clin. Sci.* **44:**425.

Crane, M. F., Malathi, P., Caspary, W. F., and Ramaswany, K., 1970, A new transport system as the basis for the kinetic advantage contributed to absorption by brush border digestive enzymes, *Gastroenterology* **58:**1038.

Crane, R. K., 1975, The physiology of the intestinal absorption of sugars, *ACS Symp. Ser.* **15:**2.

Crane, R. K., Forstner, G., and Eichholz, A., 1965, Studies on the mechanism of the intestinal absorption of sugars. X, *Biochim. Biophys. Acta* **109:**467.

Dahlqvist, A., and Nordstrom, C., 1966, The distribution of disaccharidase activities in the villi and crypts of the small-intestine mucosa, *Biochim. Biophys. Acta* **113:**624.

Dahlqvist, A., and Thompson, D. L., 1963, The digestion and absorption of sucrose by the intact rat, *J. Physiol. (London)* **167**:193.

Davenport, H. W., 1961, *Physiology of the Digestive Tract,* Yearbook Publishers, Chicago.

Davies, J., 1955, Permeability of the rabbit placenta to glucose and fructose, *Am. J. Physiol.* **181**:532.

Davies, J., 1957, Differential permeability of the rabbit placenta to various sugars, *Am. J. Physiol.* **188**:21.

Dodds, C., Fairweather, F. A., Miller, A. L., and Rose, C. F. M., 1959, Blood sugar response of normal adults to dextrose, sucrose and liquid glucose, *Lancet* **1**:485.

Duncan, D. L., 1955, The physiological effects of lactose, *Nutr. Abstr. Rev.* **25**:309.

Durand, A. N. A., Fisher, N., and Adams, M., 1968, The influence of type of dietary carbohydrate, *Arch. Pathol.* **85**:318.

Fajans, S. S., Floyd, J. C., Knoff, R. F., and Conn, J. W., 1967, Effect of amino acids and proteins on insulin secretion in man, *Recent Prog. Horm. Res.* **23**:617.

Fernstrom, J. D., 1977, Effects of the diet on brain neurotransmitters, *Metabolism* **26**:207.

Fredrickson, D. S., Levy, R. I., and Lees, R. S., 1967, Fat transport in lipoproteins—an integrated approach to mechanisms and disorders, *N. Engl. J. Med.* **276**:273.

Froesch, E. R., 1972, Fructose metabolism in adipose tissue, *Acta Med. Scand. Suppl.* **542**:37.

Ginsberg, V., and Hers, H. G., 1960, On the conversion of fructose to glucose by guinea pig intestine, *Biochim. Biophys. Acta* **38**:427.

Glueck, C. J., Sheel, D., Fishback, J., and Steiner, P., 1972, Estrogen-induced pancreatitis in patients with previously covert familial Type V hyperlipoproteinemia, *Metab. Clin. Exp.* **21**:567.

Goldwater, W. H., and Stetten, D. W., 1947, Studies in fetal metabolism, *J. Biol. Chem.* **169**:723.

Gracey, M., Burke, V., and Oshin, A., 1972, Intestinal transport of fructose, *Biochim. Biophys. Acta* **266**:397.

Gray, G. M., and Ingelfinger, F. J., 1966, Intestinal absorption of sucrose in man: Interrelations of hydrolysis and monosaccharide product absorption, *J. Clin. Invest.* **45**:388.

Grimble, R. F., 1975, Effect of sucrose on recovery of serum albumin concentrations of protein deficient rats to normal, *Nutr. Abstr. Int.* **12**:331.

Hahn, P., Koldovsky, O., Melichar, V., and Novak, M., 1961, Interrelationship between fat and sugar metabolism in infant rats, *Nautre (London)* **192**:1296.

Hausberger, F. X., 1958, Action of insulin and cortisone on adipose tissue, *Diabetes* **7**:211.

Hausberger, F. X., and Milstein, S. W., 1965, Dietary effects on lipogenesis in adipose tissue, *J. Biol. Chem.* **214**:483.

Havel, R. J., 1957, Early effects of fasting and of carbohydrate ingestion on lipids and lipoproteins of serum in man, *J. Clin. Invest.* **36**:855.

Higgins, H. L., 1919, The rapidity with which alcohol and some sugars may serve as nutrients, *Am. J. Physiol.* **41**:258.

Higginson, J., and Pepler, W. J., 1954, Fat intake, serum cholesterol concentration and atherosclerosis in South African Bantu, *J. Clin. Invest.* **33**:1366.

Himsworth, H. P., 1933, The physiological activation of insulin, *Clin. Sci.* **1**:1.

Holdsworth, C. D., and Dawson, A. M., 1964, The absorption of monosaccharides in man, *Clin. Sci.* **27**:371.

Horecker, B. L., Lang, K., and Takagi, Y., 1969, *Metabolism, Physiology and Clinical Uses of Pentoses and Pentitols,* Springer-Verlag, Berlin and New York.

Isselbacher, K. J., and McCarthy, E. A., 1960, Effects of alcohol on the liver: Mechanism of impaired galactose utilization, *J. Clin. Invest.* **39**:999.

Jansen, G. R., Hutchinson, C. F., and Zanetti, N. E., 1966, Studies on lipogenesis *in vivo*, *Biochem. J.* **99**:323.

Jeanrenaud, B., and Hepp, D., 1970, *Adipose Tissue: Regulation and Metabolic Functions,* Academic Press, New York.

Jelinek, B., Katayama, M. C., and Harper, A. E., 1952, The inadequacy of potato starches as dietary carbohydrates for the albino rat, *Can. J. Med. Sci.* **30**:447.

Jos, J., Frezal, J., Rey, J., and Lamy, M., 1967, Histochemical localization of intestinal disaccharidases: Application to peroral biopsy specimens, *Nature (London)* **213**:516.

Jourdan, M. H., 1972, The influence of different amounts and concentrations of glucose on the oral glucose tolerance test, *Guy's Hosp. Rep.* **121**:155.

Kaufmann, N. A., Poznanski, R., Blondheim, S. A., and Stein, Y., 1966, Changes in serum lipid levels of hyperlipemic patients following the feeding of starch, sucrose, and glucose, *Am. J. Clin. Nutr.* **18**:261.

Kekki, M., and Nikkila, E. A., 1971, Plasma triglyceride turnover during use of oral contraceptives, *Metab. Clin. Exp.* **20**:878.

Kessler, J. I., 1963, Effect of diabetes and insulin on the activity of myocardial and adipose tissue lipoprotein lipase of rats, *J. Clin. Invest.* **42**:362.

Kupke, I., and Lamprecht, W., 1967, Über die Biosynthese von Lipiden aus Fructose in Leber. 1. Einbau von uniform markierter [^{14}C]Fructose in Leberlipiden, *Hoppe-Seyler's Z. Physiol. Chem.* **348**:17.

Langworthy, C. F., and Deuel, H. J., 1922, Digestibility of raw rice, arrowroot, canna, cassava, taro, treefern and potato starches, *J. Biol. Chem.* **52**:251.

Lawes, J. B., and Gilbert, J. H., 1852, Composition of foods in relation to respiration and feeding of animals, *Br. Assoc. Adv. Sci.* Ref. 323.

Llewellyn, A. F., 1966, Variations in the composition of the skin surface lipid associated with dietary carbohydrate, *Proc. Nutr. Soc.* **26**:ii.

Macdonald, I., 1966, Influence of fructose and glucose on serum lipid levels in men and pre- and post-menopausal women, *Am. J. Clin. Nutr.* **20**:345.

Macdonald, I., 1972, Relationship between dietary carbohydrates and fats in their influence on serum lipid levels, *Clin. Sci.* **43**:265.

Macdonald, I. (ed.), 1973a, *Effects of Carbohydrates on Lipid Metabolism* (*Progress in Biochemical Pharmacology,* Vol. 8), Phiebig, White Plains, N. Y.

Macdonald, I., 1973b, Effects of dietary carbohydrates on serum lipids, *Prog. Biochem. Pharmacol.* **8**:216.

Macdonald, I., and Taylor, J., 1973, Differences in body weight loss on diets containing either sucrose or glucose syrups, *Guy's Hosp. Rep.* **122**:155.

Macdonald, I., and Turner, L. J., 1968, Serum fructose levels after sucrose or its constituent monosaccharides, *Lancet* **1**:841.

Macdonald, I., Keyser, A., and Pacey, D., 1978, Some effects, in men, of varying the load of glucose, sucrose, fructose or sorbitol on various metabolites in blood, *Am. J. Clin. Nutr.* **31**:1305.

Maruhama, Y., 1970, Conversion of ingested carbohydrate ^{14}C into glycerol and fatty acids of serum triglyceride in patients with myocardial infarction, *Metab. Clin. Exp.* **19**:1085.

Maruhama, Y., and Macdonald, I., 1973, Incorporation of orally administered glucose-U-^{14}C and fructose-U-^{14}C into the triglyceride of liver, plasma and adipose tissue of rats, *Metab. Clin. Exp.* **22**:1205.

McMichael, H. B., Webb, J., and Dawson, A. M., 1967, The absorption of maltose and lactose in man, *Clin. Sci.* **33**:135.

Miller, D., and Crane, R. K., 1961, Localization of disaccharide hydrolysis in isolated brush border portion of intestinal epithelial cells, *Biochim. Biophys. Acta* **52**:293.

Miller, G. J., and Miller, N. E., 1975, Plasma high density lipoprotein concentration and development of ischaemic heart disease, *Lancet* **1**:16.

Moskowitz, H. R., 1974, The psychology of sweetness, in: *Sugars in Nutrition* (H. L. Sipple and K. W. McNutt, eds.) pp. 38–64, Academic Press, New York.

Nestel, P. J., Carroll, K. F., and Havenstein, N., 1970, Plasma triglyceride response to carbohydrates, fats and calorie intake, *Metab. Clin. Exp.* **19**:1.

Ockerman, P. A., 1964, Glucose-6-phosphatase in human jejunal mucosa, *Clin. Chim. Acta* **9**:151.

Oppenheimer, S., 1912, Über Milchsäurebildung in der künstlich durchströmten Leber, *Biochem. Z.* **45**:30.

Owen, D. E., Morgan, A. P., Kemp, H. G., Sullivan, J. M., Herrara, M. G., and Cahill, G. T., 1967, Brain metabolism during fasting, *J. Clin. Invest.* **46**:1589.

Pigman, W., 1957, *The Carbohydrates,* Academic Press, New York.

Randle, P. J., Garland, P. B., Hales, C. N., and Newsholme, E. A., 1963, The glucose fatty-acid cycle, *Lancet* **1**:785.

Rebello, T., and Macdonald, I., 1974, The effect of a high-carbohydrate diet on skin lipogenesis in the rat, *Proc. Nutr. Soc.* **33**:52A.

Reboud, J. P., Marchis-Mouren, G., Cozzone, A., and Desnvelle, P., 1966, Variations in the biosynthesis rate of pancreatic amylase and chymotrypsinogen in response to a starch-rich or protein-rich diet, *Biochem. Biophys. Res. Commun.* **22**:94.

Reiser, S., Michaelis, D. E., and Hallfrisch, J., 1975, Effects of sugars on leucine and lysine uptake by intestinal cells from rats fed sucrose and starch diets, *Proc. Soc. Exp. Biol. Med.* **150**:110.

Renold, A. E., and Cahill, G. F. (eds.), 1965, *Handbook of Physiology,* Section 5; *Adipose Tissue,* Williams and Wilkins, Baltimore.

Richter, C. P., and Duke, J. R., 1970, Cataracts produced in rats by yogurt, *Science* **168**:1372.

Rosensweig, N. S., and Herman, R. H., 1968, Control of jejunal sucrase and maltase activity by dietary sucrose or fructose in man, *J. Clin. Invest.* **47**:2253.

Samols, E., and Holdsworth, D., 1968, Disturbances in carbohydrate metabolism: Liver disease, in: *Carbohydrate Metabolism and Its Disorders,* Vol. 2 (F. Dickens, P. J. Randle, and W. J. Whelan, eds.), pp. 289–336, Academic Press, London.

Schlierf, G., Reinheimer, W., and Stossberg, V., 1971, Diurnal patterns of plasma triglycerides and free fatty acids in normal subjects and in patients with endogenous (type IV) hyperlipoproteinemia, *Nutr. Metab.* **13**:80.

Schonfeld, G., Weidman, S. W., Witztum, S. L., and Bowen, R. M., 1976, Alterations in levels and interrelationships of plasma apolipoproteins induced by diet, *Metabolism* **25**:261.

Schonthal, H., Al-Hujaj, M., and Elbrechter, J., 1972, Zur Therapie der Gicht, *Dtsch. Med. Wochenschr.* **97**:1195.

Shallenberger, R. S., 1971, The theory of sweetness, in: *Sweetness and Sweetners* (G. G. Birch, L. F. Green, and C. B. Coulson, eds.), pp. 42–48, Applied Science, London.

Simon, L. G., 1907, L'activité diastasique de la salive mixte chez l'homme normal et au cours des maladies, *J. Physiol. Pathol. Gén.* **9**:261.

Sokoloff, L., 1959, Metabolism of the central nervous system *in vivo,* in: *Handbook of Physiology,* Section 1, *Neurophysiology,* Vol. 1 (A. W. Magown, ed.), pp. 1843–1964, American Physiological Society, Williams and Wilkins, Baltimore.

Starling, E. H., 1906, *Recent Advances in the Physiology of Digestion,* Constable, London.

Stone, M. C., and Dick, T. B. S., 1973, Prevalence of hyperlipoprotinemias in a random sample of men and in patients with ischaemic heart disease, *Br. Heart J.* **35**:954.

Stovin, V., and Macdonald, I., 1973, Some effects of diet with oral contraceptive on carbohydrate:lipid metabolism in the baboon, *Proc. Nutr. Soc.* **32**:34A.

Sun, D. C. H., and Shay, H. 1961, An evaluation of the starch tolerance test in pancreatic insufficiency, *Gastroenterology* **40**:379.

Sutherland, E. W., Oye, I., and Butcher, R. W., 1965, The action of epinephrine and the role of the adenyl cyclase system in hormone action, *Recent Prog. Horm. Res.* **21**:623.

Swan, D. C., Davidson, P., and Albrink, M., 1966, Effect of simple and complex carbohydrates on plasma non-esterified fatty acids, plasma sugar and plasma insulin during oral carbohydrate tolerance tests, *Lancet* **1**:60.

Togel, O., Brezina, E., and Durig, A., 1913, Über die kohlenhydratsparende Wirkung des Alkoholes, *Biochem. Z.* **1**:296.

Trowell, H. C., 1975, Diabetes mellitus and obesity, in: *Refined Carbohydrate Foods and Disease* (D. P. Burkitt and H. C. Trowell, eds.), pp. 227–249, Academic Press, London.

Uram, J. A., Friedman, L., and Kline, O. L., 1958, Influence of diet on glucose tolerance, *Am. J. Physiol.* **192**:521.

Villar-Palasi, C., and Sols, A., 1957, Phosphorylation du fructose par la phosphorofructokinase du muscle, *Bull. Soc. Chim. Biol.* **39** (Suppl. 2):71.

Weber, G., Singhal, R. L., Stamm, N. B., Lea, M. A., and Fisher, E. A., 1966, Synchronous behaviour pattern of key glycolytic enzymes: Glucokinase, phosphofructokinase and pyruvate kinase. *Adv. Enzyme Regul.* **4**:59.

West, K. M., and Kalbfleisch, J. M., 1971, Influence of nutritional factors on prevalence of diabetes, *Diabetes* **20**:99.

Wood, P. D. S., Stern, M. P., Silver, A., Reaven, G. M., and Groeben, J., 1972, Prevalence of plasma lipoprotein abnormalities in a free-living population of the Central Valley, California, *Circulation* **45**:114.

Woods, H. F., and Alberti, K. G. M. M., 1972, Dangers of intravenous fructose. *Lancet* **2**:1354.

Wynn, V., and Doar, J. W. H., 1969, Some effects of oral contraceptives on carbohydrate metabolism, *Lancet* **2**:761.

Ylikahri, R. H., Kahonen, M. T., and Hassinen, I., 1972, Modification of metabolic effects of ethanol by fructose, *Acta Med. Scand. Suppl.* **542**:141.

Zakim, D., 1973, Influence of fructose on hepatic synthesis of lipids, *Prog. Biochem. Pharmacol.* **8**:161.

15. Further Reading

Berdanier, C. D. (ed.), 1976, *Carbohydrate Metabolism,* Hemisphere Publishing Corp., Washington, D.C.

Dickens, F., Randle, P. J., and Whelan, W. J. (eds.), 1968, *Carbohydrate Metabolism and Its Disorders,* Vols. 1 and 2, Academic Press, London.

Sipple, H. L., and McNutt, K. W. (eds.), 1974, *Sugars in Nutrition,* Academic Press, New York.

Yudkin, J., Edelman, J., and Hough, L. (eds.), 1971, *Sugar,* Butterworths, London.

4

Suppliers of Energy: Fat

Roslyn B. Alfin-Slater and Lilla Aftergood

1. Introduction

Lipids are a heterogenous group of hydrophobic compounds which have two properties in common: They are soluble in organic solvents and they have an important physiological role as structural components of living cells. Lipids have been further identified as "molecules, synthesized by biological systems which have as a major part of their structure long aliphatic hydrocarbon chains unbranched or branched, which may form carbocyclic rings and which may contain unsaturated linkages" (Davenport and Johnson, 1971).

Among the edible lipids are the fats and oils, which are arbitrarily differentiated by their melting points since their other characteristics are quite similar. Both are predominantly triglycerides, are major dietary components for animals (including humans), and are important storage forms of energy in animals and seeds. It is generally considered that oils are liquid at room temperature, whereas fats are solid.

The special suitability of fats as a source of energy is due to certain of their unique properties. First, since they are hydrophobic, and also, since they yield approximately 9 kcal/g as compared with the approximately 4 kcal/g contributed by proteins and carbohydrates, they provide the most efficient way of storing calories. The oxidative breakdown of fatty acids produces a number of high-energy bonds, which translates into a potentially available source of calories. Also, since the storage of hydrophobic proteins and carbohydrates per se is impractical and inefficient, after satisfying the specific metabolic needs of the organism, these nutrients enter the pool of metabolic precursors which are ultimately converted to, and stored as, fat. This process is now readily reversible and therefore provides adipose tissue deposits which are available for future energy needs.

Roslyn B. Alfin-Slater and Lilla Aftergood • Environmental and Nutritional Sciences, School of Public Health, University of California, Los Angeles, California 90024.

The importance of fat in the diet over and above its content of essential fatty acids and fat-soluble vitamins is well documented. Fat brings flavor, satiety, and palatability to food and also provides a concentrated source of energy, especially when the bulk of a diet needs to be limited. Within the body, fat insulates and protects various organs and is involved in the regulation of nutrients crossing the cell membrane.

Several topics related to the subject of lipids, such as the interrelationships of fat and carbohydrate metabolisms, their role as carriers of fat-soluble vitamins and contributors of essential fatty acids, and the metabolism of cholesterol and its role in health and disease, will be discussed elsewhere. This presentation will be limited to a discussion of triglycerides (neutral fat) and their component fatty acids, which include fatty acids of various chain lengths, of various degrees of saturation, and of different spatial configuration (geometric isomers). The contribution of these lipids to nutritional status, their digestion, absorption, and transport, and the factors which affect these processes, their deposition in tissues, their metabolism, and their effect on tissue functions will also be reviewed.

2. Distribution of Lipids in the Diet

Fats and oils are present in varying amounts in many foods. The principal sources of fat in the diet are meats, dairy products, poultry, fish, nuts, salad oils, shortenings, and vegetable seeds from which various oils are expressed to be used for cooking and processing. Most naturally occurring vegetables and fruits contain little if any fat. At present, fats and oils provide more than 40% of the caloric needs in the United States. Of this total dietary fat, approximately 14% is as table spreads, 12% as shortenings, and 6% as cooking and salad oils. Since 1945, there has been a shift away from the more saturated fats in, and derived from, animal products toward the consumption of a greater proportion of unsaturated fats of vegetable origin. This change in food patterns may be a result of recommendations of various health agencies as well as the availability of food products containing larger amounts of unsaturated fatty acids resulting from improved methodology in food technology.

3. Triglycerides: Physical and Chemical Properties

Triglycerides are esters of glycerol with three fatty acids. The number of possible combinations between the three hydroxyl groups of the glycerol molecule and the large number of identified fatty acids is obviously of considerable magnitude.

Fatty acids occur as chains of methylene groups (varying in number and in degree and position of unsaturation) with a terminal carboxyl group. The melting point of the fatty acid depends on its chain length, degree of unsaturation, and spatial configuration; the melting point of a triglyceride is, in turn, a function of its fatty acid composition. For example, the more saturated and the longer the fatty acid chain, the higher the melting point. Saturated fatty acids with more than ten carbons are usually solid at room temperature, whereas the unsaturated C_{18} fatty acid with two double bonds is liquid at room temperature. The unsaturation of a fat, that is, the number of double bonds present in the chain, is generally characterized by the iodine value (IV), since iodine has a quantitative affinity for double bonds. The IV is the number of grams of iodine which will react with 100 g of fat.

In milk fat (and, consequently, butter) there is a relatively larger concentration of the short- or medium-chain triglycerides (6—12 carbons). On the other hand, in most edible oils, the predominant fatty acids consist of chains of 16 or 18 carbon atoms, either saturated or unsaturated to various degrees. Double bonds in fatty acids normally occur in a nonconjugated position; that is, they are of the divinylmethane type: $—CH=CH—CH_2—CH=CH—$. Two such unsaturated fatty acids, linoleic (9,12-octadecadienoic) and arachidonic (5,8,11,14-eicosatetraenoic) are the so-called "essential" fatty acids (EFA). EFA are required by many animal species. Their specific needs in nutrition have been established, and being unable to be synthesized by the body, they must be obtained from food (see Chapter 8).

Isomerism of fatty acids affects not only their properties but also those of the triglycerides in which the isomers are present. Isomers are compounds of the same composition but different molecular structure. Two important types of isomerism may occur in dietary fatty acids, namely, positional and geometric. Positional isomers are fatty acids having the same number of carbon atoms and double bonds but with difference in location of the double bond; for example, vaccenic acid, a minor component of tallow and butterfat, is an 11-octadecenoic acid (C_{18} acid with the double bond between positions 11 and 12) and is a positional isomer of oleic acid, 9-octadecenoic acid (C_{18} acid with a double bond between positions 9 and 10). The position of the double bonds also affects the melting point of the fatty acid, albeit to a limited extent. During hydrogenation, there may be a shift in the location of a double bond in the fatty acid chain.

Geometric, or *cis–trans,* isomerism involves the spatial configuration of the carbon chain around the double bond. If the carbon chains are on the same side of the double bond, the isomer is called *cis,* whereas if the carbon chains are on the opposite sides of the double bond, the isomer is referred to as the *trans* isomer; for example, elaidic (*trans* $C_{18:1}$) and oleic (*cis* $C_{18:1}$) acids are geometric isomers. Obviously, as the number of double bonds in a fatty acid increases, so does the possible number of geometric isomers. The geometric configuration of the atoms at the double bond has an appreciable effect upon the melting point of the fatty acid; *trans* isomers have higher melting points than do their *cis* counterparts.

In general, *cis* isomers occur naturally in food fats and oils. Small amounts of *trans* isomers occur in fats from ruminants, but most *trans* isomers result from industrial hydrogenation processes. The metabolic and nutritional implications of these *trans* isomers in food products are now being studied.

4. Digestion and Absorption

The mechanisms of fat digestion and absorption have been comprehensively reviewed (Ockner and Isselbacher, 1974; Senior, 1964; Westergaard and Dietschy, 1974). Fat from meats, dairy products, and vegetable oils is ingested into the lumen of the gastrointestinal tract in the form of mixed triglycerides. The processes by which water-insoluble triglycerides are catabolized and resynthesized in an aqueous medium by means of interaction with water-soluble enzymes involve the presence of emulsifying agents so that digestion, absorption, and resynthesis may occur.

No digestion of fat occurs in the mouth. In the stomach, fats are released from the food mixture as a result of the partial digestion of carbohydrates and proteins. Little digestion of fat occurs in the stomach, although it has been shown recently that in humans, glands located near the pharynx secrete a lipase that acts in the stomach to hydrolyze long-chain triglycerides (Hamosh *et al.*, 1975). However, the quantitative significance of this enzyme activity is questionable, although it may aid in emulsification. The gastric lipase has a lower molecular weight and a lower pH optimum than does pancreatic lipase, which is the principal enzyme concerned with fat digestion. Gastric lipase is active in infants (Sickinger, 1975) but has only about 10% of its maximum activity in adults.

The chyme is delivered intermittently into the duodenum. Food entering the stomach and the duodenum stimulates the secretion of enteric hormones such as cholecystokinin, enterogastrone, pancreozymin, secretin, and enterocrinin (Sheehy and Floch, 1964). These hormones cause the gallbladder to contract and to secrete stored bile into the duodenum, and they also stimulate the pancreas to secrete pancreatic lipase. Pancreatic lipase is activated by conjugated bile acids due to their surfactant effect and to their ability to lower the pH from 8–9 to 6–7, which is optimal for lipase activity. Peristaltic action in the stomach and upper jejunum breaks down the triglyceride droplets into minute particles. As a result, the relative surface of the triglyceride particles is increased by a factor of 10,000 (Sickinger, 1975). Pancreatic lipase attacks the glycerol–fatty acid ester bonds stepwise at positions 1 and 3, resulting eventually in the release of the 2-monoglyceride and fatty acids.

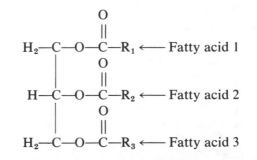

The rate of hydrolysis of triglycerides by pancreatic lipase is dependent upon the chain length and the extent of unsaturation of the constituent fatty acids; unsaturated fatty acid triglycerides are hydrolyzed more rapidly than are saturated fatty acids (Hofmann and Borgström, 1965). Simultaneously, cholesteryl esters and phospholipids are hydrolyzed by a series of enzymes to yield free cholesterol from its ester, lysolecithin from the phospholipid lecithin (phosphatidyl choline), and fatty acids from both substrates:

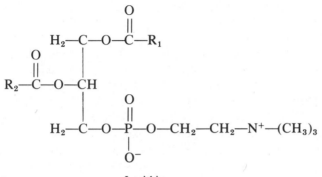

Lecithin

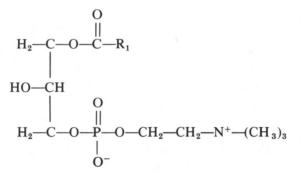

Lysolecithin

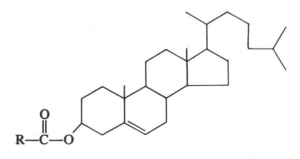

Cholesteryl ester

The emulsified particles are acted upon at the oil–water interface, and the products of the lipolysis are released into the aqueous environment. Only small amounts of these products would be dissolved in the aqueous phase were it not for the presence of phospholipids and bile acids.

The two primary bile acids, cholic and chenodeoxycholic acids, are synthesized from cholesterol by the liver. They are conjugated with taurine or glycine before being secreted into bile. In the intestinal lumen, a portion of the primary bile acid pool is converted to deoxycholic and lithocholic acids. The bile acid pool in humans equals approximately 2–3 g. The primary and secondary bile acids are stored in the gallbladder between meals and are delivered into the lumen during the ingestion of a meal.

Bile acids and their salts have both a hydrophilic and a hydrophobic region, which is responsible for their detergent properties. Above certain concentrations, bile salt molecules aggregate to form macromolecular complexes called micelles (Hofman and Small, 1967). These micelles have the capability of solubilizing less polar compounds, that is, digestive products of lipolysis such as fatty acids, monoglycerides, cholesterol, and lecithin. In this way, water-insoluble compounds such as fat acquire aqueous solubility. Furthermore, bicarbonate secreted by the pancreas provides the proper pH in the duodenum for triglyceride hydrolysis and the concurrent solubilization of the products. The water solubility of monoglycerides and fatty acids is slight, and micellar solubilization is therefore necessary. The critical micellar concentration is dependent upon the degree of saturation of the fatty acids. Unsaturated fatty acids and monoglycerides with an unsaturated fatty acid require a lower bile acid concentration for micellar solubilization than do saturated fatty acids and their monoglycerides. Bile acids also stimulate water secretion from the small and large intestine (Mekhijian *et al.*, 1971).

Above a critical micellar concentration, bile salts tend to inhibit the activity of pancreatic lipase (Borgström *et al.*, 1974). However, another enzyme, namely, colipase, prevents this inhibition by binding to the surface of fat droplets in the presence of bile salts and by providing an attachment site for lipase (Patton and Carey, 1979). It has been suggested that the complex of colipase–bile salts–pancreatic lipase is the enzymatic unit that interacts with the lipid emulsion (Sari *et al.*, 1975; Masoro, 1977).

Direct evidence for the existence of the postprandial mixed micelle has been confirmed by Mansbach *et al.* (1975). Because there is virtually no interaction of bile salts with triglyceride, significant quantities of triglyceride are not incorporated into the mixed micelles. However, long-chain fatty acids are readily incorporated, especially in the presence of monoglyceride.

It has now been established that the multimolecular aggregate, that is, the mixed bile salt–fatty acid–monoglyceride micelle, is stable in water, carries a net negative charge, and is in dynamic equilibrium with its more polar components, each of which may exist in monomeric aqueous solution at relatively low concentration (Ockner and Isselbacher, 1974).

The medium-chain-length fatty acids (6–12 carbon atoms) derived from triglycerides are relatively polar compounds with relatively high water solubility. Thus, they do not require micellar formation for solubilization and further metabolism. In contrast, saturated and unsaturated long-chain fatty acids and cholesterol are much less polar, and consequently, over 96% of long-chain fatty acids are present in micelles rather than in solution in the aqueous phase. In the absence of bile salt micelles in the intestinal lumen, there is a marked decrease in absorption of long-chain fatty acids, although the absorption of short- and medium-chain fatty acids is not affected.

Within the lumen, phosphoglycerides and cholesteryl esters are also hydrolyzed, and free fatty acids, monoglycerides, and lysophosphoglycerides (the compounds with increased polarity) diffuse into bile salt–lecithin–cholesterol micelles delivered into the lumen in the bile (Masoro, 1977). These micelles dissociate at the microvillus border of the intestinal mucosal cell. Fatty acids, monoglycerides, and cholesterol diffuse through the intestinal wall. The released conjugated bile salts participate in the solubilization of additional lipids and eventually travel to the ileum, from where they are absorbed and recirculated via liver and bile (enterohepatic circulation). However, some bile salts are normally lost to the colon, where they undergo bacterial degradation to form a large number of different metabolites, most of which are excreted via the feces (Borgström, 1974). Each individual bile salt molecule synthesized in the liver undergoes 15–20 enterohepatic circulations before it is lost (Tyor *et al.*, 1971).

The entry of long-chain fatty acids into the microvillus membrane is a passive process. A fatty-acid-binding protein is present in the cytoplasm of the intestinal mucosa (Ockner and Isselbacher, 1974) which seems to have a greater affinity for unsaturated fatty acids since, despite the equal rates of uptake, the unsaturated fatty acids are more rapidly incorporated into intestinal lipoproteins (Ockner and Manning, 1974; Ockner *et al.*, 1972).

Luminal emulsions are also stabilized by dietary or endogenous protein, and this facilitates fat absorption similar to that occurring with detergent-stabilized emulsions (Meyer *et al.*, 1976).

In the intestinal mucosal cells, the long-chain fatty acids are activated to combine with coenzyme A (CoA) by an enzyme which has a high affinity for long-chain fatty acids but is relatively inactive toward medium-chain fatty acids. These activated thioesters are then esterified to form triglycerides, preferably via the monoglyceride pathway (Johnston, 1968), which accounts for approximately 85% of newly synthesized triglycerides. The enzymes required for this process are localized in the microsomal portion of the cell, that is, the endoplasmic reticulum.

Unlike absorption, the resynthesis of monoglycerides and fatty acids to triglycerides is an energy-consuming process. In the monoglyceride synthesis pathway, fatty acids activated by a monoglyceride transacylase are transformed chiefly into β-monoglycerides. The L-α-glycerophosphate pathway is an alternate for triglyceride synthesis. The glycerol skeleton in this case is derived either from glycolysis or from lipase action on monoglycerides and subsequent phosphorylation.

In the α-glycerophosphate pathway, glycerol is first phosphorylated to α-

glycerophosphate, which, in turn, reacts with two molecules of activated fatty acids, yielding phosphatidic acid, which then is converted to either triglycerides or phospholipids. However, monoglyceride actually inhibits the α-glycerophosphate pathway of triglyceride synthesis (Polheim *et al.,* 1973).

It has been suggested that the selection of one or the other pathway for triglyceride resynthesis depends upon the location on the microvillus itself (tip versus crypt cell) and subsequently upon the availability of the type of fatty acid (Ockner and Isselbacher, 1974). But there is no apparent preference for oleyl-CoA over palmityl-CoA in the acylation of 2-monoglyceride, suggesting that saturated and unsaturated fatty acyl-CoA molecules generally are utilized to a similar extent (Johnston and Rao, 1965).

The final products of fat absorption, chylomicrons and a significant amount of very-low-density lipoproteins (VLDLs), are formed in the intestinal mucosal cell (Mistilis and Ockner, 1972). Chylomicrons (0.5−1 nm in diameter) consist of over 80% triglycerides, under 10% phospholipids, about 2% cholesterol and its esters, and about 2% protein. VLDLs have a lower triglyceride and a higher phospholipid content (Westergaard and Dietschy, 1974). VLDLs are also synthesized to a large extent in the liver. They usually contain 10–15% cholesterol, 55–65% triglycerides, 15–20% phospholipids, and 5–10% protein. A typical composition of lipoprotein families is presented in Table I. VLDLs of hepatic origin differ from the VLDLs originating in the intestine. The latter contain a significantly higher amount of esterified cholesterol (Polonowski, 1976).

A primitive process, called pinocytosis, also plays a role, albeit very minor, in the absorption of dietary fat. Pinocytosis occurs following a progressive reduction in size of the emulsified fat particles; the small fat droplets come into contact with microvilli, are engulfed by invaginations of the membrane, and are thus transferred to the inner surface of the cell. A mechanism of inverse pinocytosis has been suggested for the release of chylomicrons from the mucosal cell into the intercellular spaces and eventually into the lacteals, which deliver them through lymphatic channels and subsequently into the thoracic duct leading into the blood system (Senior, 1964).

Time-sequence studies (Jersild, 1966) have pointed out that the overall digestion and absorption process is quite rapid. The entire process of fat absorption (as shown by electron-microscopic examination), resulting in the secretion of lipoproteins (''chylomicrons'') into intercellular space, took ap-

Table I. Composition of Plasma Lipoprotein Families (% of Dry Weight)[a]

Lipoprotein constituent	Chylomicrons	VLDL (pre-β)	LDL (β)	HDL (α)
Protein	1–2	5–10	25	45–55
Triglyceride	80–95	55–65	10	8
Unesterified cholesterol	1–3	10	8	14
Esterified cholesterol	2–4	5	37	3
Phospholipid	3–6	15–20	22	22

[a] From Bilheimer and Levy (1973).

proximately 12 min. The process of absorption is so efficient that most of the dietary lipid is absorbed in the first 100 cm of the small intestine (Borgström *et al.*, 1957).

A normal man excretes a relatively constant percentage of dietary fat regardless of the amount he eats. Approximately 95% of the dietary lipid is absorbed; 5% appears in feces (Kasper, 1970). This is true whether the origin of the dietary fat is animal or vegetable.

Intraluminal factors in the midportion of the intestine favor the hydrolysis and solubilization of dietary lipid, thus promoting its absorption. Although it is generally believed that fat absorption is virtually complete in the jejunum, there is evidence obtained through fat balance studies which indicates that the ileum may be in regular use as an organ of fat absorption (Wollaeger, 1973).

Chylomicron formation is a prerequisite for fat absorption. A well-functioning mechanism of the protein (apo-low-density-lipoprotein, apo-LDL) synthesis seems to be necessary for the formation of chylomicrons. In addition, small peptides associate with the lipoprotein particle after it enters lymph and are necessary for the activation of lipoprotein lipase in the capillary endothelium, thereby permitting the entry of fatty acids and 2-monoglyceride into adipose tissue (Bier and Havel, 1970).

Medium-chain triglycerides (MCTs), which contain fatty acid molecules with chain lengths varying from 6 to 12 carbon atoms, are absorbed differently from triglycerides with fatty acids of 16 and 18 carbons (Greenberger and Skillman, 1969). These medium-chain-length fatty acids are absorbed and transported directly to the liver via the portal vein without esterification. They enter the portal circulation as free fatty acids bound to plasma albumin and are subsequently oxidized in the liver.

The smaller molecular size and greater water solubility of MCTs and the fatty acids derived from them give these compounds some unique properties: more rapid hydrolysis, no requirement for bile acids for digestion and absorption, minimal reesterification, no requirement for chylomicron formation after absorption, and, subsequently, a route of transport different from that of the long-chain fatty acid triglycerides. As a result, MCTs have potential therapeutic applications in patients with lipid malabsorption, which include those with pancreatic insufficiency, cystic fibrosis, sprue, and short-bowel syndromes (Greenberger *et al.*, 1967; Wiley and Leveille, 1973). MCTs have received widespread acceptance in recent years as a means of providing calories to persons unable to absorb long-chain fatty acids.

However, MCTs have been shown to induce ketosis in several mammalian species, including humans. It has been suggested (Yeh and Zee, 1976) that this type of ketosis results from rapid oxidation of the medium-chain fatty acids. Hyperinsulinemia, hypoglycemia, and depressed lipogenesis resulting from MCT feeding appear to potentiate but not initiate ketosis.

Fat malabsorption has also been shown to occur in EFA deficiency in rats (Clark *et al.*, 1973). This is probably due to an impairment in formation of appropriate phospholipids, which then interferes with the formation of, or a delay in, the removal of newly synthesized triglyceride from the mucosa.

5. Transport

Most of the triglycerides enter the circulation as chylomicrons (Goldrick, 1971). In the circulation, these primary particles interact with other plasma lipoproteins, form smaller secondary particles, and thus become a mixture of exogenous and endogenously synthesized triglycerides, together with cholesterol (Bierman *et al.*, 1962). Phospholipids and some protein serve as stabilizers.

Approximately 20–40% of chylomicron triglycerides are removed by the liver, where they are hydrolyzed to glycerol and free fatty acids. The fatty acids enter the hepatic free fatty acid pool and are mixed with fatty acids which have been synthesized *in situ* or transported from adipose tissue. In general, the ingestion of fat (or carbohydrate) is followed by the synthesis and release of lipoproteins (VLDLs) by the liver which transport the dietary and endogenous triglycerides to muscle and adipose tissue.

It has recently been confirmed that the output of triglycerides by the liver is stimulated by the presence of free fatty acids. Also, there are sex differences in triglyceride synthesis and/or release, since livers from female rats secrete more triglycerides than do livers from male animals (Soler-Argilaga *et al.*, 1976). Phospholipids of the VLDLs in the females usually contain more stearate and less oleate and linoleate as compared to the males. This may be related to observed sex differences in hepatic phospholipid metabolism (Lyman *et al.*, 1967).

The nutritional status of the subject affects the disposition of the circulating triglycerides. In the fed animal, the excess triglycerides as chylomicrons bypass the liver and, together with VLDLs synthesized in the liver, are taken up by the adipose tissue. In the fasting state or during a low caloric intake, muscle is the major consumer of triglycerides.

The enzyme lipoprotein lipase is important in the transfer of triglycerides from chylomicrons and VLDLs, to muscle and adipose tissue. Lipoprotein lipase is produced in these tissues, and it participates in the hydrolysis of triglycerides to glycerol and free fatty acids. The fatty acids are then taken up by the tissue for subsequent oxidation or esterification, depending on the specific needs based on the nutritional status. The activity of lipoprotein lipase is stimulated by insulin (Austin and Nestel, 1968) which, in turn, is elevated during the ingestion of protein and carbohydrate, thus facilitating the transfer of triglyceride fatty acids to adipose tissue for storage. The lipoprotein lipase activity, low in adipose tissue, increases in muscle during starvation, thus releasing lipids for energy purposes.

Lipids are also transported in blood as free fatty acids, mainly oleic (Spitzer and Gold, 1962), which originate either from the hydrolysis of chylomicron triglyceride (Nestel, 1964) or, more usually, from adipose tissue. The release of free fatty acids from adipose tissue is controlled by the energy requirements of other tissues. In the fed state, they usually exist in low concentrations bound to plasma albumin. In the postabsorptive state, free fatty acid concentrations in plasma are increased and serve as a source of energy for tissues. Physical exercise and/or starvation also cause elevations in free fatty acids.

Plasma free fatty acids are also responsive to hormonal influences. A deficiency of or insensitivity of the host to insulin, epinephrine, and thyroid hormone leads to elevated levels of free fatty acids (Hales and Randle, 1963; Mueller and Horwitz, 1962; Rich *et al.*, 1959). On the other hand, prostaglandins inhibit free fatty acid mobilization (Steinberg *et al.*, 1964).

Although VLDLs are generally considered to be primarily concerned with the transport of triglycerides, other lipoproteins play minor but possibly significant roles. LDLs have been shown to be derived from VLDLs (Quarfordt *et al.*, 1970). Although not all LDLs originate in VLDLs (Bilheimer and Levy, 1973), this particular process is probably unidirectional. In the absence of circulating LDLs, a certain degree of demyelination occurs, as well as steatorrhea (Fasoli, 1973) and general disturbances in triglyceride transport. Abetalipoproteinemia, a genetic disorder, is characterized by the absence of circulating LDLs. As a result, in patients suffering from abetalipoproteinemia where LDLs as well as their protein moiety (apo-LDL) are absent from plasma, there is interference with absorption of fat. These patients are unable to form chylomicrons, and, as a result, triglycerides accumulate in the absorptive cells. In these cases, an alternate absorptive pathway allows a limited but direct entry of unesterified long-chain fatty acids into the portal vein (Kayden and Medick, 1969).

High-density lipoproteins (HDLs) also act as lipid carriers in plasma. Even though HDLs contain the least amount of lipids (approximately 50%, with less than 5% triglycerides), a congenital absence of this lipoprotein, known as Tangier disease, is associated with defective removal of triglycerides from the circulation and with excessive deposition of cholesterol esters in various tissues (Fasoli, 1973). Nichols and Smith (1965) have shown in other experiments that triglycerides are transferred from VLDLs to HDLs, associated with a reciprocal exchange of cholesteryl esters from HDLs. Thus, one of the major functions of both LDLs and HDLs appears to be the transport of triglycerides (Nichols, 1969). In addition, it is believed that HDLs are involved in transporting cholesterol from tissues to the liver for subsequent catabolism, thus contributing to its disposal (Scanu, 1978). The composition of plasma lipoprotein families is shown in Table I.

Although serum triglyceride levels (and cholesterol as well) have been shown to be lower in mentally retarded patients (Lawlor *et al.*, 1974) irrespective of diet, it has been reported recently that 27 patients given a low-fat diet (30% fat) for one week showed a 41% increase in triglyceride levels; cholesterol levels were not affected (Ginsberg *et al.*, 1976). On the other hand, in acromegaly, serum triglyceride levels are usually higher, and cholesterol levels lower, than in the normal population (Nikkilä and Pelkonen, 1975). A possible seasonal variation of plasma triglycerides has been reported in a healthy population (Fuller *et al.*, 1974); a highly significant decrease in plasma triglycerides in winter as compared with summer and a less marked decrease in spring as compared with autumn were observed. These changes could not be explained in terms of changes in weight or diet.

6. Utilization of Fats by Tissues

6.1. Heart

The heart muscle appears to be unique in its fuel requirements. Since it must depend on energy sources that are available even during fasting, it preferentially utilizes fat. The low respiratory quotient (0.70) of the fasting heart muscle shows its dependence on fat oxidation (Goodale *et al.*, 1959). Some workers (Rothlin and Bing, 1961; Willebrands, 1964), but not all (Stein and Stein, 1963), have reported a preferential utilization of unsaturated fatty acids by this tissue.

It has been shown that animals fed rapeseed oil which is high in erucic acid (22:1) accumulate triglyceride containing erucic acid in the cardiac tissue as early as after one week of experimental feeding (Beare-Rogers and Gordon, 1976; Beare-Rogers *et al.*, 1972; Kramer, 1973). In fact, erucic acid is poorly oxidized and is preferentially incorporated (Teige and Beare-Rogers, 1973; Vasdev and Kako, 1976). This results in an inhibition of the oxidation of all dietary fatty acids coming to the heart. The inability of cardiac tissue to readily metabolize these fatty acids leads to their incorporation into triglycerides which are deposited *in situ*. A fatty acid imbalance is produced, leading to a condition similar to that of EFA deficiency. Since the amount of sphingomyelin increases in cardiac phospholipids, some alterations of the membrane lipids are suspected.

As a result of complete hydrogenation, erucic acid is converted to behenic acid (22:0). Ingestion of fat containing behenic acid did not result in the accumulation of lipids in the hearts of weanling rats (Mattson and Streck, 1974), even though feeding of similar levels of erucic acid (approximately 50% of the dietary fat) resulted in a two- to threefold increase in heart lipids.

Although studies of human patients have shown no preferential uptake of free erucic acid by myocardium (Jaillard *et al.*, 1973), a possible cumulative effect after long-term administration of low doses cannot be overlooked. Rapeseed oil with low erucic acid content has been developed in Canada (Downey *et al.*, 1969).

6.2. Adipose Tissue

The mature fat cell (in adipose tissue) consists of cytoplasm stretched over a droplet of triglyceride which occupies up to 99% of the volume (Goldrick, 1971). In the cytoplasmic compartment, triglycerides are hydrolyzed to free fatty acids and glycerol and are resynthesized by the esterification of free fatty acids with α-glycerophosphate. The mobilization of lipid via the release of free fatty acids from the fat droplet into the extracellular fluid occurs as the need arises.

In fed subjects under conditions of caloric balance, the rate-limiting factor in esterification of fatty acids to form triglycerides is the availability of α-glycerophosphate, The rate of esterification also depends upon the levels of plasma insulin available to promote the metabolism of carbohydrate leading to formation of this precursor. During the height of insulin availability, triglycerides from lipoproteins are incorporated into the fat cell, and, at the same time, carbohydrate is converted into fat.

Lipoprotein lipase functions at the endothelial cell surface in adipose tissue *in vivo;* it is derived from a precursor that exists in the fat cell (Robinson *et al.,* 1975). Two forms of this enzyme (*a* and *b*) differing in molecular weight have been identified. The proportion of the total activity that is contributed by the enzyme with the higher molecular weight is greater in rats in a fed state, suggesting that this is the one which functions at the cell surface (Schotz and Garfinkel, 1972).

In times of caloric deprivation, lipolysis in adipocytes is effected by lipases, which eventually produce glycerol and free fatty acids. The lipase that converts tri- to diglycerides is the hormone-sensitive lipase which also requires activation by cyclic AMP (cAMP). The hormones epinephrine and ACTH accelerate lipolysis by stimulating the production of cAMP, whereas caffeine and theophylline are lipolytic due to an inhibition of the breakdown of cAMP. The relationship between the changes in adipose cell cAMP and the effect of lipolytic hormones has been well documented, but it is also possible that other factors, such as possible changes in the intracellular calcium, may be involved (Siddle and Hales, 1975).

The increased release of free fatty acids from adipose tissue into the bloodstream in fasting individuals is due mainly to a decreased esterification resulting from a decreased glucose uptake into adipose tissue. A feedback hypothesis relating lipolysis to the synthesis of prostaglandins (PGs) has been proposed (Shaw and Ramwell, 1968). If this is true, diets which increase the levels of PG precursors in adipose tissue should lower rates of lipolysis. It has been reported that lipolysis in adipose tissue of rats fed high levels of linolenic acid (18:3) is greater than in rats given high levels of linoleic acid (18:2) (Larking and Nye, 1975). It has been suggested that there is competition by 18:3 for the enzymes involved in the elongation and desaturation of 18:2 to arachidonic acid (20:4) and other precursors of PGE_1 and PGE_2, resulting in a decreased production of PGs.

The rate of fatty acid release from stored adipose tissue triglycerides during periods of caloric deficiency and stress is determined by the relative rates of hydrolysis of these triglycerides and of reesterification of fatty acids, and by transport from the cell whereby fatty acids can combine with albumin. Free fatty acids from adipose tissue are utilized in many organs and supply a considerable portion of the metabolic energy of heart and skeletal muscle.

Adipose tissue is an important source of energy, especially in the newborn and in infants (Shiff *et al.,* 1966). The lipid content (98% triglyceride) of adipose tissue accounts for approximately 40% of adipose tissue weight in the newborn and increases to 75% with increasing age in the adult (Baker, 1969). The fatty acid composition of adipocytes also changes with age (Birkbeck, 1970). The concentration of linoleic acid in adipose tissue is usually high in the infant and thereafter diminishes. Eventually sex differences in fatty acid composition of adipose tissue develop; that is, lower oleic and higher stearic acid concentrations have been reported in normal adult males as compared with normal adult females (Heffernan, 1964). Inhabitants of the United States have higher levels of myristic, palmitic, stearic, and oleic acids and lower levels of palmitoleic, linolenic, and linolenic acids in adipose tissue than their Japanese counterparts. In addition, whereas the Japanese have a higher content of polyunsaturated acids regardless of age, linoleic acid seems to decrease with age in the United States population (Insull *et al.,* 1969).

In studies with rats, physical training has been found to be associated with a significant increase in the enzyme glyceride synthetase in muscle and adipose tissue (Askew *et al.,* 1975a,b). In contrast to glyceride synthesis, no increase in adipose tissue lipogenic potential was noted in response to training, indicating that the physically trained rat may have an enhanced ability to store but not synthesize fatty acids.

The proportion of total energy derived from free fatty acid is considerably elevated during exercise. The initial fall in plasma free fatty acids is accompanied by an increase in the rate of efflux from plasma (Carlson and Pernow, 1961); this is followed by an elevation associated with an increased turnover, indicating an increase in free fatty acid mobilization from adipose tissue. The sudden increase of free fatty acids at the end of exercise reflects the continued influx, even though its utilization by muscle falls abruptly. A rise in plasma glycerol concentration during exercise has also been reported (Carlson *et al.,* 1963). There has been a suggestion that exercising muscle may utilize some fatty acids in preference to others (i.e., oleate rather than palmitate) (Wood *et al.,* 1965).

A relationship between the composition of dietary fat and the fatty acid composition of adipose tissue has been demonstrated in humans, although the adipose tissue in adults changes only slowly in response to changes in the diet (Hirsch *et al.,* 1960). In the young infant, the composition of the body fat is profoundly influenced by the type of fat being fed (Widdowson *et al.,* 1975).

An excessive accumulation of adipose tissue in the body leads to obesity. The etiology of obesity may be viewed as an anomaly of normal weight regulation. Excessive weight gain during early infancy has been associated with a greater incidence of obesity during later years. It has been reported that adipose tissue develops through stages of hyperplasia first and hypertrophy later (Cioffi and Speranza, 1972; Hirsch and Han, 1969). It has been suggested that the number of fat cells which thereafter regulate fat storage becomes fixed before adulthood is reached (Knittle, 1972).

6.3. Skin

The skin occupies a rather unique position among the organs. It serves as a transition surface between the exterior and the internal milieu. Its lipids are very specific. The structural entity is provided by phospholipids and sterols; however, there is also a good supply of triglycerides, assuring sufficient materials to provide the energy necessary to produce heat to maintain temperature control. In fact, the lipid of the product of sebaceous glands (sebum) contains approximately 60% triglycerides. The uniqueness of skin lipids is expressed primarily by the unusual types of fatty chains synthesized, the wide range of chain lengths, and unique patterns of unsaturation. Branched-chain fatty acids, odd-numbered-chain fatty acids, as well as fatty acids with up to 30 carbon atoms in length are easily found among skin lipids. It has been suggested (Nicolaides, 1974) that such compounds may pose metabolic problems to potential pathogens—blocking their entry and, in fact, causing their destruction—and thus contribute to the survival of only compatible microorganisms.

6.4. Liver

The liver is the major source of endogenously synthesized triglycerides. In hepatectomized animals, practically no synthesis of triglycerides occurs (Borgström and Olivecrona, 1961). Hepatic triglycerides are the immediate precursors of triglycerides contained in plasma VLDLs. The turnover rate and pool size of hepatic triglycerides are considerably greater (about three times) than those of plasma triglycerides. Only about 3% of newly synthesized triglyceride is secreted from the liver into the plasma. During fasting, extrahepatic removal of triglyceride is minimal (Nestel, 1967). Minor changes in hepatic triglyceride turnover are reflected by major changes in plasma triglyceride concentration (Farquhar *et al.*, 1965). In general, the accumulation of triglycerides in the liver ("fatty liver") appears to be due to a defect in the transport system for fatty acids rather than to a defect in the fatty acid synthesizing mechanism.

The consumption of alcohol leads to an increase in both plasma and liver triglycerides. In fact, it has been shown that a single dose of alcohol leads to an accumulation in the liver of fat derived from adipose fatty acids. When alcohol was consumed for many days, the fatty acids found in the liver resembled those derived from endogenous synthesis; however, when fats were fed along with the alcohol, much of the fat in the liver was of dietary origin (Lieber *et al.*, 1965). It has been suggested that alcohol-fed subjects have a diminished utilization of triglyceride by extrahepatic tissues and oxidation of fat in the liver (Jones *et al.*, 1963). Also, an increase in fatty acid esterification within the liver is a major factor in the pathogenesis of alcohol-induced hyperlipemia.

6.5. Central Nervous System

In the central nervous system, brain lipids contain primarily long-chain fatty acids which are inserted enzymatically into their respective positions in phospholipids with a high degree of specificity (Ramsey and Nicholas, 1972). Fatty acid elongation correlates closely with the period of rapid myelination (in the rat). The incorporation of acetate into long-chain fatty acids of the adult rat brain is low as compared to analogous fractions in the weanling rat (Dhopeshwarkar *et al.*, 1969). However, linoleic acid administered orally can also be incorporated into brain lipids, thus crossing the blood–brain barrier (Dhopeshwarkar *et al.*, 1971). A large group of even- and odd-numbered α-OH fatty acids with 20–26 carbon atoms occurs in brain cerebrosides (Fulco and Mead, 1961).

The predominant fatty acids in human fetal brain lipids are palmitic, stearic, and oleic. In the adult brain, oleic acid is the major fatty acid of phospholipid fractions. Linoleic acid is practically absent in the fetal brain but represents 4.5% of total fatty acids of the adult brain (Eichberg *et al.,* 1969; Hansen and Clausen, 1968).

Changes in brain lipid content are very often associated with a variety of disease states (Ramsey and Nicholas, 1972). One of the pathological states involving an imbalance of neutral lipids is Refsum's disease, which is characterized by the accumulation of 3,7,11,15-tetramethylhexadecanoic acid (phytanic acid) of exogenous origin. Apparently, α-oxidation, usually quite active in the brain, and necessary for the conversion of phytanic to pristanic acid, is blocked. The methyl group on the β position of phytanic acid prevents its degradation by β-oxidation unless prior α-oxidation occurs. Wolman's disease is characterized by elevated levels of triglycerides and cholesterol in the central nervous system, evidently related to an enzymatic defect of lipolysis in the spleen and liver. Possible changes in the synthesis of $C_{20:1}$ fatty acid may influence a predisposition to multiple sclerosis. These and many other pathological states involving lipids in the brain emphasize the important role of fats in the function of this tissue.

Gaucher's disease is a hereditary disorder in which the main clinical finding is hepatosplenomegaly. In infants, there is an involvement of the central nervous system. Neurological manifestations are caused by a disturbance in the metabolism of brain gangliosides, leading to an accumulation of some minor gangliosides (Svennerholm, 1966). The white matter shows a decrease in all lipid classes (Banker *et al.,* 1962).

In general, a common abnormality in all infantile cases of hereditary disorders of lipid metabolism is that the fatty acid pattern of the brain is similar to that found normally in the immature brain (Schettler and Kahlke, 1967b). In Niemann–Pick disease, sphingomyelin deposits are found in most organs and tissues. For instance, total brain sphingomyelins were increased tenfold as compared to the normal infant brain (Rouser *et al.,* 1965).

In Tay–Sachs disease, the infantile amaurotic family idiocy, there is accumulation in the brain of a ganglioside which occurs normally only in trace amounts (Schettler and Kahlke, 1967a).

7. Fat as an Energy Source

Triglycerides are the best storage form of energy because of difficulties associated with the deposition of excess protein or carbohydrate. Proteins are usually involved in functions other than those of energy production, and carbohydrate storage is impractical. Storage of fat is most efficient (until it becomes exaggerated and results in obesity).

The very rapid fractional turnover rate of plasma free fatty acids in a man (approximately 20 times each hour) results in sufficient lipid to provide for his basal requirements. However, a substantial amount of fuel is derived from lipids other than free fatty acids. These may be fatty acid esters associated with the circulating lipoproteins or the endogenous stores of lipid in other tissues. In the fasting, resting man, about 25% of the energy requirement is met by the oxidation of free fatty acids, 15% is derived from glucose metabolism, and the remainder is provided by endogenous stores of esterified fatty acid (Nestel, 1967).

7.1. Fatty Acid Oxidation

The end products of reactions catalyzed by lipases and phospholipases are mainly free fatty acids. The major energy pathway subsequently involves oxidation of these fatty acids through β-oxidation, catalyzed by a series of enzymes located in the mitochondrion. Formation of a thioester between the CoA and the long-chain fatty acid molecule constitutes the activation necessary for the process to proceed. This activation requires Mg^{++} and three ATP-dependent synthetases with specificities varying according to the chain length of the fatty acid. A GTP (guanosine triphosphate) synthetase has also been identified and may play a role.

The β-oxidation cycle consists of an α,β-dehydrogenation, hydration of the *trans*-α,β-unsaturated acyl-CoA, followed by the dehydrogenation of $L(+)$-β-hydroxyacyl-CoA to yield β-ketoacyl-CoA. Finally, a cleavage takes place under the influence of a thiolase to form acetyl-CoA and an acyl-CoA with two carbon atoms less than the original fatty acid. This reaction is practically irreversible, and the new acyl-CoA product reenters the β-oxidation cycle to be further degraded. Unsaturated fatty acids require additional enzymes to allow the reactions to go to completion. Assuming that the oxidation of fatty acids proceeds by the successive removal of two-carbon fragments from the carboxyl end of the compound, yielding acetyl-CoA molecule, the overall reaction for the oxidation of stearic acid would be

$$CH_3(CH_2)_{16}COOH + 9\,CoASH + 4\,O_2 \rightarrow 9\,CH_3COSCoA + H_2O + 8\,H_2$$

The net yield of energy from the oxidation of one molecule of stearic acid amounts to 146 molecules of ATP with an approximate ΔF of 8000 cal/mol of ATP. (The change in free energy, ΔF, is the energy that becomes available to be utilized.)

The major product of the β-oxidation of fatty acids is acetyl-CoA, which either can be completely oxidized to carbon dioxide and water via the citric acid cycle or else can enter other metabolic processes. In the liver, acetoacetate produced by the condensation of two molecules of acetyl-CoA can lead to the production of ketone bodies (acetoacetic acid, β-hydroxybutyric acid, and acetone) during fatty acid oxidation. An active carbohydrate metabolism which assures the availability of the receptor molecule, oxaloacetate, prevents the accumulation of acetoacetate in the liver.

Originally, the formation of ketones was thought to indicate an abnormal metabolism produced during diabetic ketoacidosis as a result of altered carbohydrate metabolism. However, it is now known that under certain conditions, ketones can be a valuable source of energy. Skeletal muscle can oxidize ketone bodies, and this is of more importance in physically trained than in untrained individuals (Askew *et al.,* 1975b). The brain can utilize ketone bodies as precursors for lipid synthesis as well as for energy purposes (Patel, 1975).

Also, during fasting, circulating ketone bodies exert an inhibitory influence on the rate of ketogenesis (Balasse and Neef, 1975), thereby preventing an uncontrolled hyperketonemia during starvation.

7.2. Fatty Acid Synthesis

The equilibria of certain reactions involved in the oxidation of fatty acids (in particular, the 3-ketoacyl-CoA thiolase reaction) favor the breakdown of fatty acids and thus preclude the idea that fatty acid synthesis is a simple reversal of oxidation. Furthermore, biotin, not required for oxidative breakdown of fatty acids, is required for fatty acid synthesis.

The oxidative and synthetic pathways have many similarities but many differences as well. The main (*de novo*) pathway of biosynthesis takes place in the cytoplasm. The first step involves the formation of malonyl-CoA by the carboxylation of acetyl-CoA, followed by a series of reactions catalyzed (in liver) by a multienzyme complex, and eventually leading to the production of saturated fatty acids containing predominantly 16 or 18 carbon atoms. Only one molecule of acetate is incorporated directly into the fatty acid molecule, and this is always located at the methyl end of the chain. The elongation proceeds through a stepwise addition of two-carbon units derived from malonyl-CoA. The protein-attached acyl thioester is the preferred substrate for all reactions. The overall energy-consuming reaction leading to the synthesis of palmitic acid is as follows:

$$1 \text{ Acetyl-CoA} + 7 \text{ Malonyl-CoA} + 14 \text{ NADPH} + 14 \text{ H}^+ \rightarrow 1 \text{ Palmitic acid}$$

$$+ 7 \text{ CO}_2 + 8 \text{ CoA} + 14 \text{ NADP}^+ + 6 \text{ H}_2\text{O}$$

Liver mitochondria have an enzyme system capable of carrying out the elongation of long-chain acyl-CoAs by the addition of acetate. And the microsomal fraction of liver contains an enzyme system which catalyzes the elongation of acyl-CoAs containing ten or more carbon atoms by the addition of two-carbon units derived from malonyl-CoA.

The proper functioning of the biosynthetic mechanism depends upon many factors, among others, the availability of a transfer mechanism for acetyl-CoA from the mitochondrion into the cytoplasm and the availability of reducing cofactors (NADPH), which, in turn, depends upon the nutritional status of the organism. NADPH is provided by the well-functioning carbohydrate metabolism.

8. Fatty Acid Isomerism

The process of hydrogenation of vegetable oils containing large amounts of polyunsaturated fatty acids in their natural *cis–cis* form has resulted in the formation of unsaturated fatty acid isomers with a *trans* configuration. These *trans* fatty acids are incorporated into various lipid formations. Recently, it was shown that when rats were fed a diet containing 15% of calories as elaidic acid (*trans* 18:1), serum triglycerides (followed by the phospholipids) contained the greatest amount of *trans* fatty acid (Schrock and Connor, 1975). The triglycerides of HDLs contained fewer *trans* fatty acids than did the triglycerides of other lipoproteins.

Other workers (Alfin-Slater *et al.*, 1976) have shown that when fats containing approximately 45% *trans* monounsaturated fatty acids and 6% *trans,trans*-linoleic acid are fed to rats for extended periods of time, phospholipids rather than other lipid fractions, in tissues such as liver, plasma, red blood cells and heart preferentially accumulated *trans* fatty acids. Since a mixture of phosphoglycerides (phospholipids) plays a significant role in cell membranes, providing a hydrophobic–hydrophilic bridge active in the transfer of nutrients, changes in the composition of these compounds may affect metabolic processes. Although they are oxidized to carbon dioxide as readily as their *cis* counterparts (Anderson and Coots, 1967), the *trans* isomers of linoleic acid cannot function as an EFA and, in fact, they actually interfere with the conversion of *cis,cis*-linoleic acid to arachidonic acid (Privett and Blank, 1964). In animals, *trans* fatty acids, when fed as a sole source of fat in the diet, have been shown to intensify the symptoms of an EFA deficiency and, in fact, resemble saturated fatty acids in many metabolic reactions (Takatori *et al.*, 1976). Feeding of hydrogenated fat (containing *trans* isomers) to rats has been reported to induce an accumulation of lipids in certain organs and depositions of *trans* isomers in certain lipid fractions (Egwim and Kummerow, 1972). However, a margarine fat containing up to 35% *trans* isomers has been fed to rats for 75 generations with no deleterious effects (Alfin-Slater *et al.*, 1973).

The *trans* isomers of oleic and linoleic acid are present in shortenings and margarines, and it has been estimated that at least 15% of the fat consumed in the American diet is derived from fatty acids having the *trans* configuration (Kummerow, 1974; Meyer, 1974). These acids are often not metabolically equivalent to their *cis* isomers, and, therefore, it is important to elucidate the metabolic activity of these isomers not only as concerns their tissue deposits, but also with respect to any possible interferences with the metabolism of other fatty acids and/or possible changes in tissue function which may result.

9. Conclusions

Hypercholesterolemia and hypertriglyceridemia have been implicated as risk factors in atherosclerosis and coronary heart disease, and epidemiological and laboratory evidence has suggested that dietary fat may be important in the regulation of these lipid parameters. It has, therefore, been suggested (American Heart Association Council Statement, 1972) that the total fat intake should be reduced to less than 35% total calories and saturated fatty acids to less than 10%, while the monounsaturated and the essential polyunsaturated fatty acids should each achieve at least 10% calories. Although some dietary trials have achieved successful results in reducing hyperlipidemia, whether this reduction will affect the incidence of atherosclerosis and coronary heart disease remains to be proved (Ahrens, 1976).

Dietary fat has also been implicated as a causative factor in obesity, gallbladder disease, diabetes, and colon cancer. Research in these areas is in progress.

Reductions in dietary fat levels automatically result in a substitution of the displaced calories by carbohydrate, and elevated carbohydrate intakes are obviously not without undesirable effects either. Furthermore, the requirements for polyunsaturated fatty acids have yet to be quantitated, and the long-range effects of large amounts of polyunsaturated fatty acids need to be evaluated, especially as they affect the requirements for other nutrients. Furthermore, the effects of the ingestion of fatty acids modified by food processing need evaluation as well. Obviously, there is still much to learn about the contribution of fat to the health and well-being of individuals and populations.

10. References

Ahrens, E. H., 1976, The management of hyperlipidemia: Whether, rather than how, *Ann. Intern. Med.* **85**:87.

Alfin-Slater, R. B., Wells, P., Aftergood, L., and Melnick, D., 1973, Dietary fat composition and tocopherol requirement. IV. Safety of polyunsaturated fats, *J. Am. Oil Chem. Soc.* **50**:479.

Alfin-Slater, R. B., Aftergood, L., and Whitten, T., 1976, Nutritional evaluation of *trans* fatty acids, *J. Am. Oil Chem. Soc.* **53**:468A.

American Heart Association Council Statement, 1972, Diet and coronary heart disease, *J. Am. Med. Assoc.* **222**:1647.

Anderson, R. L., and Coots, R. H., 1967, The catabolism of the geometric isomers of uniformly ^{14}C-labelled Δ^9-octadecenoic acid and uniformly ^{14}C-labelled $\Delta^{9,12}$-octadecadienoic acid by the fasting rat, *Biochim. Biophys. Acta* **144**:525.

Askew, E. W., Dohm, G. L., Doub, W. H., Jr., Houston, R. L., and Van Notta, P. A., 1975a, Lipogenesis and glyceride synthesis in the rat, *J. Nutr.* **105**:190.

Askew, E. W., Dohm, G. L., and Huston, R. L., 1975b, Fatty acid and ketone body metabolism in the rat: Response to diet and exercise, *J. Nutr.* **105**:1422.

Austin, W., and Nestel, P. J., 1968, The effect of glucose and insulin *in vitro* on the uptake of triglyceride and on lipoprotein lipase activity in fat pads from normal, fed rats, *Biochim. Biophys. Acta* **164**:59.

Baker, G. L., 1969, Human adipose tissue composition and age, *Am. J. Clin. Nutr.* **22**:829.

Balasse, E. O., and Neef, M. A., 1975, Inhibition of ketogenesis by ketone bodies in fasting humans, *Metabolism* **24**:999.

Banker, B. G., Miller, J. Q., and Crocker, A. C., 1962, The cerebral pathology of infantile Gaucher's disease, in: *Cerebral Sphingolipidoses* (S. M. Aronson and B. W. Volk, eds.), pp. 73–99, Academic Press, New York.

Beare-Rogers, J. L., and Gordon, E., 1976, Myocardial lipids and nucleotides of rats fed olive oil or rapeseed oil, *Lipids* **11**:287.

Beare-Rogers, J. L., Nera, E. A., Craig, B. M., 1972, Cardiac lipids in rats and gerbils fed oils containing C_{22} fatty acids, *Lipids* **7**:548.

Bier, D. M., and Havel, R. J., 1970, Activation of lipoprotein lipase by lipoprotein fractions of human serum, *J. Lipid Res.* **11**:565.

Bierman, E. L., Gordis, E., and Hamlin, J. T., 1962, Heterogeneity of fat particles in plasma during alimentary lipemia, *J. Clin. Invest.* **41**:2254.

Bilheimer, D. W., and Levy, R. S., 1973, Origin and fate of lipoproteins, *Adv. Exp. Med. Biol.* **38**:39.

Birkbeck, J. A., 1970, The fatty acid composition of depot fat in childhood, *Acta Paediatr. Scand.* **59**:505.

Borgström, B., 1974, Bile salts—their physiological functions in the gastrointestinal tract. *Acta Med. Scand.* **196**:1.

Borgström, B., Erlanson, C., and Sternby, B., 1974, Further characterization of two colipases from porcine pancreas, *Biochem. Biophys. Res. Commun.* **59**:902.

Borgström, B., and Olivecrona, T., 1961, The metabolism of palmitic acid-1-C^{14} in functionally hepatectomized rats, *J. Lipid Res.* **2:**263.

Borgström, B., Dahlquist, A., Lundh, G., and Sjövall, J., 1957, Studies of intestinal digestion and absorption in the human, *J. Clin. Invest.* **36:**1521.

Carlson, L. A., and Pernow, B., 1961, Studies on blood lipids during exercise, *J. Lab. Clin. Med.* **58:**673.

Carlson, L. A., Ekelund, L. G., and Orö, L., 1963, Studies on blood lipids during exercise, *J. Lab. Clin. Med.* **61:**724.

Cioffi, L. A., and Speranza, A., 1972, Physiological and psychological components of the body weight control in the obese, *Bibl. Nutr. Dieta* **17:**154.

Clark, S. B., Ekkers, T. E., Singh, A., Balint, J. A., Holt, P. R., and Rodgers, J. B., Jr., 1973, Fat absorption in EFA deficiency, *J. Lipid Res.* **14:**581.

Davenport, J. B., and Johnson, A. R., 1971, The nomenclature and classification of lipids, in: *Biochemistry and Methodology of Lipids* (A. R. Johnson and J. B. Davenport, eds.) pp. 1–28, Wiley–Interscience, New York.

Dhopeshwarkar, G. A., Maier, R., and Mead, J. F., 1969, Incorporation of [1–^{14}C]acetate into the fatty acids of the developing rat brain, *Biochim. Biophys. Acta* **187:**6.

Dhopeshwarkar, G. A., Subramanian, C., and Mead, J. F., 1971, Fatty acid uptake by the brain, *Biochim. Biophys. Acta* **231:**8.

Downey, R. K., Craig, B. M., and Youngs, C. G., 1969, Breeding rapeseed for oil and meal quality, *J. Am. Oil. Chem. Soc.* **41:**475.

Egwim, P. O., and Kummerow, F. A., 1972, Incorporation and distribution of dietary elaidate in the major lipid classes of rat heart and plasma lipoproteins, *J. Nutr.* **102:**783.

Eichberg, J., Hauser, G., and Karnovsky, M. L., 1969, Lipids of nervous tissue, in: *The Structure and Function of Nervous Tissue,* Vol. 3 (G. H. Bourne, ed.), pp. 185–287, Academic Press, New York.

Farquhar, J. W., Gross, R. C., Wright, P. W., and Reaven, J. M., 1965, Kinetics of plasma lipoprotein triglyceride in man and the dog, *J. Clin. Invest.* **44:**1046.

Fasoli, A., 1973, Biological significance of serum lipoproteins, *Adv. Exp. Med. Biol.* **38:**23.

Fulco, A. J., and Mead, J. F., 1961, The biosynthesis of lignoceric, cerebronic and nervonic acids, *J. Biol. Chem.* **236:**2416.

Fuller, J. H., Grainger, G. L., Jarrett, R. J., and Keen, H., 1974, Possible seasonal variations of plasma lipids in a healthy population, *Clin. Chim. Acta* **52:**305.

Ginsberg, H., Olefsky, J. M., Kimmerling, G., Crapo, P., and Reaven, G. M., 1976, Induction of hypertriglyceridemia by a low-fat diet, *J. Clin. Endocrinol. Metab.* **42:**729.

Goldrick, R. B., 1971, Deposition and mobilization of lipids in man, in: *Biochemistry and Methodology of Lipids* (A. R. Johnson and J. B. Davenport, eds.), pp. 501–514, Wiley–Interscience, New York.

Goodale, W. T., Olson, R. E., and Hackel, D. B., 1959, The effect of fasting and diabetes mellitus on myocardial metabolism in man, *Am. J. Med.* **27:**212.

Greenberger, N. J., and Skillman, T. G., 1969, Medium chain triglycerides, *N. Engl. J. Med.* **280:**1045.

Greenberger, N. J., Ruppert, R. D., and Tzagournis, M., 1967, Use of MCT in malabsorption, *Ann. Intern. Med.* **66:**727.

Hales, C. N. T., and Randle, P. J., 1963, Effects of low-carbohydrate diet and diabetes mellitus on plasma concentrations of glucose, nonesterified fatty acid, and insulin during oral glucose tolerance tests, *Lancet* **1:**790.

Hamosh, M., Klaeveman, H. L., Wolf, R. O., and Scow, R. O., 1975, Pharyngeal lipase and digestion of dietary TG in man, *J. Clin. Invest.* **55:**908.

Hansen, J. B., and Clausen, J., 1968, The fatty acids of the human fetal brain, *Scand. J. Clin. Lab. Invest.* **22:**231.

Heffernan, A. G. A., 1964, Fatty acid composition of adipose tissue in normal and abnormal subjects, *Am. J. Clin. Nutr.* **15:**5.

Hirsch, J., and Han, P. W., 1969, Cellularity of rat adipose tissue: Effects of growth, starvation, and obesity, *J. Lipid Res.* **10:**77.

Hirsch, J., Farquhar, J. W., Ahrens, E. H., Jr., Peterson, M. L., and Stoffel, W., 1960, Studies of adipose tissue in man, *Am. J. Clin. Nutr.* **8**:499.

Hofmann, A. F., and Borgström, B., 1965, Hydrolysis of long-chain monoglycerides in micellar solution by pancreatic lipase, *Biochim. Biophys. Acta* **70**:317.

Hofmann, A. F., and Small, D. M., 1967, Detergent properties of bile salts: Correlation with physiological function, *Annu. Rev. Med.* **18**:333.

Insull, W., Jr., Lang, P. D., Hsi, B. P., and Yoshimura, S., 1969, Studies of arteriosclerosis in Japanese and American men, *J. Clin. Invest.* **48**:1313.

Jaillard, J., Sezille, G., Dewailly, P., Fruchart, J. C., and Bertrand, M., 1973, Étude expérimentale chez l'homme et chez l'animal du métabolisme de l'acide érucique, *Nutr. Metab.* **15**:336.

Jersild, R. A., Jr., 1966, A time sequence study of fat absorption in the rat jejunum, *Am. J. Anat.* **118**:135.

Johnston, J. M., 1968, Mechanism of fat absorption, in: *Handbook of Physiology,* Section 6, *The Alimentary Canal,* American Physiological Society, Vol. 3 (C. F. Code and W. Heidel, eds.), pp. 1353–1375, Williams and Wilkins, Baltimore.

Johnston, J. M., and Rao, G. A., 1965, Triglyceride biosynthesis in the intestinal mucosa, *Biochim. Biophys. Acta* **106**:1.

Jones, D. P., Losowsky, M. S., Davidson, C. S., and Lieber, C. S., 1963, Low plasma lipoprotein lipase activity as a factor in the pathogenesis of alcoholic hyperlipemia, *J. Clin. Invest.* **42**:945.

Kasper, H., 1970, Fecal fat excretion, diarrhea, and subjective complaints with highly dosed oral fat intake, *Digestion* **3**:321.

Kayden, H. J., and Medick, M., 1969, The absorption and metabolism of short and long chain fatty acids in puromycin-treated rats, *Biochim. Biophys. Acta* **176**:37.

Knittle, J. L., 1972, Obesity in childhood: A problem in adipose tissue cellular development, *J. Pediatr.* **81**:1048.

Kramer, J. K. G., 1973, Changes in liver lipid composition of male rats fed rapeseed oil diets, *Lipids* **8**:641.

Kummerow, F. A., 1974, Current studies on relation of fat to health, *J. Am. Oil Chem. Soc.* **51**:255.

Larking, P. W., and Nye, E. R., 1975, Effect of dietary lipids on lipolysis in rat adipose tissue, *Br. J. Nutr.* **33**:291.

Lawlor, T., O'Hara, F., and Birtwistle, D. T., 1974, Serum cholesterol and serum TG in mental retardation, *Postgrad. Med.* **50**:140.

Lieber, C. S., Jones, D. P., and di Carli, L. M., 1965, Effects of prolonged ethanol intake, *J. Clin. Invest.* **44**:1009.

Lyman, R. L., Tinoco, J., Bouchard, P., Sheehan, G., Ostwald, R., and Miljanich, P., 1967, Sex differences in the metabolism of phosphatidyl cholines in rat liver, *Biochim. Biophys. Acta* **137**:107.

Mansbach, C. M., Cohen, R. S., and Leff, P. B., 1975, Isolation and properties of the mixed lipid micelles present in intestinal content during fat digestion in man, *J. Clin. Invest.* **56**:781.

Masoro, E. J., 1977, Lipids and lipid metabolism, *Ann. Rev. Physiol.* **39**:301.

Mattson, F. H., and Streck, J. A., 1974, Effect of the consumption of glycerides containing behenic acid on the lipid contents of the heart of weanling rats, *J. Nutr.* **104**:483.

Mekhijian, H. S., Philips, S. F., and Hofmann, A. F., 1971, Colonic secretion of water and electrolytes induced by bile acids, *J. Clin. Invest.* **50**:1569.

Meyer, J. H., Stevenson, E. A., and Watts, H. D., 1976, The potential role of protein in the absorption of fat, *Gastroenterology* **70**:232.

Meyer, W. H., 1974, *Food Fats and Oils,* Institute of Shortening and Edible Oils, Washington, D. C.

Mistilis, S. P., and Ockner, R. K., 1972, Effects of ethanol on endogenous lipid and lipoprotein metabolism in small intestine, *J. Lab. Clin. Med.* **80**:34.

Mueller, P. S., and Horwitz, D., 1962, Plasma FFA and blood glucose responses to analogues of norepinephrine in man, *J. Lipid Res.* **3**:251.

Nestel, P. J., 1964, Relationship between plasma triglycerides and removal of chylomicrons, *J. Clin. Invest.* **43:**943.

Nestel, P. J., 1967, Lipoprotein transport, in: *Newer Methods of Nutritional Biochemistry,* Vol. 3 (A. A. Albanese, ed.), pp. 243–302, Academic Press, New York.

Nichols, A. V., 1969, Functions and interrelationships of different classes of plasma lipoproteins, *Proc. Natl. Acad. Sci. U.S.A.* **64:**1128.

Nichols, A. V., and Smith, L., 1965, Effect of very low density lipoproteins on lipid transfer in incubated serum, *J. Lipid Res.* **6:**206.

Nicolaides, N., 1974, Skin lipids: Their biochemical uniqueness, *Science* **186:**19.

Nikkilä, E. A., and Pelkonen, R., 1975, Serum lipids in acromegaly, *Metabolism* **24:**829.

Ockner, R. K., and Isselbacher, K. J., 1974, Recent concepts of intestinal fat absorption, *Rev. Physiol. Biochem. Pharmacol.* **71:**107.

Ockner, R. K., and Manning, J. A., 1974, Fatty acid-binding protein in small intestine, *J. Clin. Invest.* **54:**326.

Ockner, R. K., Pittman, J. P., and Yager, J. L., 1972, Differences in the intestinal absorption of saturated and unsaturated long chain fatty acids, *Gastroenterology* **62:**981.

Patel, M. S., 1975, The metabolism of ketone bodies in developing human brain, *J. Neurochem.* **25:**905.

Patton, J. S., and Carey, M. C., 1979, Watching fat digestion, *Science* **204:**145.

Polheim, D., David, J. S. K., Schultz, F. M., Wylie, M. B., and Johnston, J. M., 1973, Regulation of triglyceride biosynthesis in adipose and intestinal tissue, *J. Lipid Res.* **14:**415.

Polonowski, J., 1976, Some aspects of the metabolism of the lipoproteins, *Biochimie* **58:**971.

Privett, O., and Blank, M. L., 1964, Studies on the metabolism of linoelaidic acid in the EFA-deficient diet, *J. Am. Oil Chem. Soc.* **41:**292.

Quarfordt, S. N., Frank, A., Shames, D. M., Berman, M., and Steinberg, D., 1970, VLDL triglyceride transport in Type IV hyperlipoproteinemia and the effects of carbohydrate-rich diets, *J. Clin. Invest.* **49:**2281.

Ramsey, R. B., and Nicholas, H. J., 1972, Brain lipids, *Adv. Lipid Res.* **10:**143.

Rich, C., Bierman, E. L., and Schwartz, I. L., 1959, Plasma nonesterified fatty acids in hyperthyroid states, *J. Clin. Invest.* **38:**275.

Robinson, D. S., Cryer, A., and Davies, P., 1975, Role of clearing factor lipase in transport of plasma TG, *Proc. Nutr. Soc.* **34:**211.

Rothlin, M. E., and Bing, R. J., 1961, Extraction and release of individual free fatty acids by the heart and fat depots, *J. Clin. Invest.* **40:**1380.

Rouser, G., Galli, C. and Kritchevsky, G., 1965, Lipid class composition of normal human brain and variations in metachromatic leucodystrophy, Tay–Sachs, Niemann–Pick, chronic Gaucher's and Alzheimer's diseases, *J. Am. Oil Chem. Soc.* **42:**404.

Sari, H., Entressangles, B., and Desnuelle, P., 1975, Interactions of co-lipase with bile salt micelles, *Eur. J. Biochem.* **58:**561.

Scanu, A. M., 1978, Plasma lipoproteins and coronary heart disease, *Ann. Clin. Lab. Sci.* **8:**79.

Schettler, G., and Kahlke, W., 1967a, Gangliosides, in: *Lipids and Lipidoses* (G. Schettler, ed.), pp. 213–259, Springer-Verlag, New York.

Schettler, G., and Kahlke, W., 1976b, Gaucher's disease, in: *Lipids and Lipidoses* (G. Schettler, ed), pp. 260–287, Springer-Verlag, New York.

Schotz, M. C., and Garfinkel, A. S., 1972, Effect of nutrition on species of lipoprotein lipase, *Biochim. Biophys. Acta* **270:**472.

Schrock, C. G., and Connor, W. E., 1975, Incorporation of the dietary *trans* fatty acid into the serum lipids, the serum lipoproteins and adipose tissue, *Am. J. Clin. Nutr.* **28:**1020.

Senior, J. R., 1964, Intestinal absorption of fats, *J. Lipid Res.* **5:**495.

Shaw, J. E., and Ramwell, P. W., 1968, Release of prostaglandin from rat epididymal fat pad on nervous and hormonal stimulation, *J. Biol. Chem.* **243:**1498.

Sheehy, T. W., and Floch, M. H., 1964, *The Small Intestine,* Harper and Row, New York.

Shiff, D., Stern, L., and Leduc, J., 1966, Chemical thermogenesis in newborn infants, *Pediatrics* **37:**577.

Sickinger, K., 1975, Clinical aspects and therapy of fat malassimilation, in: *The Role of Fats in Human Nutrition* (A. J. Vergroesen, ed.), pp. 115–209, Academic Press, New York.

Siddle, K., and Hales, C. N., 1975, Hormonal control of adipose tissue lipolysis, *Proc. Nutr. Soc.* **34**:233.

Soler-Argilaga, C., Wilcox, H. G., and Heimberg, M., 1976, The effect of sex on the quantity and properties of the VLDL secreted by the liver *in vitro, J. Lipid Res.* **17**:139.

Stein, O., and Stein, Y., 1963, Metabolism of fatty acids in the isolated perfused rat heart, *Biochim. Biophys. Acta* **70**:517.

Spitzer, J. J., and Gold, M., 1962, Effect of catecholamines of the individual free fatty acids of plasma, *Proc. Soc. Exp. Biol. Med.* **110**:645.

Steinberg, D., Vaughan, M., Nestel, P. J., Strand, O., and Bergström, S., 1964, Effects of the prostaglandins on hormone-induced mobilization of free fatty acids, *J. Clin. Invest.* **43**:1533.

Svennerholm, L., 1966, The patterns of gangliosides in mental and neurological disorders, *Biochem. J.* **98**:20P.

Takatori, T., Phillips, F. C., Shimasaki, H., and Privett, O. S., 1976, Effects of dietary saturated and *trans* fatty acids on tissue lipid composition and serum LCAT activity in the rat, *Lipids* **11**:272.

Teige, B., and Beare-Rogers, J. L., 1973, Cardiac fatty acids in rats fed marine oils, *Lipids* **8**:584.

Tyor, M. P., Garbutt, J. T., and Lack, L., 1971, Metabolism and transport of bile salts in the intestine, *Am. J. Med.* **51**:614.

Vasdev, S. C., and Kako, K. J., 1976, Metabolism of erucic acid in the isolated perfused rat heart, *Biochim. Biophys. Acta* **431**:22.

Westergaard, H., and Dietschy, J. M., 1974, Normal mechanisms of fat absorption and derangements induced by various gastrointestinal diseases, *Med. Clin. North Am.* **58**:1413.

Widdowson, E. M., Dauncey, M. J., Gairdner, D. M. T., Jonxis, J. A. P., and Pelikan-Filipkova, M., 1975, Body fat of British and Dutch infants, *Br. Med. J.* **1**:653.

Wiley, J. H., and Leveille, G. A., 1973, Metabolic consequences of dietary MCT in the rat, *J. Nutr.* **103**:829.

Willebrands, A. F., 1964, Myocardial extraction of individual nonesterified fatty acids, esterified fatty acids and acetoacetate in the fasting human, *Clin. Chim. Acta* **10**:435.

Wollaeger, E. E., 1973, Role of the ileum in fat absorption, *Mayo Clin. Proc.* **48**:836.

Wood, P., Schlierf, G., and Kinsell, L., 1965, Changes in human plasma free fatty acids during exercise, *Clin. Res.* **13**:134.

Yeh, Y. Y., and Zee, P., 1976, Relation of ketosis to metabolic changes induced by acute MCT feeding in rats, *J. Nutr.* **106**:58.

5

Suppliers of Energy: Carbohydrate–Fat Interrelationships

Dale R. Romsos and Steven D. Clarke

1. Introduction

Carbohydrate and fat are major sources of energy in the diets of many people. Carbohydrate and fat make nearly equal contributions to the energy content of the United States diet, 46 and 42%, respectively (FNB, 1974). To compare carbohydrate and fat as energy sources in a diet, replacement of one with the other must be on an equal-energy basis rather than on a weight basis. Unfortunately, reports still periodically appear in which the differences in energy values of fat and carbohydrate are not considered when the diets are formulated. The nutrient-to-energy ratio is thus altered, and it is very difficult to interpret the results.

The gross energy value of fat as determined in a bomb calorimeter is approximately twice the energy value of carbohydrate. These differences in energy values are related to the amount of oxygen and hydrogen relative to the amount of carbon in the molecule (Maynard and Loosli, 1962). The gross energy value of glucose ($C_6H_{12}O_6$, 3.76 kcal/g) is slightly less than the energy value of sucrose ($C_{12}H_{22}O_{11}$, 3.96 kcal/g) because there is relatively less carbon in a gram of glucose. Likewise, the gross energy value of an unsaturated fat from plants (9.33 kcal/g) is slightly less than the energy value of a more saturated fat such as lard (9.48 kcal/g) because the hydrogen-to-carbon ratio is lower in the unsaturated fat.

There are marked differences in the gross energy values of fat and carbohydrate, but these two sources of dietary energy appear to be oxidized with

Dale R. Romsos • Department of Food Science and Human Nutrition, Michigan State University, East Lansing, Michigan 48824.   ***Steven D. Clarke*** • Department of Food Science and Nutrition, Ohio State University, Columbus, Ohio 43210.

approximately equal energetic efficiency in animals. Oxidation of either glucose or palmitate and conservation of the energy in the form of ATP retains approximately 25% of the molecule's energy (Hochachka, 1974; Prusiner and Poe, 1968). The exact fraction of the energy that is conserved in ATP is not known with certainty because the energy content of the terminal pyrophosphate bond of ATP has not been determined *in vivo*. Regardless, carbohydrate and fat appear to be oxidized with equal efficiencies.

A complete understanding of the interrelationships of carbohydrate and fat as dietary sources of energy is much more complex than a simple examination of their gross energy values and efficiency of oxidation. A consideration of the energy costs for digestion, absorption, transport, and storage of the energy is also necessary. Current information in this area has been summarized by others (Baldwin, 1968, 1970; Baldwin and Smith, 1971; Blaxter, 1971; Milligan, 1971); only one example will be presented here to illustrate that the energy cost of certain metabolic processes is dependent upon the source of dietary energy. After a meal, a portion of the dietary carbohydrate and fat is stored. The energy cost of storing glucose as glycogen is approximately 5% of the molecule's energy, whereas storage of glucose as fat represents a heat loss of approximately 10–20% (Baldwin, 1968). Less than 5% of the energy in a molecule of triglyceride is lost when it is taken up and stored in adipose tissue. Thus, as one would predict, the energy costs for storage of energy are somewhat greater when carbohydrates are consumed than when fats are consumed. But heat generated during these processes may partially reduce the need for thermoregulatory heat production in animals living at temperatures below their thermoneutral zone. Conversely, for animals living at temperatures above their thermoneutral zone, heat generated during metabolism of food would be expected to increase energy expenditure for thermoregulation. Consequently, environmental conditions may interact, under certain conditions, to either partially negate or to exacerbate predicted differences in energy efficiency.

Various tissues require either glucose or fatty acids as a major energy source (Krebs, 1972). Erythrocytes, leukocytes, lymphocytes, renal medulla, brain, skeletal muscle in severe exercise, retina, and intestinal mucosa depend primarily upon glucose as the major source of energy. Fatty acids serve as a major energy source for liver, kidney cortex, cardiac muscle, and skeletal muscle except in severe exercise. A metabolic product of fatty acids, ketone bodies, can supply energy to cardiac muscle, renal cortex, skeletal muscle, and brain.

Because animals are capable of *de novo* synthesis of carbohydrate and fat from appropriate precursors, our specific dietary need for these suppliers of energy is minimal. Dietary fat serves as a carrier for fat-soluble vitamins and provides certain fatty acids that are essential nutrients. Except for these needs, which humans can satisfy by consuming approximately 200 kcal of appropriate food fats, there is no specific requirement for fat as a nutrient in the diet (FNB, 1974). When high-carbohydrate diets are fed, a portion of the dietary carbohydrate is converted to fatty acids before utilization by tissues which preferentially utilize fatty acids.

In many species (as discussed later), a dietary need for glucose cannot be demonstrated; however, as already mentioned, glucose is an essential supplier of energy for certain tissues. A normal postabsorptive man utilizes approximately 180–225 g glucose/24 hr, 125 g of which is utilized by the brain (Cahill, 1976; Felig, 1975; Felig and Wahren, 1974; Hultman and Nilsson, 1975). When the level of circulating ketone bodies is sufficiently elevated, ketone bodies can spare glucose utilization by the brain. In the postabsorptive state, glucose is produced by two different mechanisms: glycogenolysis and gluconeogenesis (Hultman and Nilsson, 1975). The glycogen content of the liver varies with diet but averages approximately 200–350 mmol glycosyl units/kg liver in the postabsorptive state. When humans were fed a carbohydrate-free diet for 1–10 days, the values averaged less than 75 mmol glycosyl units/kg liver; these values increased to 300–600 units within 1–2 days after feeding a high-carbohydrate diet. Glucose output by the liver decreased by 65% when the carbohydrate-free diet was fed; the decrease was attributed to reduced glycogenolysis, with the rate of gluconeogenesis remaining relatively constant. A compensatory reduction in peripheral glucose consumption occurred to maintain circulating glucose levels near normal. The three- to fivefold increase in circulating ketone bodies observed when the carbohydrate-free diet was fed may have spared glucose oxidation by the brain. The activity of glucokinase, a rate-limiting enzyme in hepatic glucose metabolism, was also depressed when the carbohydrate-free diet was fed. While there is no specific requirement for carbohydrate in the diet, current recommendations suggest that our diets include 200–400 kcal of digestible carbohydrate/day (FNB, 1974).

2. Growth

The dietary fat-to-carbohydrate ratio affects the energy intake of certain animals. While many animals are able to maintain approximately the same energy intake when the dietary fat-to-carbohydrate ratio is altered, this is not true for all animals. For example, certain strains of rodents consume more energy when carbohydrate in the diet is replaced by fat (Lemonnier *et al.*, 1971; Schemmel *et al.*, 1970). Obesity-prone rodents are less able to regulate their intake when fed the high-fat diets. Whether the ability of humans to regulate their energy intake over long periods of time is altered by the fat and carbohydrate content of the diet has not been clearly established.

The influence of the fat and carbohydrate content of the diet on *ad libitum* food intake and growth of male infants has been reported (Fomon *et al.*, 1976). A combination of infant formula and strained foods were fed. Each diet contained 9% of energy from protein; one diet contained 62% of energy from carbohydrate and 29% of energy from fat, whereas the other diet contained 34 and 57% of energy from carbohydrate and fat, respectively. Energy density was similar for the two diets. These diets were fed to 36 infants from 8 to 112 days of age. Energy intakes per kilogram and rates of gain in length and weight were similar for infants fed the two diets. More extensive studies on the interrelationship of dietary fat and carbohydrate have been conducted with laboratory animals.

If dietary carbohydrate is totally replaced by neutral fat, rats (Goldberg, 1971), chickens (Brambila and Hill, 1966; Renner, 1964; Renner and Elcombe, 1964), and dogs (Romsos *et al.*, 1976) will grow at a rate at least equal to that of animals fed high-carbohydrate diets. But when fatty acids, rather than triglycerides, replace all the carbohydrate in the diet, energy intake and growth rate are depressed. Addition of glycerol or glucose to the fatty acid diet will usually improve growth. Thus, it appears that many animals do not require any more dietary carbohydrate to maintain a normal rate of growth than that supplied as glycerol in neutral fat.

Over thirty years ago, Forbes and co-workers (1946a,b) fed rats diets containing various fat-to-carbohydrate ratios and noted that fat increased the efficiency of dietary energy utilization. When growing rats were pair-fed diets containing 2, 10, or 30% fat (fat increased at the expense of carbohydrate on an equalized energy basis), metabolizable energy intake was not influenced, but heat production decreased 6% and energy retention increased 15% with the increased level of fat in the diet. When these diets were fed to adult rats at either a maintenance level or above maintenance, similar results were obtained. Heat increments of the diets were determined by measuring the difference in heat production of rats fed each diet at maintenance or above maintenance. The heat increments, which represented the energy required to utilize the nutrients, were equivalent to 31, 26, and 16%, respectively, of the gross energy of diets containing 2, 10, and 30% fat. Wood and Reid (1975) also have reported that rats pair-fed a high-fat diet are energetically more efficient than rats fed a high-carbohydrate diet.

This effect of dietary fat on efficiency of energy utilization has been clearly demonstrated in poultry (Carew and Hill, 1964; Carew *et al.*, 1964; Polin and Wolford, 1976). Chicks were fed diets containing 2.5, 12.5, and 22.5% corn oil substituted on an equalized energy basis for glucose. Metabolizable energy intake was equal, but tissue energy gain per unit metabolizable energy intake increased as the level of dietary corn oil increased. The energy value of the added fat was 24% greater than could be explained by its metabolizable energy value. Similarly, replacement of carbohydrate with fat increased the feed efficiency of turkeys (Jensen *et al.*, 1970). The improved efficiency was approximately 32% more than expected from the metabolizable energy value of the added fat. These effects of the fat-to-carbohydrate ratio on energy efficiency have been related to the lower heat increment associated with dietary fat. It has been suggested that replacement of dietary carbohydrate with fat might improve performance of chickens during heat stress (Fuller and Mora, 1973). An increase in the fat-to-carbohydrate ratio did improve feed efficiency in the heat-stressed birds.

Hartsook and Hershberger (1971; Hartsook *et al.*, 1973) employed the body balance technique in an extensive experiment to further define the interaction of dietary fat and carbohydrate in whole-animal energy metabolism. They fed nine different diets to rats for a six-week period. Rats were pair-fed;

Table I. Effect of the Dietary Fat-to-Carbohydrate Ratio on
Metabolized Energy and Heat Production in Rats[a]

Parameter	Dietary fat-to-carbohydrate ratio	Dietary protein (%)		
		10	40	70
Metabolized energy[b]	1.4	1615	1635	1595
	0.2	1675	1630	1515
Heat production[b]	1.4	1440	1455	1475
	0.2	1540	1445	1350

[a] Estimated from data presented by Hartsook and Hershberger (1971).
[b] kcal/rat per six weeks. Rats were fed isocaloric amounts of the diets.

thus, differences in gross energy intake did not influence the results. Metabolized energy was altered by the fat-to-carbohydrate ratio (Table I). When rats were fed low-protein diets, a high fat-to-carbohydrate ratio depressed the amount of energy which was metabolized; at intermediate protein levels in the diet, the fat-to-carbohydrate ratio did not influence the fraction of the dietary energy which was metabolized; but at high-protein levels, the metabolized energy was depressed when the fat-to-carbohydrate ratio was low. Thus, metabolism of fat and carbohydrate is related to their ratio in the diet as well as to the protein content of the diet. A similar phenomenon was observed when the effects of the fat-to-carbohydrate ratio were related to heat production (Table I). Heat production was obtained by subtracting energy stored as body weight gain from metabolized energy. At low protein intake, heat production was greatest when the diets with a low fat-to-carbohydrate ratio were fed, whereas the reverse was true when high-protein diets were fed. The authors summarized their findings by stating that the fat-to-carbohydrate ratio had the most important effect on energy metabolism when dietary protein levels were extremely low or extremely high. Only minimal effects of the dietary fat-to-carbohydrate ratio were observed when intermediate levels of protein were fed.

The influence of the dietary fat-to-carbohydrate ratio on heat production in humans was evaluated by Swift *et al.* (1959). They fed diets of equal protein and energy content, but one contained only 9.8% of energy from fat, while the other contained 60.1% of energy from fat. During the ten-day period, no significant differences in heat production, as measured with a respiration calorimeter on day 3 and again on day 10, were recorded. Metabolizable energy intake also was not influenced by the diet fed. Thus, the influence of the dietary fat-to-carbohydrate ratio on heat production which has been noted in rats and chickens was not confirmed in this short-term study with four human subjects.

In summary, dietary fat is generally utilized more efficiently than dietary carbohydrate; however, the differences are relatively small in most situations. The utilization also depends upon the dietary protein content. Likewise, other factors, such as environmental temperature, may modulate efficiency of dietary fat and carbohydrate utilization.

3. Obesity

Obesity is a syndrome which has a variety of causes, all of which result in an absolute or relative increase in body fat mass. Many factors may contribute to the positive energy balance during the development of obesity (Bray and Campfield, 1975); however, this discussion will be restricted to the possible influence of the dietary fat-to-carbohydrate ratio on the development and control of obesity.

Certain strains of rats become very obese when fed a high-fat diet but are able to maintain a reduced body fat content when fed a high-carbohydrate diet (Schemmel *et al.*, 1970). Other strains of rats retain similar levels of body fat regardless of whether a high-fat or a high-carbohydrate diet is fed. Likewise, in genetically obese (*ob/ob*) mice, marked differences in response to a high-fat or high-carbohydrate diet are observed (Genuth, 1976; Lemonnier *et al.*, 1971). When the obese mice consumed a high-fat diet, a 50% increase in body weight resulted, whereas when lean littermates consumed a high-fat diet, their body weight increased only 11% (Lemonnier *et al.*, 1971). Part of the difference in response of rodents to the high-fat diet can be explained on the basis of their ability to regulate energy intake. Rodents fed *ad libitum* consumed more energy when fed the high-fat diet than when fed the high-carbohydrate diet. But even when obese mice are pair-fed they gain more body energy when fed high-fat diets than when fed high-carbohydrate diets (Lin *et al.*, 1979). This difference in energy efficiency is not observed in pair-fed lean mice. Diet composition did not influence maintenance energy requirements of obese mice but rather improved the utilization of energy consumed above maintenance. Adult dogs fed a high-fat diet also accumulate more body fat than dogs fed a high-carbohydrate diet (Romsos *et al.*, 1978).

Additional studies are needed to delineate the role of the dietary fat-to-carbohydrate ratio in the development and maintenance of obesity in animal models with the propensity to become obese. The role of the dietary fat-to-carbohydrate ratio in the development of obesity in humans is not clear (Bray *et al.*, 1972). The possibility of interactions of genetic factors in humans with the regulation of energy intake by fat and carbohydrate and with the metabolic efficiency of dietary fat and carbohydrate utilization is an unexplored area.

More attention has been focused on the effects of the dietary fat-to-carbohydrate ratio on weight reduction than on the influence of this ratio on the development of obesity. Initial weight loss is usually greater when a high-fat, low-carbohydrate hypocaloric diet is fed than when the weight-reduction diet contains greater quantities of carbohydrate. But, these low-carbohydrate diets also promote increased water excretion (Azarnoff and Shoeman, 1976; Bray, 1976; Howard, 1975).

Moderately obese young men were fed 1800-kcal diets varying in fat and carbohydrate to compare their efficacies during weight reduction (Young *et al.*, 1971). The diets contained 115 g protein and 30, 60, or 104 g carbohydrate. Fat and carbohydrate were exchanged on an equal energy basis. Two or three men were fed each diet for 9 weeks. Calculated weight lost as fat was greatest for men fed the low-carbohydrate diet; these men lost approximately 900 kcal more each day than men fed the equal-caloric, high-carbohydrate diet. It would appear difficult to attribute such a large difference in energy expenditure to diet composition alone. A switch-back design, where each subject was fed each hypocaloric diet, would have reduced the possibility that the observed results represent individual differences rather than true responses to diet composition.

In another study the energy–nitrogen balance method was used to quantify loss of protein, fat, and water in grossly-obese subjects fed a 800-kcal ketogenic diet (5% of energy from carbohydrate) or a 800-kcal mixed diet (45% of energy from carbohydrate) (Yang and Van Itallie, 1976). Each subject consumed each diet for a ten-day period. The experimental design provided a five-day stabilization period before and after each experimental diet to reduce possible carry-over effects from the preceding regimen. Weight loss was greater when the 800-kcal ketogenic diet was fed than when the 800-kcal mixed diet was fed (Table II). The differences in weight loss between the ketogenic diet and the mixed diet were due almost entirely to differences in water balance. Fat loss, as well as protein loss, was similar regardless of whether the ketogenic or mixed diet was fed. Thus, the wide ratio of fat to carbohydrate used in this study did not alter the amount of energy lost during a ten-day period. As the authors pointed out, results from this relatively short-term study cannot necessarily be applied to longer-term situations.

Hegsted *et al.* (1975) examined the influence of diet composition on weight reduction in rats. They fed adult obese rats restricted amounts of diets which varied in protein, fat, and carbohydrate content. Carcass analysis at the end of the 52 days of restricted feeding indicated that loss in body weight and in body fat was similar for all diets. The high-protein diets (42% of energy) did spare body protein; however, the fat-to-carbohydrate ratio did not influence body protein.

Table II. Influence of Diet Composition on Composition of Weight Lost during Short-Term Weight Reduction[a]

	Treatment[b]	
	800-kcal ketogenic diet	800-kcal mixed diet
Body weight loss (g/day)	467 ± 51	278 ± 32
Composition of weight loss (g/day)		
Water	286	103
Fat	163	165
Protein	18	10

[a] From Yang and Van Itallie (1976).
[b] The ketogenic diet contained 70% of energy from fat and 5% from carbohydrate, whereas the mixed diet contained 30% of energy from fat and 45% from carbohydrate. Each of six obese subjects was maintained on each treatment for a ten-day period.

In summary, there is no clear evidence that loss of body energy during weight reduction is directly influenced by the dietary fat-to-carbohydrate ratio. It is, however, conceivable that alterations in the dietary fat-to-carbohydrate ratio may influence satiety of the diet and consequently subject compliance. Carefully controlled, long-term studies are needed to resolve this latter issue.

Dietary carbohydrates may influence body protein loss during weight reduction. Flatt and Blackburn (1974) have discussed the involvement of glucose and insulin in body protein sparing during energy deprivation. In the past, exogenous glucose has been given to reduce the need for gluconeogenesis and thereby spare tissue protein during energy deprivation. Flatt and Blackburn (1974) postulate that carbohydrate, by stimulating insulin release, may lead to a reduction in fat mobilization and thus not effectively spare body protein. Nitrogen balance was maintained during weight reduction of obese, diabetic adults with a daily protein supplement of 1.2–1.4 g/kg body weight, in the absence of dietary carbohydrate or fat (Bistrian *et al.*, 1976). Baird *et al.* (1974) were also able to maintain nitrogen balance during weight reduction of obese patients given 30 g of amino acids/day. But when a small amount (30–45 g) of carbohydrate was added to their diets, the minimum intake of amino acids required for nitrogen balance was reduced to about 15 g/day, suggesting that carbohydrates spare protein during energy deprivation. These patients had been on diets of 800–900 kcal for four weeks and then fasted for two weeks before they were given the amino acids and carbohydrate combinations. In another shorter-term study (Yang and Van Itallie, 1976), obese subjects on an 800-kcal diet containing 50 g of protein were unable to maintain nitrogen balance regardless of whether a high or a low fat-to-carbohydrate ratio was utilized to supply the remaining energy in the diet. When 1000-kcal diets were fed to obese women for eight days, nitrogen balance was less negative when increased amounts of carbohydrate were present, despite the lower protein content of these higher-carbohydrate diets (Hood *et al.*, 1970). Thus, the role of the fat-to-carbohydrate ratio on protein sparing during weight reduction has not been completely resolved. Undoubtedly, the length and degree of energy deprivation play an important role. Provided the level of protein is high (1.2–1.4 g/kg body weight), it would appear that neither carbohydrate nor fat is needed to minimize protein loss during weight reduction.

Others have studied the influence of carbohydrates and of the fat-to-carbohydrate ratio on nitrogen retention of postoperative, depleted, or septic patients receiving hypocaloric infusions of nutrients. Results are variable and likely reflect, in part, the patient population studied and the level of amino acids infused. Addition of glucose to infusion mixtures and concomitant reduction of amino acid intake from 90 to 70 g/day depressed the nitrogen balance of nutritionally depleted patients (Blackburn *et al.*, 1973a,b). Although it was suggested that these results were attributable to the addition of glucose, differences in nitrogen intake cannot be ignored. When nonobese surgical patients received 90 g amino acids plus 100 g glucose daily they lost less nitrogen than when they received amino acids alone (Elwyn *et al.*, 1978). These results suggest that carbohydrates elicit a protein-sparing effect. But others have shown that glucose (150 g/day) did not affect nitrogen balance in obese surgical patients infused with 150 g amino acids/day (Freeman *et al.*, 1977). Likewise, infusion of either glucose (150 g/day) or lipid (50 g/day) into postoperative patients receiving 1 g of protein/kg body weight did not influence the nitrogen balance (Greenberg *et al.*, 1976). Greenberg *et al.* suggest that endogenous fat mobilization is adjusted to meet the energy deficit when hypocaloric diets are given, independent of the source of nonprotein energy. Consequently, the major determinant of protein sparing in the absence of total caloric replacement is protein itself (Greenberg *et al.*, 1976; Freeman *et al.*, 1977). In patients receiving total parenteral nutrition, administration of 40–50 kcal/kg body weight as predominantly fat or carbohydrates failed to influence nitrogen balance when infused with 1–2 g protein/kg body weight (Bark *et al.*, 1976; Jeejeebhoy *et al.*, 1976). Changing the source of nonprotein energy did result in a temporary alteration in nitrogen balance during the first several days of each infusion (Jeejeebhoy *et al.*, 1976). Replacement of glucose with lipid reduced the nitrogen retention. Similar changes have been noted in animals; however, these changes are transitory.

Kasper *et al.* (1973, 1975) have proposed that high-fat diets may alter metabolic efficiency of humans. They fed normal-weight individuals formula diets containing a constant amount of protein and carbohydrate but increasing amounts of fat. The volunteers gained weight at a rate much slower than expected from their energy intake, which approached 7000 kcal/day in several instances. Weight gain was reported to be less when a fat high in linoleic acid (corn oil) was fed than when a fat lower in linoleic acid (olive oil) was fed. The authors further reported that when the daily intake of fat reached 300–400 g, the subjects reported a marked sensation of heat extending over the entire body and a marked tendency toward sweating. Hirsch and Van Itallie (1973) have criticized the design and authors' interpretations of this study. Clearly, the results of these studies need to be documented with measurements of heat loss and water balance. The entire subject of gluttony and thermogenesis needs further and careful detailed investigation.

The ratio of fat to carbohydrate in diets may alter energy metabolism during fasting. Rats previously fed a high-fat diet for 14 days lost more body weight and more body fat during a 120-hr fast than rats previously fed a high-carbohydrate diet (Suzuki *et al.*, 1975). Changes in carcass protein were not influenced by the diet fed. The influence of hypothyroidism, produced by oral administration of propylthiouracil, on the weight loss in rats was also evaluated. When the hypothyroid agent was added to the high-fat diet, weight loss during fasting was reduced to the rate observed when the high-carbohydrate diet was fed. Additions of propylthiouracil to the high-carbohydrate diet did not influence weight loss during fasting. It was speculated that the high-fat diet might accelerate thyroid function (Suzuki and Fuwa, 1971). Several parameters of thyroid function have been reported to be altered by feeding a high-fat diet; however, these results must be interpreted with reservation, because the diets were not formulated on an equal-energy basis (Yoshimura *et al.*, 1972). Consequently, it is not possible to attribute the results to an alteration in the fat-to-carbohydrate ratio.

Composition of the fat previously fed may also influence energy metabolism during fasting. Rats previously fed a diet containing safflower oil lost more than twice as much carcass fat during a three-day fast as did rats previously fed a diet containing lard (Anderson and Boggs, 1975). In agreement with the greater rate of fat catabolism, rats previously fed safflower oil also exhibited increased circulating ketone levels. The mechanism(s) whereby composition of the previous diet fed alters energy metabolism during fasting is not clear. Rates of adipose tissue lipolysis (Pawar and Tidwell, 1968) and the rate of fatty acid oxidation (Dupont and Mathias, 1969) may be influenced by the saturation of the fatty acids. Whether the composition of the previous diet alters energy metabolism of fasting humans has not been established.

4. Diabetes

Diabetes is characterized by alterations in the metabolism of glucose; these alterations may contribute to the development of certain diabetic complications (Anderson, 1974, 1975). Traditionally, diabetics have been treated with rather low-carbohydrate, high-fat diets. More recently, it has been recommended that the carbohydrate content of the diet be increased to at least 45% of energy (Albrink, 1974). High-carbohydrate diets have been associated with improved glucose tolerance in normal-weight, obese, and mildly diabetic individuals. Anderson *et al.* (1973) fed liquid diets containing 20–80% of energy as carbohydrate or fat to normal young men. As the carbohydrate content of the diet increased, a gradual improvement in the oral glucose tolerance was observed. An increase in the proportion of energy from carbohydrate, as well as weight reduction, also improved glucose tolerance of obese subjects (Jourdan *et al.*, 1974). Similarly, diabetics requiring less than 30 units of insulin per day, or requiring sulfonylureas, have been shown to have lower fasting plasma glucose levels and a significant improvement in glucose tolerance when high-carbohydrate diets (75% of energy from carbohydrate) are consumed. In fact, insulin or sulfonylurea therapy was not needed in some subjects after they had been on this diet for several weeks (Kiehm *et al.*, 1976). Further, their serum triglyceride levels were not adversely altered after consuming this high-carbohydrate diet containing a high portion of polysaccharides and dietary fiber. But, diabetics requiring more insulin (40–55 units per day) did not respond positively to the high-carbohydrate diets. These latter subjects may have had very limited residual capacity to produce and secrete insulin.

Deterioration of glucose tolerance associated with an elevated ratio of fat to carbohydrate in the diet appears to be related to an increase in energy from fat rather than to a decrease in energy from carbohydrate. For example, Anderson and Herman (1975) demonstrated that a low-carbohydrate diet (57 g/day) did not impair glucose tolerance of normal men provided carbohydrate was replaced with protein rather than fat.

Changes in glucose tolerance with alterations in the dietary fat-to-carbohydrate ratio reflect the capacity of tissues to clear glucose from the circulation. The liver is quantitatively the most important site for glucose uptake following an oral glucose tolerance test (Felig, 1975). Reduced capacity of liver to phosphorylate glucose after consumption of high-fat, low-carbohydrate diets (Romsos and Leveille, 1974) is consistent with the observed impairment in glucose tolerance. Extra hepatic tissues also exhibit reduced ability to clear glucose from the circulation after a high-fat diet is consumed. Glucose oxidation to carbon dioxide and conversion to glyceride-glycerol by adipose tissue from humans fed a high-fat diet is less than when the same subjects consumed a high-carbohydrate diet (Salans *et al.*, 1974). An alteration in the dietary fat-to-carbohydrate ratio may influence adipocyte-membrane-associated events (Ip *et al.*, 1976, 1977; Lavau *et al.*, 1979) as well as intracellular glucose metabolism (Romsos and Leveille, 1974).

Although the impact that the dietary fat-to-carbohydrate ratio exerts at the cellular level in diabetics remains to be completely elucidated, some diabetics clearly improve their glucose tolerance if fed relatively high levels of complex carbohydrates. But, total energy intake must still be closely regulated or high-carbohydrate diets may be poorly tolerated (West, 1973).

5. Blood Lipids

Epidemiological studies of the incidence and distribution of coronary heart disease have revealed that the disease rate is correlated with both dietary saturated fat and sucrose separately, with $r = 0.82$ and 0.80, respectively. However, the interaction of the two dietary factors is revealed, with a correlation of $r = 0.92$ (McGandy *et al.*, 1967) between sucrose and fat together and the incidence of arteriosclerosis. From such observations has originated a massive search for a dietary component in the etiology and control of degenerative vascular disease. One focus relates to changes in the concentrations of serum triglycerides. Collectively, evidence from both human and animal investigations now indicates that levels of serum triglycerides vary with the amount and type of dietary carbohydrate and/or fat (Antonis and Bersohn, 1961; Ginsberg *et al.*, 1976; Romsos and Leveille, 1974).

Simply raising the proportion of nonsucrose carbohydrate from 40 to 55% of the total calories caused blood triglycerides to be elevated by 41% over a seven-day period in both normal and hyperlipemic men (Ginsberg *et al.*, 1976). Schreibman and Ahrens (1976) have also reported that dextrose exchanged isocalorically for polyunsaturated fat in a liquid formula increased plasma triglyceride levels in hyperlipemic patients. When healthy subjects are fed high-carbohydrate diets, blood triglyceride concentrations appear to gradually decline and achieve basal levels after 17–32 weeks (Antonis and Bersohn,

1961). Therefore, under most normal conditions, accelerated triglyceride production associated with carbohydrate consumption evidently is transient. Ultimately, the mechanisms of peripheral tissue clearance will "catch up," thereby normalizing serum triglyceride concentrations (Cahlin *et al.*, 1973). If the high-carbohydrate diet is rich in dietary fiber, even the transient hyperlipemia often observed when refined carbohydrates are consumed may not occur (Albrink *et al.*, 1979). The authors suggest that the insulin-stimulating potential of the diet may be a critical determinant of the magnitude of carbohydrate-induced lipemia.

Unlike normolipemic patients, the hyperlipoproteinemic action of carbohydrate in the form of sucrose may not be transient in hypertriglyceridemic patients (Cahlin *et al.*, 1973). The defect in carbohydrate and triglyceride metabolism of these subjects may be at the level of triglyceride synthesis, triglyceride clearance, or both (Kimmerling *et al.*, 1976). Insulin resistance in obesity and adult-onset diabetes may impair triglyceride clearance and thus maintain the hypertriglyceridemic state. All of these conditions may be accentuated by a synergistic action of carbohydrate and fat.

Antar and co-workers concluded that starch and sucrose (40% of dietary calories) resulted in similar serum concentrations of cholesterol, phospholipids, and triglycerides when the proportion of polyunsaturated and saturated fatty acids was 10 and 9% of dietary calories, respectively (Birchwood *et al.*, 1970). However, when the amount of saturated and polyunsaturated fatty acids was changed to 18 and 2% of calories, respectively, the sucrose and saturated fat acted synergistically to cause a dramatic rise in serum cholesterol, phospholipid, and triglyceride concentrations in both normal and hyperlipoproteinemic patients (Antar *et al.*, 1970). Elevations in triglycerides in lipemic patients was far more dramatic (+141 mg/dl) than in the normal subjects (+60 mg/dl). Mechanisms responsible for the associative effect of dietary energy sources remain to be elucidated.

Dietary sucrose or fructose stimulates the rate of hepatic fatty acid and triglyceride synthesis (Romsos and Leveille, 1974) and may impair triglyceride clearance in rats (Bruckdorfer *et al.*, 1972; Toppings and Mayes, 1971). Humans given 800 kcal of sucrose in place of 160 kcal of protein, 290 kcal of fat, and 350 kcal of starch in a 2500-kcal diet exhibited increased rates of hepatic lipid synthesis (Cahlin *et al.*, 1973). Unfortunately, this experiment is confounded by the change not only in carbohydrate but in fat and protein as well.

The composition of the dietary fat may also influence lipid synthesis in the liver. Polyunsaturated fatty acids are more effective inhibitors of fatty acid synthesis in rodent liver than are saturated fatty acids (Allmann and Gibson, 1965; Bartley and Abraham, 1972; Clarke *et al.*, 1976; Musch *et al.*, 1974). In addition to differentially affecting hepatic lipid synthesis, the composition of the dietary fat may alter the rate of peripheral triglyceride clearance (Bagdade *et al.*, 1970; Nestel and Barter, 1973). Collectively, the data now available indicate that dietary carbohydrates and fats, as well as their ratio in the diet, can alter blood lipid levels. Whether these alterations are clinically significant in degenerative vascular disease relative to the regulation of energy balance and obesity remains to be ascertained.

6. Exercise

In the resting state, skeletal muscle derives virtually all of its energy from the oxidation of fatty acids; glucose uptake accounts for less than 10% of the total oxygen consumed by muscle (Felig and Wahren, 1975). During exercise, both fats and carbohydrates contribute to the energy supply of muscle. The intensity of the exercise affects the percentage contribution of fat and carbohydrate utilized by muscle. During mild exercise, proportions of fat and carbohydrate utilized are similar to the resting muscle. As the intensity of exercise increases, the proportion of carbohydrate utilized increases until, at maximum activity, nearly all the energy is derived from carbohydrate.

Physical training alters the relative amounts of fat and carbohydrate utilizing during exercise. Animals adapted to exercise have an increased capacity to use both fatty acids and ketone bodies during exercise of submaximal intensity (Askew *et al.*, 1975; Mole *et al.*, 1971). Palmitate oxidation was doubled in leg muscles of trained rats relative to the rates observed in sedentary animals. Physical training also appears to increase the capacity of animals to mobilize fatty acids from adipose tissue stores, which increases the availability of substrate for muscle oxidation (Askew and Hecker, 1976; Askew *et al.*, 1972). The increased utilization of fatty acid by muscle spares glucose utilization (Rennie *et al.*, 1976).

During vigorous exercise, work performance can be maintained as long as muscle glycogen levels are maintained above a critical level (Consolazio and Johnson, 1971). Exhaustion occurs when this critical level of muscle glycogen is reached. A number of studies have now been conducted to examine the influence of the dietary fat-to-carbohydrate ratio on muscle glycogen levels and work performance. In an earlier study, subjects were fed a diet extremely high in fat (less than 5% of energy from carbohydrate) for several days or a diet in which 90% of the energy was derived from carbohydrate (Astrand, 1967). Work time on a bicycle ergometer was three times as long when the diet very high in carbohydrate had been fed. Work time for subjects fed a normal mixed diet was somewhat less than for those fed the high-carbohydrate diet but considerably better than when subjects consumed the high-fat diet. Subjects fed the high-fat diet for less than a week probably had not fully adapted to the diet; an abrupt switch to this diet would cause temporary loss of electrolytes and water. This metabolic shift could have influenced work performance. Rats fed a high-fat diet for four weeks were able to exercise as long as those fed high-carbohydrate diets (Huston *et al.*, 1975).

More recently, dietary alterations have been utilized to increase muscle glycogen levels prior to an athletic event (Slovic, 1975; Williams, 1976). The most pronounced effect of diet on muscle glycogen levels is observed when the muscle glycogen content is first depleted by heavy exercise, followed by a diet very high in carbohydrate. An overshoot in muscle glycogen levels occurs; levels reach values several times higher than normally observed. Bergstrom *et al.* (1972) demonstrated this phenomenon very clearly when they exercised one leg of their subjects and rested the other leg. Following the exercise, glycogen was almost totally depleted in the exercised leg, while the glycogen content of the unexercised leg remained unchanged. A high-carbohydrate diet was then fed for several days, and the glycogen content of the exercised leg increased rapidly to levels severalfold above normal. Only a very small increase occurred in the glycogen content of the unexercised leg. Thus, glycogen loading only occurred in the previously depleted muscle. The ability to perform vigorous exercise was correspondingly altered. The overshoot in muscle glycogen requires dietary carbohydrate; a high-fat and high-protein diet will not elevate muscle glycogen levels.

Concomitant with glycogen storage, 2.7 times as much water is also stored. Thus, depending on the extent of glycogen loading, body weight may be increased, which might adversely affect physical performance. The beneficial as well as possible adverse effects of this added water need to be considered. Athletes should not undertake such a program of glycogen loading without careful medical supervision.

Flatt has presented a hypothesis which relates physical activity and the dietary fat-to-carbohydrate ratio in control of energy balance (Flatt, 1978). He suggests that weight maintenance may be facilitated by adjusting physical activities and diet composition to balance the fuel mix oxidized. This intriguing biochemical approach to the integration of the dietary fat to carbohydrate ratio in overall energy balance is amenable to testing.

7. References

Albrink, M. J., 1974, Dietary and drug treatment of hyperlipidemia in diabetes, *Diabetes* 23:913.

Albrink, M. J., Newman, T., and Davidson, P. C., 1979, Effect of high- and low-fiber diets on plasma lipids and insulin, *Am. J. Clin. Nutr.* 32:1486.

Allmann, D. W., and Gibson, D. M., 1965, Fatty acid synthesis during early linoleic acid deficiency in the mouse, *J. Lipid Res.* 6:51.

Anderson, J. W., 1974, Alterations in metabolic fate of glucose in the liver of diabetic animals, *Am. J. Clin. Nutr.* 27:746.

Anderson, J. W., 1975, Metabolic abnormalities contributing to diabetic complications. I. Glucose metabolism in insulin-insensitive pathways, *Am. J. Clin. Nutr.* 28:273.

Anderson, J. W., and Herman, R. H., 1975, Effects of carbohydrate restriction on glucose tolerance of normal men and reactive hypoglycemic patients, *Am. J. Clin. Nutr.* 28:748.

Anderson, J. W., Herman, R. H., and Zakim, D., 1973, Effect of high glucose and high sucrose diets on glucose tolerance of normal men, *Am. J. Clin. Nutr.* 26:600.

Anderson, R. L., and Boggs, R. W., 1975, Gluconeogenic and ketogenic capacities of lard, safflower oil and triundecanoin in fasting rats, *J. Nutr.* 105:185.

Antar, M. A., Little, J. A., Lucas, C., Buckley, G. C., and Csima, A., 1970, Interrelationship between the kinds of dietary carbohydrate and fat in hyperlipoproteinemic patients. III. Synergistic effect of sucrose and animal fat on serum lipids, *Atherosclerosis* 11:191.

Antonis, A., and Bersohn, I., 1961, The influence of diet on serum triglycerides, *Lancet* 1:3.

Askew, E. W., and Hecker, A. L., 1976, Adipose tissue cell size and lipolysis in the rat: Response to exercise intensity and food restriction, *J. Nutr.* 106:1351.

Askew, E. W., Dohm, G. L., Huston, R. L., Sneed, T. W., and Dowdy, R. P., 1972, Response of rat tissue lipases to physical training and exercise, *Proc. Soc. Exp. Biol. Med.* **141**:123.

Askew, E. W., Dohm, G. L., and Huston, R. L., 1975, Fatty acid and ketone body metabolism in the rat: Response to diet and exercise, *J. Nutr.* **105**:1422.

Astrand, P. O., 1967, Diet and athletic performance, *Fed. Proc. Fed. Am. Soc. Exp. Biol.* **26**:1772.

Azarnoff, D. L., and Shoeman, D. W., 1976, Diet and drugs in obesity control, in: *Lipid Pharmacology,* Vol. II (R. Paoletti and C. J. Glueck, eds.), pp. 161—181, Academic Press, New York.

Bagdade, J. D., Hazzard, W. R., and Carlin, J., 1970, Effect of unsaturated dietary fat on plasma lipoprotein lipase activity in normal and hyperlipidemic states, *Metabolism* **19**:320.

Baird, I. M., Parson, R. L., and Howard, A. N., 1974, Clinical and metabolic studies of chemically defined diets in the management of obesity, *Metabolism* **23**:645.

Baldwin, R. L., 1968, Estimation of theoretical calorific relationships as a teaching technique. A review, *J. Dairy Sci.* **51**:104.

Baldwin, R. L., 1970, Metabolic functions affecting the contribution of adipose tissue to total energy expenditure, *Fed. Proc. Fed. Am. Soc. Exp. Biol.* **29**:1277.

Baldwin, R. L., and Smith, N. E., 1971, Application of a simulation modeling technique in analyses of dynamic aspects of animal energetics, *Fed. Proc. Fed. Am. Soc. Exp. Biol.* **30**:1459.

Bark, S., Holm, I., Hakansson, I., and Wretlind, A., 1976, Nitrogen-sparing effect of fat emulsion compared with glucose in the postoperative period, *Acta Chir. Scand.* **142**:423.

Bartley, J. C., and Abraham, S., 1972, Hepatic lipogenesis in fasted refed rats and mice, *Biochim. Biophys. Acta* **280**:258.

Bergstrom, J., Hultmann, E., and Roch-Norlund, A. E., 1972, Muscle glycogen synthetase in normal subjects. Basal values, effect of glycogen depletion by exercise and of a carbohydrate-rich diet following exercise, *Scand. J. Clin. Lab. Invest.* **29**:231.

Birchwood, B. L., Little, J. A., Antar, M. A., Lucas, C., Buckley, G. C., Csima, A., and Kallos, A., 1970, Interrelationship between the kinds of dietary carbohydrate and fat in hyperlipoproteinemic patients. II. Sucrose and starch with mixed saturated and polyunsaturated fats, *Atherosclerosis* **11**:183.

Bistrian, B. R., Blackburn, G. L., Flatt, J. P., Sizer, J., Scrimshaw, N. S., and Sherman, M., 1976, Nitrogen metabolism and insulin requirements in obese diabetic adults on a protein-sparing modified fast, *Diabetes* **25**:494.

Blackburn, G. L., Flatt, J. P., Clowes, G. H. A., O'Donnell, T., and Hensle, T. E., 1973a, Protein-sparing therapy during periods of starvation with sepsis or trauma, *Ann. Surg.* **177**:588.

Blackburn, G. L., Flatt, J. P., Clowes, G. H. A., and O'Donnell, T., 1973b, Peripheral intravenous feeding with isotonic amino acid solutions. *Am. J. Surg.* **125**:447.

Blaxter, K. L., 1971, Methods of measuring the energy metabolism of animals and interpretation of results obtained, *Fed. Proc. Fed. Am. Soc. Exp. Biol.* **30**:1436.

Brambila, S., and Hill, F. W., 1966, Comparison of neutral fat and free fatty acids in high lipid—low carbohydrate diets for the growing chicken, *J. Nutr.* **88**:84.

Bray, G. A., 1976, *The Obese Patient,* W. B. Saunders, Philadelphia.

Bray, G. A., and Campfield, L. A., 1975, Metabolic factors in the control of energy stores, *Metabolism* **24**:99.

Bray, G. A., Davidson, M. B., and Drenick, E. J., 1972, Obesity: A serious symptom, *Ann. Intern. Med.* **77**:779.

Bruckdorfer, K. R., Khan, I. H., and Yudkin, J., 1972, Fatty acid synthetase activity in the liver and adipose tissue of rats fed with various carbohydrates, *Biochem. J.* **129**:439.

Cahill, G. F., Jr., 1976, Protein and amino acid metabolism in man, *Circulation Res.* **38**:109.

Cahlin, E., Jonsson, J., Persson, B., Stakeberg, H., Bjorntorp, P., Gustafson, A., and Schersten, T., 1973, Sucrose feeding in man. Effects on substrate incorporation into hepatic triglycerides and phospholipids *in vitro* and on removal of intravenous fat in patients with hyperlipoproteinemia, *Scand. J. Clin. Lab. Invest.* **32**:21.

Carew, L. B., Jr., and Hill, F. W., 1964, Effect of corn oil on metabolic efficiency of energy
    utilization by chicks, *J. Nutr.* **83**:293.
Carew, L. B., Jr., Hopkins, D. T., and Nesheim, M. C., 1964, Influence of amount and type of
    fat on metabolic efficiency of energy utilization by the chick, *J. Nutr.* **83**:300.
Clarke, S. D., Romsos, D. R., and Leveille, G. A., 1976, Specific inhibition of hepatic fatty acid
    synthesis exerted by dietary linoleate and linolenate in essential fatty acid adequate rats,
    Lipids **11**:485.
Consolazio, C. F., and Johnson, H. L., 1971, Measurement of energy cost in humans, *Fed. Proc.
    Fed. Am. Soc. Exp. Biol.* **30**:1444.
Dupont, J., and Mathias, M. M., 1969, Bio-oxidation of linoleic acid via methylmalonyl-CoA,
    Lipids **4**:478.
Elwyn, D. H., Gump, F. E., Iles, M., Long, C. L., and Kinney, J. M., 1978, Protein and energy
    sparing of glucose added in hypocaloric amounts to peripheral infusions of amino acids,
    Metabolism **27**:325.
Felig, P., 1975, The liver in glucose homeostasis in normal man and in diabetes, in: *Diabetes: Its
    Physiological and Biochemical Basis* (J. Vallance-Owen, ed.), pp. 93–123, University Park
    Press, Baltimore.
Felig, P., and Wahren, J., 1974, Protein turnover and amino acid metabolism in the regulation of
    gluconeogenesis, *Fed. Proc. Fed. Am. Soc. Exp. Biol.* **33**:1093.
Felig, P., and Wahren, J., 1975, Fuel homeostasis in exercise, *N. Engl. J. Med.* **293**:1078.
Flatt, J. P., 1978, The biochemistry of energy expenditure, in: *Recent Advances in Obesity
    Research,* II (G. A. Bray, ed.), pp. 211–228, Newman Publishing, London.
Flatt, J. P., and Blackburn, G. L., 1974, The metabolic fuel regulatory system: Implications for
    protein-sparing therapies during caloric deprivation and disease, *Am. J. Clin. Nutr.* **27**:175.
FNB (Food and Nutrition Board), 1974, *Recommended Dietary Allowances,* National Academy
    of Sciences, Washington, D.C.
Fomon, S. J., Thomas L. N., Filer, L. J., Jr., Anderson, T. A., and Nelson, S. E., 1976, Influence
    of fat and carbohydrate content of diet on food intake and growth of male infants, *Acta
    Paediatr. Scand.* **65**:136.
Forbes, E. B., Swift, R. W., James, W. H., Bratzler, J. W., and Black, A., 1946a, Further
    experiments on the relation of fat to economy of food utilization. I. By the growing albino
    rat, *J. Nutr.* **32**:387.
Forbes, E. B., Swift, R. W., Thacker, E. J., Smith, V. F., and French, C. E., 1946b, Further
    experiments on the relation of fat to economy of food utilization. II. By the mature albino
    rat, *J. Nutr.* **32**:397.
Freeman, J. B., Steginck, L. D., Wittine, M. F., Danney, M. M. and Thompson, R. G., 1977,
    Lack of correlation between nitrogen balance and serum insulin levels during protein sparing
    with and without dextrose, *Gastroenterology* **73**:31.
Fuller, H. L., and Mora, G., 1973, Effect of diet composition on heat increment feed intake and
    growth of chicks subjected to heat stress, *Poult. Sci.* **52**:2029.
Genuth, S. M., 1976, Effect of high fat vs. high carbohydrate feeding on the development of
    obesity in weanling *ob/ob* mice, *Diabetologia* **12**:155.
Ginsberg, H., Olefsky, J. M., Kimmerling, G., Crapo, P., and Reaven, G. M., 1976, Induction of
    hyperglycemia by a low fat diet, *J. Clin. Endocrinol. Metab.* **42**:729.
Goldberg, A., 1971, Carbohydrate metabolism in rats fed carbohydrate-free diets, *J. Nutr.* **101**:693.
Greenberg, G. R., Marliss, E. B., Anderson, G. H., Langer, B., Spence, W., Tovee, E. B., and
    Jeejeebhoy, K. N., 1976, Protein-sparing therapy in postoperative patients. Effects of added
    hypocaloric glucose or lipid, *N. Engl. J. Med.* **294**:1411.
Hartsook, E. W., and Hershberger, T. V., 1971, Interactions of major nutrients in whole-animal
    energy metabolism, *Fed. Proc. Fed. Am. Soc. Exp. Biol.* **30**:1466.
Hartsook, E. W., Hershberger, T. V. and Nee, J. C. M., 1973, Effects of dietary protein content
    and ratio of fat to carbohydrate calories on energy metabolism and body composition of
    growing rats, *J. Nutr.* **103**:167.
Hegsted, D. M., Gallagher, A., and Hanford, H., 1975, Reducing diets in rats, *Am. J. Clin. Nutr.*
    28:837.

Hirsch, J., and Van Itallie, T. B., 1973, The treatment of obesity, *Am. J. Clin. Nutr.* 26:1039.

Hochachka, P. W., 1974, Regulation of heat production at the cellular level, *Fed. Proc. Fed. Am. Soc. Exp. Biol.* 33:2162.

Hood, C. E. A., Goodhart, J. M., Fletcher, R. F., Gloster, J., Bertrand, P. V., and Crooke, A. C., 1970, Observations on obese patients eating isocaloric reducing diets with varying proportions of carbohydrates, *Br. J. Nutr.* 24:39.

Howard, A. N., 1975, Dietary treatment of obesity, in: *Obesity: Pathogenesis and Management* (J. T. Silverstone, ed.), pp. 123–153, Publishing Sciences Group, Acton, Mass.

Hultman, E., and Nilsson, L. H., 1975, Factors influencing carbohydrate metabolism in man, *Nutr. Metab.* 18:45.

Huston, R. L., Weiser, P. C., Dohm, G. L., Askew, E. W., and Boyd, J. B., 1975, Effects of training, exercise and diet on muscle glycolysis and liver gluconeogenesis, *Life Sci.* 17:369.

Ip, C., Tepperman, H. M., DeWitt, J., and Tepperman, J., 1977, The effect of diet fat on rat adipocyte glucose transport, *Horm. Metab. Res.* 9:218.

Ip, C., Tepperman, H. M., Holohan, P., and Tepperman, J., 1976, Insulin binding and insulin response of adipocytes from rats adapted to fat feeding, *J. Lipid Res.* 17:588.

Jeejeebhoy, K. N., Anderson, G. H., Nakhooda, A. F., Greenberg, G. R., Sanderson, I., and Marliss, E. B., 1976, Metabolic studies in total parenteral nutrition with lipid in man, *J. Clin. Invest.* 57:125.

Jensen, L. S., Schumaier, G. W., and Latshaw, J. D., 1970, "Extra caloric" effect of dietary fat for developing turkeys as influenced by calorie–protein ratio, *Poult. Sci.* 49:1697.

Jourdan, M., Goldbloom D., Margen, S., and Bradfield, R. B., 1974, Differential effects of diet composition and weight loss on glucose tolerance in obese women, *Am. J. Clin. Nutr.* 27:1065.

Kasper, H., Thiel, H., and Ehl, M., 1973, Response of body weight to a low-carbohydrate, high-fat diet in normal and obese subjects, *Am. J. Clin. Nutr.* 26:197.

Kasper, H., Schonborn, J., and Rabast, U., 1975, Behavior of body weight under a low-carbohydrate, high-fat diet, *Am. J. Clin. Nutr.* 28:801.

Kiehm, T. G., Anderson, J. W., and Ward, K., 1976, Beneficial effects of a high carbohydrate, high fiber diet on hyperglycemic diabetic men, *Am. J. Clin. Nutr.* 29:895.

Kimmerling, G., Javorski, W. C., Olefsky, J. M., and Reaven, G. M., 1976, Locating the site of insulin resistance in patients with nonketotic diabetes mellitus, *Diabetes* 25:673.

Krebs, H. A., 1972, Some aspects of the regulation of fuel supply in omnivorous animals, in: *Advances in Enzyme Regulation* (G. Weber, ed.), pp. 397–420, Pergamon Press, New York.

Lavau, M., Fried, S. K., Susini, C. and Freychet, P., 1979, Mechanisms of insulin resistance in adipocytes of rats fed a high-fat diet, *J. Lipid Res.* 20:8.

Lemonnier, D., Winand, J., Furnelle, J., and Christope, J., 1971, Effect of a high-fat diet on obese–hyperglycaemic and non-obese Bar Harbor mice, *Diabetologia* 7:328.

Lin. P. Y., Romsos, D. R., Vander Tuig, J. G., and Leveille, G. A., 1979, Maintenance energy requirements, energy retention and heat production of young obese (ob/ob) and lean mice fed a high-fat or a high-carbohydrate diet. *J. Nutr.* 109:1143.

Maynard, L. A., and Loosli, J. K., 1962, *Animal Nutrition,* McGraw Hill, New York.

McGandy, R. B., Hegsted, D. M., and Stare, F. J., 1967, Dietary fats, carbohydrates and atherosclerotic vascular disease, *N. Engl. J. Med.* 277:186.

Milligan, L. P., 1971, Energetic efficiency and metabolic transformations, *Fed. Proc. Fed. Am. Soc. Exp. Biol.* 30:1454.

Mole, P. A., Oscai, L. B., and Holloszy, J. O., 1971, Adaptation of muscle to exercise. Increase in levels of palmityl-CoA synthetase, carnitine palmityltransferase, and palmityl-CoA dehydrogenase, and in the capacity to oxidize fatty acids, *J. Clin. Invest.* 50:2323.

Musch, K., Ojakian, M. A., and Williams, M. A., 1974, Comparison of α-linolenate and oleate in lowering activity of lipogenic enzymes in rat liver: Evidence for a greater effect of dietary linolenate independent of food and carbohydrate intake, *Biochim. Biophys. Acta* 337:343.

Nestel, P. J., and Barter, P. J., 1973, Triglyceride clearance during diets rich in carbohydrate or fats, *Am. J. Clin. Nutr.* 26:241.

Pawar, S. S., and Tidwell, H. C., 1968, Effect of prostaglandin and dietary fats on lipolysis and esterification in rat adipose tissue *in vitro, Biochim. Biophys. Acta* 164:167.

Polin, D., and Wolford, J. H., 1976, Various types of diets, sources of energy, and positive energy balance in the induction of fatty liver hemorrhagic syndrome, *Poult. Sci.* **55**:325.

Prusiner, S., and Poe, M., 1968, Thermodynamic considerations of mammalian thermogenesis, *Nature* **220**:235.

Renner, R., 1964, Factors affecting the utilization of "carbohydrate-free" diets by the chick, *J. Nutr.* **84**:322.

Renner, R., and Elcombe, A. M., 1964, Factors affecting the utilization of "carbohydrate-free" diets by the chick, *J. Nutr.* **84**:327.

Rennie, M. J., Winder, W. W., and Holloszy, J. O., 1976, A sparing effect of increased plasma fatty acids on muscle and liver glycogen content in the exercising rat, *Biochem. J.* **156**:647.

Romsos, D. R., and Leveille, G. A., 1974, Effect of diet on activity of enzymes involved in fatty acid and cholesterol synthesis, *Adv. Lipid Res.* **12**:97.

Romsos, D. R., Belo, P. S., Bennink, M. R., Bergen, W. G., and Leveille, G. A., 1976, Effects of dietary carbohydrate, fat and protein on growth, body composition and blood metabolite levels in the dog, *J. Nutr.* **106**:1452.

Romsos, D. R., Hornshuh, M. J., and Leveille, G. A., 1978, Influence of dietary fat and carbohydrate on food intake, body weight and body fat of adult dogs, *Proc. Soc. Exp. Biol. Med.* **157**:278.

Salans, L. B., Bray, G. A., Cushman, S. W., Danforth, E., Glennon, J. A., Horton, E. S., and Sims, E. A. H., 1974, Glucose metabolism and the response to insulin by human adipose tissue in spontaneous and experimental obesity. Effects of dietary composition and adipose cell size, *J. Clin. Invest.* **53**:848.

Schemmel, R., Mickelsen, O., and Gill, J. L., 1970, Dietary obesity in rats: Body weight and body fat accretion in seven strains of rats, *J. Nutr.* **100**:1041.

Schreibman, P. H., and Ahrens, E. H., 1976, Sterol balance in hyperlipidemic patients after dietary exchange of carbohydrate for fat, *J. Lipid Res.* **17**:97.

Slovic, P., 1975, What helps the long distance runner run? *Nutr. Today* **10**:18.

Suzuki, H., and Fuwa, H., 1971, Interaction of dietary fat and thyroid function with hepatic and renal gluconeogenesis of rats, *J. Nutr.* **101**:919.

Suzuki, H., Goshi, H., and Sugisawa, H., 1975, Effects of previous feeding of a high-carbohydrate or a high-fat diet on changes in body weight and body composition of fasted rats, *J. Nutr.* **105**:90.

Swift, R. W., Barron, G. P., Fisher, K. H., Cowan, R. L., Hartsook, E. W., Hershberger, T. V., Keck, E., King, R. P., Long, T. A., and Berry, M. E., 1959, The utilization of dietary protein and energy as affected by fat and carbohydrate, *J. Nutr.* **68**:281.

Toppings, D. L., and Mayes, P. A., 1971, The concentration of fructose, glucose and lactate in splanchnic blood vessels of rats absorbing fructose, *Nutr. Metab.* **13**:331.

West, K. M., 1973, Diet therapy of diabetes: An analysis of failure, *Ann. Intern. Med.* **79**:425.

Williams, M. H., 1976, *Nutritional Aspects of Human Physical and Athletic Performance*, C. C. Thomas, Springfield, Ill.

Wood, J. D., and Reid, J. T., 1975, The influence of dietary fat on fat metabolism and body fat deposition in meal-feeding and nibbling rats, *Br. J. Nutr.* **34**:15.

Yang, M. U., and Van Itallie, T. B., 1976, Composition of weight lost during short-term weight reduction. Metabolic responses of obese subjects to starvation and low-calorie ketogenic and nonketogenic diets, *J. Clin. Invest.* **58**:722.

Yoshimura, M., Hori, S., and Yoshimura, H., 1972, Effect of high-fat diet on thermal acclimation with special reference to thyroid activity, *Jpn. J. Physiol.* **22**:517.

Young, C. M., Scanlan, S. S., Im, H. S., and Lutwak, L., 1971, Effect on body composition and other parameters in obese young men of carbohydrate level of reduction diet, *Am. J. Clin. Nutr.* **24**:290.

6

Energetics and the Demands for Maintenance

Keith A. Crist, R. L. Baldwin, and Judith S. Stern

1. Introduction

Maintenance, the physiological steady state in which there is no net change in body energy content, has long been a focal point in the study of nutritional energetics. Early workers used balance studies and direct heat measurements on animals to estimate the energy required as food to support daily requirements but could only speculate on the nature of "vital processes" involved. A brief review of fundamental concepts and techniques in nutritional energetics is presented below. Specific partitioning of energy input and identification of fixed and variable costs of maintenance are recent advances dependent upon current knowledge of intermediary metabolism. Students in nutrition need an overall concept of energy expenditure to appreciate this variability. For example, dietary protein is used more efficiently as a source of muscle protein than for liver glycogen or storage triglyceride. Proteins differing in amino acid composition differ in energy yield as ATP when oxidized; this reflects the different pathways by which individual amino acids are metabolized.

If body weight is to be maintained, energy provided for performance of vital processes must be regulated to keep pace with tissue demands. Food intake is controlled to supply this energy, analogous to the control of cardiac output to regulate blood pressure. This chapter will focus on identification of specific energy-requiring processes which maintain adult body weight and efficiency of nutrient energy utilization.

Keith A. Crist and R. L. Baldwin • Department of Animal Science, University of California, Davis, California 95616. *Judith S. Stern* • Department of Nutrition, University of California, Davis, California 95616.

2. Measurement of Energy Exchange

2.1. Heat as a Measure of Metabolism

The energy required to maintain a steady state, grow, or do work is derived from chemical transformation of absorbed nutrients. Such transformations obey the first law of thermodynamics. Energy can be converted quantitatively only to heat; other transformations occur at less than 100% efficiency (second law of thermodynamics). Determination of heat production is thus a convenient as well as theoretically sound measure of metabolism. Nutritional energetics is based on general statements of the first and second laws of thermodynamics.

The first law of thermodynamics is the law of conservation of energy and holds that energy can be transferred or transformed but neither created nor destroyed. Based on early studies of animal heat production by Lavoisier and Laplace (discussed below), the law was formulated by Robert J. Mayer, a physician, in 1842. Mayer demonstrated the conversion of kinetic energy to heat by shaking water and observing the increase in temperature. In terms of living systems, the first law implies that any change in energy (or heat content) must be accounted for as an input, such as food, or an output, such as heat, work, body energy gain, or waste products. Stated another way, any input of food energy increases body energy content. Despite claims of some fad diets, this energy remains unless balanced by an equivalent output. No exceptions to this law have been demonstrated. If inputs and outputs are known, energy retained can be determined by difference, an important assumption in energy balance studies. Experimental validation of the equivalence between heat and work in a biological system was provided by Atwater and Benedict (1903).

An additional principle, formulated by Hess in 1840, is the law of constant heat sums. This was later included in the more general first law formulated by Mayer. The law of Hess states that heat generated in a net energy transformation is independent of path. The work of Rubner, combining direct and indirect calorimetry to determine resting heat production in dogs, validated this concept for animals. Consider, for example, the several metabolic paths by which a molecule of fatty acid may be oxidized to CO_2 and water. Heat production is the same, whether converted to acetate and other intermediates before terminal oxidation or as heat produced during direct oxidation in a bomb calorimeter.

The second law of thermodynamics states that all forms of energy are quantitatively convertable to heat, that heat is the lowest energy form, and that the driving force of all energy transactions is the tendency to reach the lowest energy form, heat. Nutritionists depend on this law for all of their measurements because energy transformations to heat are quantitative and heat is easily determined.

The calorie has classically been the unit of measurement for heat in nutrition and is defined as the amount of heat required to raise the temperature of 1 g of water from 14.5 to 15.5°C. Adoption of the International System of

Units will mean discarding this unit of heat for a more generalized unit of energy, the joule. The joule (J) can be derived from the primary units of mass (kg), length (m), and time (sec), each explicitly defined in the International System, and is equal to $(1 \text{ kg·m}^2)/\text{sec}^2$. This international calorie is equal to 4.185 J. Although the transition is in progress, arguments of Kleiber in favor of retaining the calorie and those in favor of the joule are worth repeating.

As stated by Kleiber (1975), energy is an abstraction and can be measured only when it is transformed. When this transformation is to heat, the calorie is the logical unit, since a change in water temperature is the basis of measurement. When data of this type, meant to express heat, are reported as joules, the presentation is less clear (Kleiber, 1972). The joule is preferred because of its generality, but a strong argument for the calorie as a fundamental unit in nutrition can be made as well. Heat is the most degraded form of energy and can be obtained with 100% efficiency. It is therefore a logical common denominator on which other forms of energy can be based. Kleiber took exception to the suggestion that electrical units are the reference standard for calibration of calorimeters, justifying the change to joules. While electrical energy is used as input, the caloric effect is measured and then converted to joules. He further stated that as long as chemical energy is measured as heat of combustion and no accuracy is gained by conversion of this heat to joules, the result is more clearly stated in calories. Finally, it is well recognized that conversion of standard tables will result in a large amount of work and familiarization.

Arguments in favor of the joule were expressed by Ames (1970) and Harper (1970). The calorie, being dependent on the specific heat of water, is a temperature-dependent unit. The nutritional calorie defined at 17°C has a value slightly lower than the 15°C calorie of 4.184 J. [The benzoic acid used to calibrate calorimeters is standardized against the thermochemical (17°C) calorie.] Therefore, the unit is defined with less precision, although the difference is too small to affect measurements of animal heat. Common nutritional constants can be expressed in joules with a minimum of inconvenience. The present physiological fuel values of 4 kcal/g for protein and carbohydrate and 9 kcal/g for fat are average values. Equivalent expressions in International System Units are 17 kJ/g for carbohydrate and protein and 38 kJ/g for fat. These values are no less accurate. Energy requirements are normally rounded to the nearest 50 kcal. Values in kilojoules can be rounded to the nearest 100 kJ to similarly reflect individual variability. The difference between the two expressions is no more than 12 kcal. This difference is negligible and should pose no problem in acceptance of the joule (Harper, 1970).

2.2. Direct Calorimetry

Measurement of animal heat production originated from studies by Crawford (1788) and by Lavoisier and Laplace (in Kleiber, 1975). Lavoisier's calorimeter is illustrated in Fig. 1. The mixture of ice and water prevents heat flow by forming a constant temperature layer. Heat produced by the animal melts the ice [80 cal (335 J)/g water melted], and the collected water is weighed.

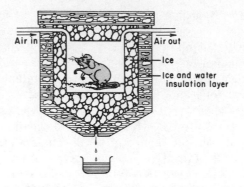

Fig 1. Lavoisier's calorimeter. Melted ice is collected and weighed to calculate animal heat production.

Modern calorimeters which directly measure heat production are designed to measure heat either transferred across the walls or absorbed by a cooling system. In gradient-layer calorimetry, the subject is surrounded by a solid heat-flow metering wall. Heat flow across the wall is uniform and independent of the way in which it is transferred to the wall, whether by radiation, conduction, convection, or condensation of water vapor (Benzinger, 1969). Additional apparatus can be added to measure evaporative heat loss completely, but, in so doing, the response time is reduced to that of respiration chambers used for indirect calorimetry. Complete measurement of evaporative heat loss is important, as it represents 20–30% of total heat loss in humans. The latter heat-sink calorimeters are similar to Lavoisier's in having an insulated wall, but heat produced is removed by a cooling stream of air or water to maintain an isothermal chamber. These are less expensive, can be more efficient, and are less restrictive on the life-style of the subject. A more detailed discussion of direct calorimetry, including a description of the Atwater calorimeter, can be found in Garrow (1974).

2.3. Indirect Calorimetry

The demonstration by Lavoisier and Laplace that CO_2 production was related to the amount of ice melted in their calorimeter marked the beginning of indirect calorimetry. Lavoisier believed the source of this heat and CO_2 to be the slow combustion of carbon and hydrogen. This view was corrected in 1842 by Liebig, who showed that carbohydrate, fat, and protein served as energy sources.

The amount of O_2 required for combustion of these fuels is variable. During the oxidation of 1 mol of glucose, 6 mol of O_2 are required and 6 mol of CO_2 are produced: $C_6H_{12}O_6 + 6 O_2 \rightarrow 6 CO_2 + 6 H_2O$. The ratio of CO_2 produced to O_2 consumed defines the respiratory quotient, commonly referred to as RQ. Knowing that 1 mol of each of these gases at standard atmospheric pressure and temperature occupies 22.4 liters and that combustion of glucose in a bomb calorimeter yields 673 kcal (2.82 MJ) of heat, one can calculate the caloric equivalent of 1 liter of O_2 or CO_2 production for combustion of glucose [(673/6)/22.4] as 5.0 kcal (21 kJ)/liter. The average value of carbohydrates is 5.047 kcal (21.12 kJ)/liter O_2.

Oxidation of fat requires more O_2 per g-atom of carbon, since less O_2 is bound in the molecule. This leads to a lower RQ. As shown for the example of tripalmitin (Table I), combustion is associated with an RQ of 0.708 and a caloric equivalent for O_2 of 4.7 kcal (20 kJ)/liter. The average values for fat are 0.707 and 4.686 kcal (19.61 kJ)/liter, respectively.

Unlike carbohydrate and fat, protein is not completely oxidized within the cell. Of the nitrogen end products excreted in the urine of humans, urea is the most important. Determination of urinary nitrogen and the assumption that the body urea pool is constant form the bases for calculation of protein metabolism. If completely oxidized in a bomb calorimeter, protein represented by 1 g of urinary nitrogen would release 33.5 kcal (140 kJ) of heat. However, 5.4 kcal (23 kJ) appears unoxidized in excreted urea. Therefore, 1 g of urinary nitrogen represents 28.1 kcal (118 kJ) of heat produced by incomplete protein

Table I. Calculation of RQ for Combustion or Synthesis of Fat[a]

A. Elemental composition of tripalmitin

Element	g/100 g	At. wt.	g-atoms/ 100 g fat	Internal oxidation (g-atoms)	Elements requiring respiratory oxidation (g-atoms)	O_2 required (mol)
C	76	12	6.3	—	6.3	6.3
H	12	1	12.0	1.6	10.4	2.6
O	12	16	0.8	0.8	—	
						8.9

$$RQ = 6.3/8.9 = 0.708$$

$$\frac{8.9 \text{ mol } O_2 \cdot 22.4 \text{ liters/mol}}{100 \text{ g fat}} = 2.0 \text{ liters } O_2/\text{g fat}$$

$$\frac{\Delta H_c \text{ fat}}{\text{Liters } O_2/\text{g fat}} = \frac{9.4 \text{ kcal (39 kJ)/G}}{2.0} = 4.7 \text{ kcal (20kJ)/liter } O_2$$

B. Synthesis of palmitate from glucose

4 $C_6H_{12}O_6$ + O_2 . $C_{16}H_{32}O_2$ + 8 CO_2 + 8 H_2O
RQ = 8.0

[a] Adapted from Kleiber (1975).

oxidation and 5.4 kcal (23 kJ) urinary energy loss as urea. For meat protein containing 17% nitrogen, this corresponds to a heat loss of 478 kcal (2 MJ)/100 g protein, requiring 100 liters $O_2/17$ g nitrogen and liberating 83 liters $CO_2/17$ g nitrogen (Kleiber, 1975). When O_2 uptake and CO_2 production are corrected for protein oxidation, as illustrated in Table II, a nonprotein RQ (NPRQ) representing mixed carbohydrate and fat oxidation can be determined. Table III shows the quantitative relationship between the caloric value of O_2 and CO_2 and the percentage accounted for by fat oxidation for any RQ between 0.70 and 1.00. Heat production can be calculated from either O_2 or CO_2 data. However, as seen from Table III, variability in caloric values for O_2 is much lower (7%) than for CO_2 (30%). Estimation of total heat production using an average caloric value for O_2 of 4.85 kcal (20 kJ)/liter O_2 and omitting the correction for protein will introduce little error. An alternate to use of tabular data as in Table III or the 4.85 kcal/liter O_2 generalization in interpreting data from indirect calorimetry is use of the following empirical equations which are based on a large number of energy balance data:

$$\text{Energy loss} = (51.83 \text{ kJ/g carbon} \cdot \text{g carbon loss/day})$$
$$- (19.4 \text{ kJ/g urinary nitrogen} \cdot \text{g urinary nitrogen/day})$$

$$\text{Heat production} = (16.18 \text{ kJ/liter } O_2 \cdot \text{liter } O_2) + (5.02 \text{ kJ/liter } CO_2 \cdot \text{liter } CO_2)$$
$$- (5.99 \text{ kJ/g urinary nitrogen} \cdot \text{g urinary nitrogen})$$

Elevation of RQ values above 1.0 occurs during periods of fatty acid synthesis from glucose. As shown in Table I for palmitate, the theoretical RQ for this process is 8.0. Deviations of this magnitude are never observed *in vivo,* because metabolism is never completely anabolic or catabolic. The RQ can be depressed below 0.71 during periods of excessive fatty acid oxidation and ketone body excretion. However, secondary effects, including neutralization of acids formed, causing evolution of CO_2, make exact interpretations of these values impossible.

Respiration chambers, capable of measuring O_2 uptake and CO_2 production, are easier to construct than calorimeters. Early open-circuit designs, using outside air, measured only CO_2 production with carbon and nitrogen balance to compute heat produced. In closed-circuit chambers, air is circulated through absorbants to remove CO_2 and water vapor and returned to the chamber. Atwater and Benedict (1903) used a respiration calorimeter, combining closed-circuit respiration methods with direct heat determinations, to show that heat production can be determined from O_2 and CO_2 exchange with accuracy equivalent to direct heat measurement. The response time of older instrument designs has been improved by adding new ventilation systems. However, the basic problem of restrictions on normal activity remain in the most precise instruments. Less restrictive devices have been developed, but at the sacrifice of accuracy. For a detailed discussion of indirect calorimetry, see Garrow (1974).

Table II. Carbon and Oxygen Exchange in Catabolism of Meat Protein[a]

Element	Composition of meat protein (% by wt.)	At. wt.	g-atoms in 100 g protein	0.6 mol urea with 1.2 g-atoms of N (g-atoms)	100 g protein minus 0.6 mol urea (g-atoms)	Internal oxidation (g-atoms)	Remaining CH (g-atoms)	O_2 from respiration required (mol)
C	52	12	4.3	0.6	3.7		3.7	3.70
H	7	1	7.0	2.4	4.6	1.6	3.0	0.75
O	23	16	1.4	0.6	0.8	0.8	—	
N	17	14	1.2	1.2	—			
S	1	32	(0.03)					

C and O exchange: CO_2 produced—3.7 mol; O_2 consumed—4.45 mol.

RQ = 3.7/4.45 = 0.83.

100 g meat protein produces 3.7 mol or 83 liters CO_2 and consumes 4.45 mol or 100 liters O_2.

Per 1 g N in catabolized meat protein, 83/17 = 4.9 liters CO_2 is produced and 100/17 = 5.9 liters O_2 is consumed.

[a] From Kleiber (1975).

Table III. RQ and Energetic Equivalents of O_2 and CO_2 from Oxidation of
Varying Percentages of Carbohydrates and Fat[a]

RQ	O_2 (kcal/liter)	CO_2		% O_2 consumed by		% Heat produced by oxidation of	
		kcal/liter	kcal/g	Carbohydrates	Fat	Carbohydrates	Fat
0.70	4.686	6.694	3.408	0	100	0	100
0.71	4.690	6.606	3.363	1.0	99.0	1.1	98.9
0.72	4.702	6.531	3.325	4.4	95.6	4.8	95.2
0.73	4.714	6.458	3.288	7.85	92.2	8.4	91.6
0.74	4.727	6.388	3.252	11.3	88.7	12.0	88.0
0.75	4.729	6.319	3.217	14.7	85.3	15.6	84.4
0.76	4.752	6.253	3.183	18.1	81.9	19.2	80.8
0.77	4.764	6.187	3.150	21.5	78.5	22.8	77.2
0.78	4.776	6.123	3.117	24.9	75.1	26.3	73.7
0.79	4.789	6.062	3.086	28.3	71.7	29.9	70.1
0.80	4.801	6.001	3.055	31.7	68.3	33.4	66.6
0.81	4.813	5.942	3.025	35.2	64.8	36.9	63.1
0.82	4.825	5.884	2.996	38.6	61.4	40.3	59.7
0.83	4.838	5.829	2.967	42.0	58.0	43.8	56.2
0.84	4.850	5.774	2.939	45.4	54.6	47.2	52.8
0.85	4.863	5.721	2.912	48.8	51.2	50.7	49.3
0.86	4.875	5.669	2.886	52.2	47.8	54.1	45.9
0.87	4.887	5.617	2.860	55.6	44.4	57.5	42.5
0.88	4.900	5.568	2.835	59.0	41.0	60.8	39.2
0.89	4.912	5.519	2.810	62.5	37.5	64.2	35.8
0.90	4.924	5.471	2.785	65.9	34.1	67.5	32.5
0.91	4.936	5.424	2.761	69.3	30.7	70.8	29.2
0.92	4.948	5.378	2.738	72.7	27.3	74.1	25.9
0.93	4.960	5.333	2.715	76.1	23.9	77.4	22.6
0.94	4.973	5.290	2.693	79.5	20.5	80.7	19.3
0.95	4.985	5.247	2.671	82.9	17.1	84.0	16.0
0.96	4.997	5.205	2.650	86.3	13.7	87.2	12.8
0.97	5.010	5.165	2.629	89.8	10.2	90.4	9.6
0.98	5.022	5.124	2.609	93.2	6.8	93.6	6.4
0.99	5.034	5.085	2.589	96.6	3.4	96.8	3.2
1.00	5.047	5.047	2.569	100	0	100	0

[a] From Brody (1945).

Estimates of heat production of individuals in their own environment, pursuing normal activity, are largely restricted to estimates made from 24-hr activity diaries and heart-rate monitors. In an absolute sense, these methods are associated with large sources of error. The novelty of keeping a detailed minute-to-minute record degenerates rapidly to a tedious exercise. Using an observer for record-keeping works in some situations but, if extended to 24 hr, imposes some of the same restrictions one is trying to avoid. Even if given a sufficiently detailed diary, an additional 10% error is added by using standard tables of activity energy costs. Heart-rate monitors can be individually calibrated, but changes in stroke volume brought on by postural changes as well as by emotional stress alter the response curve. In the context of providing estimates of energy expenditure to determine requirements for different groups of people, these methods are much preferable to estimates made from questionnaire-based estimates of daily intakes.

3. Partition of Food Energy

3.1. Definition of Terms

Gross energy (G)—The amount of heat released when a substance is completely oxidized in a bomb calorimeter containing 25–35 atm of O_2

Metabolizable energy (ME)—The gross energy of food consumed minus energy in gaseous products of digestion and excreta. ME of a food determined at maintenance corresponds to "physiological fuel values" found in standard calorie tables.

Heat increment of maintenance (HI_m)—The sum of all energy costs to an organism associated with the digestion and assimilation of food. The similar term "specific dynamic action" was considered less precise by Kleiber (1975) and will not be used here.

Heat increment of production (HI_p)—The energy supplied above maintenance not recovered, for example, as the products of growth. In other words, HI_p is the cost of product synthesis.

Net energy for maintenance (NE_m)—Measured by the loss of body energy prevented by consuming a unit of food. This equals metabolizable energy intake (MEI) less all costs of processes involved in the digestion and assimilation of food (HI_m). In equation form, $NE_m = MEI - HI_m$ for an animal fed at maintenance.

Net energy for production (NE_p)—Energy supplied in excess of the maintenance requirement which is stored or released in a product such as milk. In equation form, $NE_p = NEI - NE_m - HI_m - HI_p$ for an individual fed above maintenance.

Basal metabolism—The minimum energy cost of vital processes in a postabsorptive state, under conditions of thermoneutrality and physical and mental rest.

3.2. Basal Metabolic Rate

Basal heat production represents the proportion of daily energy expenditure required for vital maintenance functions. Although easily defined, estimation requires mental as well as physical rest and, therefore, is subject to considerable variation. When a 1-hr representative sample is used, even mild anxiety can cause errors of 30% in estimates of daily rate (Garrow, 1974). This variability is significant due to the large amount of time spent in activities near basal energy expenditure. In studies of cadets, Edholm *et al.* (1955) found sleeping and sitting to make up 70% of daily activity and to represent 40% of energy expenditure. For people who are less active, the contribution of resting metabolism is greater and most probably greater than 50%. As a minimum energy expenditure, it is a convenient base from which to evaluate additional costs, in the adult, of productive functions such as work, gestation, or lactation. Basal metabolic rate (BMR) is not constant and varies with age, environmental temperature, nutritional status, and other factors to be discussed below. Identification of specific metabolic components is valuable in understanding the nature of these alterations and will be discussed in more detail below. While a complete list cannot be given, up to 45% of basal heat production is derived from "service" functions such as sodium transport by the kidney (7%), blood circulation (11%), respiration (7%), and nervous functions (20%) (Baldwin and Smith, 1974). These are costs unique to the maintenance of complex, as compared to unicellular, organisms. Cellular maintenance functions include costs of resynthesis of labile tissue proteins and triglycerides (10%) and ion transport for the maintenance of membrane potentials, etc. (20–30%).

Comparison of BMRs within or between species requires a common base of expression, free of variability associated with body size. The limits of this variability are easily recognized. Heat production, for example, increases with body size when the whole animal is considered, yet it is much greater per kilogram of mouse than per kilogram of human. Metabolic rate per unit of surface area appeared much more consistent and was well established by 1901 when Voit published his results comparing several species (in Kleiber, 1975). The surface area of humans can adequately be determined by the formula of DuBois and DuBois (1916):

$$S = 71.84 \ W^{0.425} + L^{0.725}$$

where S = surface area in cm^2; W = body weight in kg; and L = body length in cm. A suitable formula for other animal species depends upon an accepted definition of true body surface. This has not been agreed upon and represents a major obstacle to the use of surface area as a reference base.

The relationship between heat production and body surface area among species can best be interpreted in terms of heat transfer. In data summarized by Kleiber (1975), a 200-g rat produced 1260 kcal (5.3 MJ)/m^2 per day of heat when the difference between rectal and environmental temperature was 21.5°C. A 70-kg man with the same heat production per kilogram of body weight would produce 8827 kcal (37 MJ)/m^2 per day. Assuming no temperature-dependent difference in the man's resistance to heat flow, maintenance of temperature equilibrium would require an environmental temperature of −125°C. This example indicates why it is advantageous for large animals to have lower BMRs per kilogram than small animals. It does not indicate that heat transfer determines this rate or that more accurate determinations of true body surface would improve precision of BMR data. The lack of direct proportionality is evident from basal metabolism studies in humans in which changes in effective surface area produced by curling up or lying outstretched do not affect heat production. This is obvious, since in a neural thermal environment, heat loss is not a driving force.

Body weight is a more accurate and reproducible indication of size than is surface area. Both Kleiber (1932, 1975) and Brody (1945) published results of empirical analyses of relationships between heat production and body weight and attempted to formulate a more suitable predicative equation for BMR. They reasoned that if the logarithms of these two variables were linearly related, they could conclude that BMR is directly proportional to body weight raised to a power. Kleiber's results, covering a wide range of species at mature body weights, are presented in Fig. 2. Note that the line defined by the logarithms of body weight and heat production is linear and that the slope of the line is the familiar exponent 0.75. Based on this analysis, Kleiber proposed that the equation

$$BMR = 70 \times BW^{3/4}$$

where BW = body weight in kg, could be used on an interspecific basis to estimate heat production from body weight. As seen in Fig. 2, body weight and surface area $(BW^{2/3})$ predict heat production as well as Kleiber's equation for some homeothermic species and are preferable for many poikilothermic species (von Bertalanffy, 1957) not included in Kleiber's analysis. The significant advantage of this equation is the wide range of mature body weights among homeotherms over which it is applicable. Zeuthen (1953) found 0.7 to be a suitable exponent for some unicellular organisms and an exponent of 0.75 to be applicable to poikilotherms heavier than 40 mg. Brody's analysis (1945), in close agreement with Kleiber's, included growing as well as mature animals. Brody found that the exponent 0.73 best fit his data. Yet, there was no difference between Kleiber's and Brody's equations within the range of "a

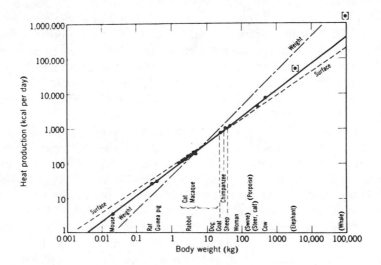

Fig. 2. Interspecies relationships between logarithms of heat production and adult body weight. From Kleiber (1975).

10-g mouse and 16-ton superelephant'' (Kleiber, 1975). The constancy of this relationship in the adult does not hold during early phases of growth. As an extreme example, use of the interspecies equation for a 1-μg embryo would predict a ''basal'' heat production of 12,450 kcal (52.1 MJ)/kg per day (Brody, 1945). The $\frac{3}{4}$ power of body weight can only be used for intraspecies comparisons after maturity. Kleiber (1932) analyzed data of Harris and Benedict (1919) to develop equations for metabolic rates in men and women, respectively, adjusted for age and stature:

$$BMR = 71.2 \times W^{3/4} [1 + 0.004(30 - a) + 0.010 (S - 43.4)]$$

$$BMR = 65.8 \times W^{3/4} [1 + 0.004(30 - a) + 0.018 (S - 42.1)]$$

where BMR = basal metabolic rate in kcal/day; W = body weight in kg; a = age in years; and s = specific stature in cm height/$W^{1/3}$.

Thonney *et al.* (1976) demonstrated differences in power functions when intraspecies data are assumed not to approximate a single population value and pointed out the error of assuming that other variables have the same relationship to body weight as does heat production. We, however, agree with Brody and Kleiber that pooling of intraspecies data to develop generalized predictive *interspecific* relationships is justified. Brody (1945), Kleiber (1975), and Thonney *et al.* (1976) are all correct in emphasizing that application of an interspecific generalization within a species is not justified.

As stated above, BMR in a given individual is influenced by several factors. These include sex, age, temperature, nutritional status, and hormones. Recent reviews of this subject are available (Bray and Atkinson, 1977; Garrow, 1974), and only major points will be discussed here.

The BMR of women is generally less than that observed in men. A 7% difference based on surface area was found (DuBois, 1936) and appears to remain between 6 and 10% from five years of age into old age (Boothby *et al.*, 1936). The different proportions of body fat in men and women, resulting in different lean body masses, do not explain this difference. In a more recent study that accounts for body composition, values for adult males and females were found to be 33.9 kcal (142 kJ)/m² per hr and 28.9 kcal (121 kJ)/m² per hour, respectively (Banerjee and Bhattacharjee, 1967). BMR correlated best with fat-free body mass and least with surface area.

BMR has generally been observed to increase rapidly after birth, reaching a maximum [about 53 kcal (220 kJ)/m² per hour] in the first year of life. It then declines until completion of growth, levels off, and gradually decreases with age until death (Brody, 1945; DuBois, 1936; Mitchell, 1962) (see Fig. 3). Garrow (1974) used the classical study of Magnus-Levy to point out alternative interpretations that depend upon the basis for calculation. Magnus-Levy's BMR had fallen 17% when calculated from difference in O_2 consumption and surface area at 26 and 76 years of age. However, if calculated from the expected decline in intracellular water as a proportion of body weight, BMR would have risen by 19%. Keys *et al.* (1973) discussed possible problems of using cross-sectional evaluations rather than long-term studies in the same

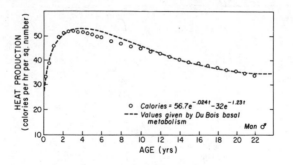

Fig. 3. The rise and decline of basal metabolism with increasing age in early life. From Brody (1945).

individuals to assess age effects and concluded that the reduction due to aging was only 1–2% per decade.

Metabolic rate in humans and other homeotherms is controlled in response to environmental temperature (Fig. 4). Within the thermoneutral range, 22–30°C for humans, no adjustment of BMR is necessary to maintain a normal body temperature of 36–38°C. Above the upper critical temperature, heat loss cannot keep pace with heat production, and metabolic rate increases 12% for each 1°C rise in body temperature. Metabolic rate (i.e., summit metabolism) increases as environmental temperature decreases and is maximal at D in Fig. 4. A further decline in environmental temperature can not be compensated for, and body temperature falls. Attempts to identify adaptive increases in BMRs of individuals living in different climates have produced conflicting results. Buskirk *et al.* (1957) found no difference in BMR among three groups of men studied for at least one month. Mean ambient temperature differed among the three locations by 59°C. Diet composition and intake were the same for all groups. Gold *et al.* (1969) found seasonal differences in metabolic rate. Consumption of O_2 was lower in the summer during periods of rest and exercise. Diet was not controlled in this group.

Overnutrition does not always produce a proportionate gain in body energy. Dogs fed 130% or more of their energy requirement showed no significant gain after one month (Share *et al.*, 1952). Similar results were obtained for dogs fed 175% of their requirement, an energy excess of 37,000 kcal (155 MJ), over 14 weeks (Janowitz and Hollander, 1955). The concept of an increase in energy expenditure in response to excess energy intake, termed "luxus consumption," is supported by results of studies of humans which, although not conclusive, are difficult to explain by other means. Excess energy is not simply eliminated in feces (Strong *et al.*, 1967). An absorption coefficient of 98% has been found after an increase of fat intake from 39 to 639 g/day (Kasper, 1970). Neither can changes in energy costs of normal activities account for this (Passmore *et al.*, 1963). Studies by Gulick on his own body weight maintenance provide a good example (Gulick, 1922). During one 41-day period, his intake

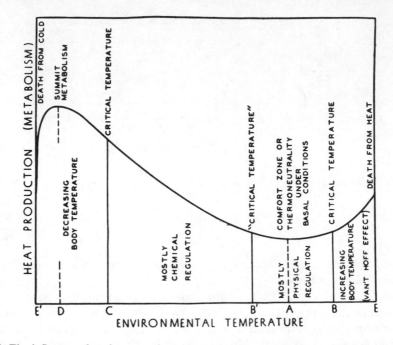

Fig. 4. The influence of environmental temperature on heat production and body temperature. From Brody (1945).

averaged 2744 kcal (11.5 MJ)/day, and he lost 510 g from an initial weight of 61.8 kg. Assuming an energy value of 6000 kcal (25 MJ)/kg of adipose tissue, the loss would have supplied 75 kcal (0.31 MJ)/day. Thus, total energy expenditure was 2810 kcal (11.8 MJ)/day. During a later 73-day period, his intake averaged 3545 kcal (14.8 MJ)/day, and he gained 2.88 kg, or 237 kcal (1 MJ)/day, as adipose tissue. This suggests an energy expenditure of 3308 kcal (13.8 MJ)/day or a value of 500 kcal (2 MJ)/day in excess of the previous daily energy expenditure. Activity was similar during both periods, and it does not seem logical to ascribe this loss totally to cost of fat deposition.

Effects of undernutrition on metabolic rate are more clearly defined and may result from a decrease in tissue mass or a lower activity per unit of tissue. The contribution of each effect is dependent upon the duration of food restriction. Short-term restriction to 1000 kcal (4.2 MJ)/day for 13 or 19 days produced a decline in BMR of 17–21% regardless of the basis of comparison (Grande et al., 1958). Lost body weight and metabolic rate returned rapidly to normal with refeeding. Semistarvation in obese patients produced a decline in O_2 consumption after two days of restriction to a 450-kcal (1.9-MJ) diet. After two weeks, O_2 consumption was reduced 21% (Bray, 1969) (Fig. 5). In chronically undernourished individuals, the loss of body tissue may account for all of the decrease (Beattie and Herbert, 1947).

Thyroid hormones most clearly affect metabolic rate, and available evidence supports a more mechanistic view than has been possible with other

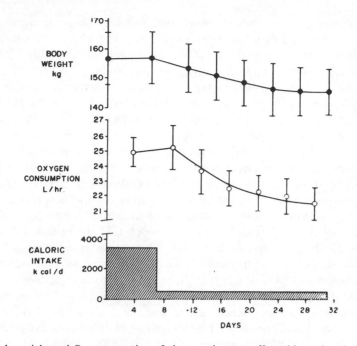

Fig. 5. Body weight and O$_2$ consumption of obese patients as affected by reduced caloric intake. From Bray (1969).

factors discussed so far. Triiodothyronine is the most active form of thyroid hormone and is produced largely by deiodination of circulating thyroxine (Larsen, 1972). Under experimental conditions in which this conversion is blocked, metabolic effects of the hormone are not observed. Triiodothyronine, however, remains effective (Bray and Hildreth, 1967). There is good evidence that triiodothyronine acts at the cell membrane by increasing ion transport. This, in turn, increases ATP utilization and the demand on oxidation metabolism (Edelman and Ismail-Beigi, 1974).

Growth hormone and the catecholamines are calorigenic as well. The action of growth hormone is not dependent upon thyroid hormone and may result from its effect on release of free fatty acids (Bray, 1969). Catecholamines may similarly act by increasing free fatty acid release. Their actions are dependent upon the presence of thyroid hormone (Goodman and Bray, 1966).

3.3. The Contribution of Various Tissues to Basal Metabolic Rate

An important goal in nutritional energetics is identification of specific processes required to maintain body weight. Those accounting for a majority of the maintenance expenditure have been identified and are discussed below, beginning with a detailed example for liver.

Expenditures of the various tissues, estimated by their O_2 consumption, must add up to the total maintenance requirement. The liver carries out a wide variety of functions in support of maintenance and accounts for 18–22% of total expenditure (Baldwin and Smith, 1974) (Table IV).

In the early postabsorptive state, liver glucose release is a major source of energy supplied to the bloodstream. Glycogenolysis accounts for 80% of glucose release; gluconeogenesis from lactate, pyruvate, glycerol, and amino acids supplies the remaining 20% (Owen *et al.*, 1978; Wahren *et al.*, 1972). Liver glycogen stores are small, about 5% of liver weight at basal and up to 10% after starvation–refeeding (Tepperman and Tepperman, 1958). A postabsorptive release of 0.86 mmol/min per 1.73 m² (Wahren *et al.*, 1972), supported completely by glycogenolysis, would deplete liver stores in approximately 6 hr. In actuality, gluconeogenesis assumes an increasingly important role, contributing 80% of the 0.51 mmol glucose/min per 1.73 m² released from liver after a three-day fast (Garber *et al.*, 1974). Glucose recycled from nonhepatic tissues as lactate (Cori cycle) accounts for 10–30% of total glucose turnover at basal (Cahill *et al.*, 1966; Reichard *et al.*, 1963). The energy cost of glucose recycling via the Cori cycle is low, requiring only 6 molecules of ATP per molecule of glucose, or an energy cost to liver of 27 kcal (113 kJ)/day, using 18 kcal (75 kJ)/mol ATP (justified below). If the maintenance requirement of resting, fasting human subjects is estimated to be 2000 kcal (8.4 MJ)/day, Cori cycle activity represents only 1% of energy expenditure (Baldwin, 1968). Heat production in Table IV from combined Cori cycle and triglyceride turnover activity suggests a net gain in liver ATP when glucose is synthesized from glycerol (derived from triglyceride degradation). This theoretical gain in ATP requires the assumption that the ATP is used for maintenance and spares oxidation of other substrates. The contribution of amino acids released from

Table IV. Energy Expenditures in Liver
Functions[a,b]

Process	kcal	Liver energy expenditure (%)
Plasma protein synthesis (18 g/day)	15.7	4.5
Liver protein synthesis (10 g/day)	8.6	2.5
Ribosomal RNA synthesis	0.15	0.045
Na^+/K^+ transport (active)	120.0	35.0
Cori cycle and triglyceride turnover	17.0	5.0
Amino acid degradation[c]	48.0	14.0
Fatty acid conversion to ketone bodies	14.4	4.2
Unaccounted	116.2	34.0

[a] From Baldwin and Smith (1974).
[b] Calculated for liver of a 70-kg man expending 340 kcal/day in liver function.
[c] Includes heat loss in glucose, ketone body, and urea synthesis.

muscle to gluconeogenesis is small but can be critically important in maintaining adequate blood glucose during starvation.

Alanine is the predominant substrate for gluconeogenesis in liver (Felig *et al.*, 1969), accounting for 41–48% of total splanchnic amino acid extraction during starvation (Owen *et al.*, 1976). However, alanine accounts for only 10% of total glucose production (Garber *et al.*, 1974) at a cost of 9 ATP/glucose. After depletion of liver glycogen, amino acids become the only significant source of new glucose synthesis.

Metabolism of 1 mol of amino acids from muscle protein to glucose, including the cost of ketone body and urea synthesis, releases 98.6 kcal (415 kJ) as heat (Baldwin, unpublished data). Assuming 110 g/mol of average amino acids from muscle protein, expenditure of 48 kcal (200 kJ) would result from utilization of 54 g of protein. This is considered reasonable for the average person living in the United States.

Ketone bodies synthesized in liver during fatty acid degredation reduce the requirement of other tissues for glucose. Estimates from humans at basal are quite variable but suggest only a minimal expenditure of liver energy (McPherson *et al.*, 1958).

Liver energy expenditures which are not associated with the supply of tissue substrates can also be identified. Structural proteins and more labile cytoplasmic proteins are constantly being degraded and resynthesized. Estimates of protein turnover (Neuberger and Richards, 1964) indicate that 8–12 g of liver protein is turned over per day. Assuming the average amino acid in protein to have a molecular weight of 110 g/mol, and that 5 ATP are hydrolyzed for each peptide bond synthesized, approximately 96 kcal (402 kJ) [5 × 19.3 kcal (81 kJ)/mol of peptide bond formed] of heat would be released/mol of peptide bond synthesized (Baldwin and Smith, 1974). Thus, 10 g of protein synthesis/day would require 8.7 kcal (36 kJ), or 2.5% of liver energy expenditure. A similar calculation for plasma proteins synthesized in liver indicates an energy expenditure of 15.7 kcal (66 kJ)/day. Energy expended in ribosomal RNA synthesis is small and based on an estimated $t_{1/2}$ for rat liver ribosomal RNA of five days (Hirsch and Hiatt, 1966), a concentration of 6 mg RNA/g wet weight of tissue, and a requirement of 4 ATP for each base in the functional RNA product (Baldwin and Smith, 1974). Transport of Na^+ and K^+ against their concentration gradients to maintain normal ion distributions and membrane potentials requires energy and represents the single largest component of liver energy expenditure (Table IV).

Ion transport contributes significantly to energy expenditures of brain (50%) and muscle (17%). These estimates are based on *in vitro* measurements but appear to be well justified (Baldwin and Smith, 1974; Milligan, 1971). Active transport of 65% of sodium reabsorbed each day by the kidney with an efficiency of approximately 2.5 sodium/pyrophosphate bond hydrolyzed requires 6.5 mol ATP/day. From the average 19 kcal (80 kJ)/mol ATP used, an expenditure of 123.5 kcal (517 kJ)/day can be calculated (Baldwin and Smith, 1974). This is equivalent to 7% of the basal energy expenditure by a 70-kg man, or on the order of 90% of energy expended by the kidney.

Consideration of additional expenditures for triglyceride resynthesis in liver and adipose tissue [21 kcal (88 kJ)/day]; protein resynthesis in muscle, liver, and intestine [167 kcal (700 kJ)/day]; and the muscular work of respiration [110 kcal (460 kJ)/day] and cardiac output (170 kcal (710 kJ)/day) leaves 30–40% of basal heat production unaccounted for. Whether this indicates major tissue expenditures not yet identified with basal requirements or the sum of a large number of minor processes cannot be determined.

3.4. Heat Increment

A portion of the ME available for maintenance is lost due to costs of digestion and assimilation of a meal. Energy lost corresponds to the HI_m and represents the difference between heat release during fasting and the ME required as food to prevent weight (body energy) loss. In Table V, the percentages of ME attributed to specific functions associated with the HI_m of a carbohydrate meal and a mixed meal are summarized. Major components of the HI_m of a carbohydrate meal are costs of absorption and assimilation or storage of glucose as glycogen and fat. These estimates are based on 1 mol ATP required/mol of glucose absorbed [17.7 kcal (74kJ) per mol ATP/673 kcal (2.8 MJ) per mol glucose], 2 mol ATP required for glycogen formation, and 15% loss of glucose energy during conversion to storage triglyceride. The HI_m for a mixed meal is lower, reflecting primarily the more efficient storage of dietary fat as storage triglyceride, and it is relatively constant over a wide range of protein intakes. These estimates agree well with observed values. The HI_m of protein used as a source of maintenance energy is much larger due to

Table V. Estimated Heat Increments of Two Meals Fed at Maintenance[a,b]

Process	Carbohydrate meal (%)	Mixed meal (%)
Bond breakage	0.5	0.3
Absorption (active transport)	2.6	1.3
Digestive secretions	2.0	2.0
Assimilation and/or storage		
Glucose as glycogen ($0.3 \times 5\%$)	1.5	1.5
Glucose as fat ($0.5 \times 15\%$)	7.5	—
Fat as fat ($0.5 \times 3\%$)	—	1.5
Theoretical HI_m	14	6.6
Observed HI_m	10–15	6–8

[a] Adapted from Baldwin and Smith (1974).
[b] Values are expressed as percentages of ME fed as carbohydrate and carbohydrate plus fat (50:50). Includes cost of synthesis of the portion of digestive proteins replenished in the period after feeding, during which heat increment is evident. Assumes, in the case of diets fed at maintenance levels, that 80% of calories provided in the diet are stored as glycogen (30%) or triglyceride (50%). Estimated energy expenditures for conversions of glucose to glycogen, glucose to storage triglyceride, and dietary triglyceride to storage triglyceride are 5%, 15%, and 3%, respectively.

extra costs of gluconeogenesis, urea synthesis (2 ATP/atom protein nitrogen), and ketone body synthesis. When these are accounted for, theoretical estimates of 15–20% are obtained (Baldwin and Smith, 1971). Excretion of unoxidized carbon skeletons, accounted for in determination of the ME, is not a component of the HI_m.

3.5. Efficiency of Energy Utilization at Maintenance

Energy for the tissue basal expenditures described above is supplied by hydrolysis of the pyrophosphate bond of ATP. ATP synthesis at basal (postabsorptive) is driven primarily by oxidation of fatty acids released from adipose tissue. Heat release during synthesis and subsequent use of this ATP provides a basis for estimating the relative efficiencies with which nutrients are used to supply energy for maintenance. From theoretical yields of ATP and heats of combustion of carbohydrate, protein, and fat listed in Table VI, energy expenditures associated with the synthesis and use of 1 mol ATP can be calculated. More ME would be required as amino acids than as glucose or long-chain fatty acids. The higher energy loss from amino acids reflects the energy lost to synthesis and excretion of urea. The average cost of synthesis in humans and other nonruminant species is 19 kcal/mol ATP (Milligan, 1971). Biochemical efficiency can be calculated using Blaxter's (1962) formula:

Biochemical efficiency

$$= 100 \times \frac{\text{Heat of combustion of fat/mol ATP formed}}{\text{Heat of combustion of nutrient/mol ATP formed}}$$

Results from this calculation represent maximal efficiencies and imply that nutrients are directly oxidized after absorption and removal from blood. Intermediate storage reduces energy available for maintenance. Baldwin (1968) estimated that the storage of glucose as glycogen reduces the available pyrophosphate bond energy by 5% and the loss from glucose storage as tripalmitin by about 15%. Storage of dietary fat as triglyceride results in a loss of only 3%.

Table VI. Estimation of Heat Energy Equivalent of High-Energy Phosphate Bonds Formed from Common Energy Sources and Utilized for Maintenance Functions[a]

Energy source	~P bonds formed (~P/mol)	ΔH_c (kcal/mol)	$\Delta H_c/$~P (kcal/bond)
Glucose	38	673	17.7
Palmitate	129	2398	18.6
Protein, 100g[b]	22.6	512	22.7

[a] Adapted from Baldwin and Smith (1974).
[b] Heat of combustion of protein minus that of urea having an equal amount of nitrogen (Milligan, 1971).

It is obvious to any nutritionist that energy released by combustion of a food in a bomb calorimeter is greater than the energy available when the same amount of food is eaten. Predicting just how much energy will be available to an individual is much less obvious. The answer depends, for example, on whether the person is growing or trying to maintain body weight, the quality of the protein, the frequency of eating, and, as discussed above, the relative efficiencies of ATP production. Ideally, a system for evaluating foods should predict with accuracy the performance of an individual based on knowledge of (1) body weight, environment, activity, and, where present, growth, pregnancy, and lactation; and (2) amount of each food in the diet and its nutritive value. No system currently in use completely satisfies this need.

The system currently in use for evaluating nutritive values of foods for human consumption was developed by W. O. Atwater and his associates at the Connecticut (Storrs) Agricultural Experiment Station (Atwater and Bryant, 1899). The nutritive values employed are called physiological fuel values (PFVs) and are essentially ME values determined at maintenance. A correction for energy losses due to gaseous products of digestion was not included but does not produce a significant error in nonruminant species. Data necessary to derive the PFVs of protein, fat, and carbohydrate are given in Table VII. Digestibility coefficients are average values compared over a wide range of standard human foods at maintenance feeding. The additional corrections required for protein reflect the cost of urea synthesis and the energy content of urea excreted in urine. Formation of 1 mol urea by the urea cycle requires 4 mol ATP or 2 mol ATP/mol amino nitrogen. Assuming protein to contain 16% nitrogen and using 19 kcal (80 kJ)/mol ATP as the average heat loss/mol ATP used, the cost of formation of urea can be calculated as 0.43 kcal (1.8 kJ)/g protein $[(0.16/14) \times 2 \times 19]$. Excretion of urea results in a loss of 5.4 kcal (23 kJ)/g nitrogen or about 0.9 kcal (3.8 kJ)/g protein. Modifications of these PFVs, which take into account digestibility results from more recent experiments with humans, were used to develop the standard tables of energy values of foods which are in use today.

A common observation among students of animal husbandry is that different animals fed precisely the same food utilize the energy available for gain with different efficiencies. When fed precisely the same amount of food as consumed by lean controls, strains of genetically obese laboratory animals

Table VII. Data for Calculation of Physiological
Fuel Value per Gram of Food (kcal/g)

	$\Delta H_c/g$	% Digestibility	Digestible energy	PFV
Protein	5.7	0.91	5.3	4.0[a]
Fat	9.4	0.95	8.9	9.0
Carbohydrate	4.15	0.97	3.98	4.0

[a] DE − [urea synthesis (0.43)] − [energy in urea (0.9)] ≈ 4.0.

gain more weight and a higher proportion of fat (Bray and York, 1971). Similarly, for humans, it seems reasonable to suggest that different individuals or the same individual at different times may not utilize dietary energy with the same metabolic efficiency (Hegsted, 1974). A change in diet composition can affect utilization as well. Samonds and Fleagle (1973) found that monkeys fed a protein-restricted diet consumed 40 kcal (170 kJ)/day more than monkeys fed a calorie-restricted diet. Both diets contained energy sufficient to maintain body weight but not to support growth. Differences in activity could account for only a small fraction of the increased consumption, suggesting that the protein-restricted group converted the extra energy to heat.

Differences in energy retention do not have to be large to produce significant long-term effects. An excess of 25 kcal (105 kJ) retained/day potentially adds up to 3–3.5 g of adipose tissue/day, or 1 kg/yr. In terms of daily food eaten, this represents one-half of an Oreo cookie or two-and-one-half french fries. If daily intake is 2500 kcal (10.5 MJ), this is an error in balance of only 1%. Similarly, small differences in heat production caused by variable efficiencies of dietary energy utilization are difficult to determine relative to daily heat production. The maximum energy converted to ATP from fat or carbohydrate is approximately 40% (assuming 8 kcal retained/pyrophosphate bond). Krebs (1964) calculated the yield from protein to be slightly lower, at 32–34%. When nutrients are recycled or stored prior to utilization, efficiency must decrease. For example, if glucose is resynthesized from pyruvate in the liver, there is a net loss of 4 ATP, 10% of the theoretical yield from glucose. Ball (1965) calculated a net cost of 0.277 kcal (1.16 kJ)/g tripalmitin formed in the rat, leading to an estimate of 15% of BMR required for triglyceride recycling; however, more recent calculations of Baldwin and Smith (1974) suggest a much lower value of 1.2%. Additional losses may occur at metabolic control points, where opposing pathways use separate enzymes to catalyze forward and reverse reactions (Katz and Rognstad, 1976). For example, phosphorylation of glucose requires ATP, while dephosphorylation is spontaneous. Cycling can occur using ATP without any change in reactants. Such is the case for the phosphorylation and dephosphorylation of glucose. Three of these "futile" cycles in liver are described in a detailed review by Katz and Rognstad (1976). They estimated that in isolated liver cells, 10% of ATP breakdown could be attributed to futile cycles. It is interesting to speculate that a low level of activity in cycles such as these could, in part, explain the eventual development of obesity in some individuals. Consistent with this speculation are observations that O_2 consumption and colonic temperature are lower in genetically obese mice over a range of ambient temperatures from 10 to 30°C (Kaplan and Leveille, 1974). This decreased O_2 consumption is evident early in development and was suggested as a useful criterion to identify the *ob/ob* genotype prior to development of obesity (Fried, 1973; Kaplan and Leveille, 1974). Heat production in response to thyroxine is reportedly less in genetically obese mice when expressed per kilogram to the 0.75 power (VanderTuig *et al.,* 1978). However, lipogenesis and, presumably, turnover are more rapid in obese than in normal rats (Sullivan *et al.,* 1977). If futile cycles can account for a significant fraction of total energy expenditure, they may help to explain why, in humans, the energy required for maintenance of some obese individuals appears no greater than that required to maintain others at a lean body weight.

4. Conclusion

Current knowledge of intermediary metabolism associated with the several tissue processes required to maintain body weight makes prediction of specific maintenance costs possible. Combination of metabolic pathways involved yields theoretical models which have been employed throughout this chapter to evaluate energy expenditure and relative efficiencies of nutrients used, for example, to supply blood glucose by gluconeogenesis in liver. Individuals differ in the efficiencies with which they utilize dietary energy. Identification of specific differences may help explain the range of body weights observed among humans and animals fed similar diets at similar intakes.

ACKNOWLEDGMENT. This work was supported, in part, by Grant AM-18899 from the National Institutes of Health.

5. References

Ames, S. R., 1970, The Joule-unit of energy, *J. Am. Diet. Assoc.* **57**:415.

Atwater, W. O., and Benedict, F. G., 1903, *Experiments on the Metabolism of Matter and Energy in the Human Body*, U.S. Dep. Agric. Off. Exp. Stn. Bull. No. 136.

Atwater, W. O., and Bryant, A. P., 1899, *The Availability and Fuel Value of Food Materials*, Conn. Storrs Agric. Exp. Stn. 12th Annu. Rep., p. 73.

Baldwin, R. L., 1968, Estimation of theoretical calorific relationships as a teaching technique. A review. *J. Dairy Sci.* **51**:104.

Baldwin, R. L., and Smith, N. E., 1971, Application of a simulation modeling technique in analysis of dynamic aspects of animal energetics, *Fed. Proc. Fed. Am. Soc. Exp. Biol.* **30**:1459.

Baldwin, R. L., and Smith, N. E., 1974, Molecular control of energy metabolism, in: *The Control of Metabolism* (J. D. Sink, ed.), pp. 17–34, The Pennsylvania State University Press, University Park and London.

Ball, E. G., 1965, Some energy relationships in adipose tissues, *Ann. N.Y. Acad. Sci.* **131**:225.

Banerjee, S., and Bhattacharjee, R. C., 1967, Interrelations of the basal metabolic rate and body composition in adult male and female medical students, *Indian J. Med. Res.* **55**:451.

Beattie, J., and Herbert, P. H., 1947, Estimation of the metabolic rate in the starvation state. *Br. J. Nutr.* **1**:185.

Benzinger, T. H., 1969, Heat regulation: Homeostasis of central temperature in man, *Physiol. Rev.* **49**:671.

Blaxter, K. L., 1962, *The Energy Metabolism of Ruminants,* Charles C. Thomas, Springfield, Ill.

Boothby, W. M., Berkson, J., and Dunn, H. L., 1936, Studies of the energy of metabolism of normal individuals: A standard for basal metabolism, with a nomogram for clinical application. *Am. J. Physiol.* **116**:468.

Bray, G. A., 1969, Effect of caloric restriction on energy expenditure in obese patients, *Lancet* **2**:397.

Bray, G. A., and Atkinson, R. L., 1977, Factors affecting basal metabolic rate, *Prog. Food Nutr. Sci.* **2**:395.

Bray, G. A., and Hildreth, S., 1967, Effect of propylthiouracil and methimazole on oxygen consumption of hypothyroid rats receiving thyroxine or triiodothyronine, *Endocrinology* **81**:1018.

Bray, G. A., and York, D. A., 1971, Genetically transmitted obesity in rodents, *Physiol. Rev.* **51**:598.

Brody, S., 1945, *Bioenergetics and Growth,* Hafner Press, New York.

Buskirk, E. R., Iampietro, P. F., and Welch, B. E., 1957, Variations in resting metabolism with changes in food, exercise and climate. *Metabolism* **6**:144.

Cahill, G. F., Jr., Herrera, M. G., Morgan, A. P., Soeldner, J. S., Steinke, J., Levy, P. L., Reichard, G. A., Jr., and Kipnis, D. M., 1966, Hormone–fuel interrelationships during fasting, *J. Clin. Invest.* **45**:1751.

Crawford, A., 1788, *Experiments and Observations on Animal Heat and the Inflammation of Combustible Bodies,* 2nd ed., J. Murray and J. Sewell, London.

DuBois, D., and DuBois, E. F., 1916, Clinical calorimetry. A formula to estimate the approximate surface area if height and weight be known, *Arch. Intern. Med.* **17**:863.

DuBois, D., 1936, *Basal Metabolism in Health and Disease,* Lea and Feibiger, Philadelphia.

Edelman, I. S., and Ismail-Beigi, F., 1974, Thyroid thermogenesis and active sodium transport, *Recent Prog. Horm. Res.* **30**:235.

Edholm, O. G., Fletcher, J. G., Widdowson, E. M., and McCance, R. A., 1955, The energy expenditure and food intake of individual men, *Br. J. Nutr.* **9**:286.

Felig, P., Owen, O. E., Wahren, J., and Cahill, G. F., Jr., 1969, Amino acid metabolism during prolonged starvation, *J. Clin. Invest.* **48**:584.

Fried, G. H., 1973, Oxygen consumption rates in litters of thin and obese hyperglycemic mice, *Am. J. Physiol.* **225**:209.

Garber, A. J., Menzel, P. H., Boden, G., and Owen, O. E., 1974, Hepatic ketogenesis and gluconeogenesis in humans, *J. Clin. Invest.* **54**:981.

Garrow, J., 1974, *Energy Balance and Obesity in Man,* American Elsevier, New York.

Gold, A. J., Zornitzer, A., and Samueloff, S., 1969, Influence of season and heat on energy expenditure during rest and exercise, *J. Appl. Physiol.* **27**:9.

Goodman, H. M., and Bray, G. A., 1966, Role of thyroid hormones in lipolysis, *Am. J. Physiol.* **210**:1053.

Grande, F., Anderson, J. T., and Keys, A., 1958, Changes in basal metabolic rate in man in semistarvation and refeeding, *J. Appl. Physiol.* **12**:230.

Gulick, A., 1922, A study of weight regulation in the adult human body during overnutrition, *Am. J. Physiol.* **60**:371.

Harper, A. E., 1970, Remarks on the Joule, *J. Am. Diet. Assoc.* **57**:416.

Harris, J. A., and Benedict, F. G., 1919, A biometric study of basal metabolism in man, *Carnegie Inst. Washginton Publ.* **279**:1.

Hegsted, D. M., 1974, Energy needs and energy utilization, *Nutr. Rev.* **32**:33.

Hirsch, C. A., and Hiatt, H. H., 1966, Turnover of liver ribosomes in fed and in fasted rats, *J. Biol. Chem.* **241**:5936.

Janowitz, H. D., and Hollander, F., 1955, The time factor in the adjustment of food intake to varied caloric requirements in the dog: A study of the precision of appetite regulation, *Ann. N.Y. Acad. Sci.* **63**:56.

Kaplan, M. L., and Leveille, G. A., 1974, Core temperature, O_2 consumption, and early detection of *ob/ob* genotype in mice, *Am. J. Physiol.* **227**:912.

Kasper, H., 1970, Faecal fat excretion, diarrhea, and subjective complaints with highly dosed oral fat intake, *Digestion* **3**:321.

Katz, J., and Rognstad, R., 1976, Futile cycles in the metabolism of glucose, in: *Current Topics in Cellular Regulation,* Vol. 10 (B. L. Horecker, and E. R. Stadtman, eds.), pp. 237–289, Academic Press, San Francisco.

Keys, A., Taylor, H. L., and Grande, F., 1973, Basal metabolism and age of adult man, *Metabolism* **22**:579.

Kleiber, M., 1932, Body size and metabolism, *Hilgardia* **6**:315.

Kleiber, M., 1972, Joules vs. calories in nutrition, *J. Nutr.* **102**:309.

Kleiber, M., 1975, *The Fire of Life,* John Wiley and Sons, New York and London.

Krebs, H. A, 1964, The metabolic fate of amino acids, in: *Mammalian Protein Metabolism,* Vol. 1 (H. N. Munro and J. B. Allison, eds.), pp. 125–176, Academic Press, New York.

Larsen, P. R., 1972, Triiodothyronine: Review of recent studies of its physiology and pathophysiology in man, *Metabolism* **21**:1073.

McPherson, H. T., Werk, E. E., Jr., Myers, J. D., and Engle, F., 1958, Studies on ketone metabolism in man. II. The effect of glucose, insulin, cortisone and hypoglycemia on splanchnic ketone production, *J. Clin. Invest.* **37**:1379.

Milligan, L. P., 1971, Energetic efficiency and metabolic transformations, *Fed. Proc. Fed. Am. Soc. Exp. Biol.* **30**:1454.

Mitchell, H. H., 1962, *Comparative Nutrition of Men and Domestic Animals,* Vol. 1, Academic Press, New York.

Neuberger, A., and Richards, F. F., 1964, Studies on turnover in the whole animal, in: *Mammalian Protein Metabolism,* Vol. 2 (H. N. Munro and J. B. Allison, eds.), pp. 243–296, Academic Press, New York.

Owen, O. E., Reichard, G. A., Jr., Boden, G., Patel, M. S., and Trapp, V. E., 1978, Interrelationships among key tissues in the utilization of metabolic substrates, in: *Advances in Modern Nutrition,* Vol. 2 (H. M. Katzen and R. J. Mahler, eds.), pp. 517–550, Halsted Press, New York.

Owen, O. R., Patel, M. S., Block, B. S. B., Krevlen, T. H., Reichle, F. A., and Mozzoli, M. A., 1976, Gluconeogenesis in normal, cirrhotic, and diabetic humans, in: *Gluconeogenesis. Its Regulation in Mammalian Species* (R. W. Hanson and A. M. Mehlaman, eds.), pp. 533–558, John Wiley and Sons, New York.

Passmore, R., Strong, J. A,, Swindells, V. E., and El Din, N., 1963, The effect of overfeeding on two fat young women, *Br. J. Nutr.* **17**:373.

Reichard, G. A., Moury, N. F., Hochella, N. J., Patterson, A. L., and Weinhouse, S., 1963, Quantitative estimation of the Cori cycle in the human, *J. Biol. Chem.* **238**:495.

Samonds, K. W., and Fleagle, J., 1973, The onset of protein as calorie deficiency in the young cebus monkey, *Fed. Proc. Fed. Am. Soc. Exp. Biol.* **32**:901 (Abstract).

Share, I., Martiniak, E., and Grossman, M. I., 1952, Effect of prolonged intragastric feeding on oral food intake in dogs, *Am. J. Physiol.* **169**:229.

Strong, J. A., Shirling, D., and Passmore, R., 1967, Some effects of overfeeding for four days in man, *Br. J. Nutr.* **21**:909.

Sullivan, A. C., Triscari, J., Stern, J. S., Hamilton, J. G., and Greenwood, M. R. C., 1977, Lipid metabolism in obese and lean Zucker rats, *Fed. Proc. Fed. Am. Soc. Exp. Biol.* **36**:1149 (Abstract).

Tepperman, J., and Tepperman, H. M., 1958, Effects of antecedent food intake pattern on hepatic lipogenesis, *Am. J. Physiol.* **193**:55.

Thonney, M. L., Touchberry, R. W., Goodrich, R. D., and Meislce, J. C., 1976, Intraspecies relationship between fasting heat production and body weight: A reevaluation of $W^{.75}$, *J. Anim. Sci.* **43**:692.

VanderTuig, J. G., Romsos, D. R., Leveille, G. A., and Pearson, A. M., 1978, Heat production, energy intake and body weight changes in response to fasting–refeeding, thyroxine administration or environmental temperature in lean and obese (*ob/ob*) mice, *Fed. Proc. Fed. Am. Soc. Exp. Biol.* **57**:676 (Abstract).

von Bertalanffy, L. 1957, Quantitative laws in metabolism and growth, *Q. Rev. Biol.* **32**:217.

Wahren, J., Felig, P., Cerasi, E., and Luft, R., 1972, Splanchnic and peripheral glucose and amino acid metabolism in diabetes mellitus, *J. Clin. Invest.* **51**:1870.

Zeuthen, E., 1953, Oxygen uptake as related to body size in organisms, *Q. Rev. Biol.* **28**:1.

7

Nutrients with Special Functions: Proteins and Amino Acids in Tissue Maintenance

Andrew J. Clifford

1. Introduction

The total protein concentration, as a percentage of total body weight, of the prenatal infant is about 8.5; it increases to about 11 at birth, further increases to a maximum of about 17.5 in the adult, and then declines with advancing age. The total amount of body protein follows a similar pattern. The amount of total-body protein present at any time represents a dynamic equilibrium between the rates of protein synthesis and protein breakdown. In order that there be a net accumulation of body protein, the set point at which equilibrium occurs between synthesis and breakdown favors synthesis. During the period of net loss of total body protein, the set point at which equilibrium occurs between synthesis and breakdown favors breakdown. Thus, there occurs over the life cycle an equilibrium set point which changes during growth, development, and aging.

There are also diurnal meal-related rhythms in body tissue protein stores. Each intake of a protein-containing meal leads to a net accumulation of liver (as well as other organs) protein. In the period between meals, there is a net depletion of liver protein. These changes in liver protein pools are caused by alterations in the balance of protein synthesis and protein breakdown. These diurnal meal-related changes in the equilibrium set points between synthesis and breakdown represent regulation of body protein pools over the short term.

Dietary protein supplies amino acids, which are essential ingredients for protein synthesis, and influences both protein synthesis and breakdown rates. Dietary proteins are not absorbed per se, but small peptides and free amino

Andrew J. Clifford • Department of Nutrition, University of California, Davis, California 95616.

acids are absorbed with great efficiency by the small intestine. The process by which dietary proteins are hydrolyzed (digestion) to small peptides and free amino acids and absorbed (absorption) into body pools of free amino acids plays a vital role in supplying these ingredients which are essential for protein synthesis and important in regulating the rate of protein synthesis and breakdown.

The intensity of body protein synthesis is also influenced by factors other than the supply of free amino acids. These other factors include the supply of nucleic acids (tRNA, mRNA) as well as initiating factors and feedback regulators. The synthesis and breakdown of nucleic acids are intimately associated with the synthesis and breakdown of tissue proteins, and, like tissue proteins, the nucleic acids show meal-related diurnal rhythms in *de novo* synthesis and reutilization. The importance of a properly functioning purine metabolism in protein synthesis is emphasized by the recent discovery that a dysfunction in purine metabolism is associated with severe deficiencies in the development of the proteins of the immune system. Since malnutrition is well known to be associated with immune dysfunction, it becomes of interest to speculate that mechanisms by which malnutrition produces impaired immunity may involve dysfunction in purine metabolism.

This chapter discusses the processes by which dietary proteins are transformed into tissue pools of free amino acids and how these amino acids, in a qualitative way, are metabolized by various organs and are involved in the synthesis and breakdown of protein. Finally, quantitative features of whole-body protein synthesis and breakdown in humans, as they are presently known, are presented.

2. Hydrolysis and Absorption of Dietary Protein

2.1. Hydrolysis

Amino acids, the building blocks of proteins, enter the free amino acid pools of intermediary metabolism from endogenous sources, *de novo* synthesis, and tissue breakdown, as well as from exogenous dietary sources. Dietary proteins per se are not absorbed from the intestine in significant amounts. Nevertheless, dietary proteins will meet nutritional needs if their amino acid quantity and quality are sufficient and if the protein is adequately hydrolyzed in the intestine and the products of hydrolysis absorbed into the portal blood. The importance of protein digestion is strikingly demonstrable in infants genetically deficient in trypsinogen who fail to grow when fed a normal diet in which milk supplies the protein (Townes, 1965). In these infants, normal growth is restored when milk protein hydrolysate supplies the protein needs. The hydrolysis of dietary protein is briefly summarized in Fig. 1.

Both brush-border and intestinal mucosal cell cytoplasm contain several different peptide hydrolases (Gray and Santiago, 1974; Heizer and Laster, 1969; Kim *et al.,* 1972, 1974; Lindberg, 1966; Lindberg *et al.,* 1968), all of

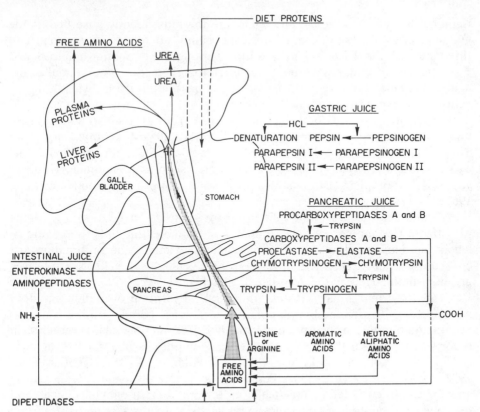

Fig. 1. Schematic outline of protein digestion and absorption.

which are aminooligopeptidases and have pH optima in the alkaline range (Heizer *et al.*, 1972; Kim *et al.*, 1974; Peters, 1973). Dipeptidases and some tripeptidases reside in the cytoplasmic compartment (Kim *et al.*, 1972, 1974; Peters, 1970), while polypeptidases reside in the brush-border compartment (Kim *et al.*, 1974). Substantial intestinal mucosal proteolysis of polypeptides of dietary origin occurs, since oligopeptides introduced into the intestinal lumen rapidly appear as the constituent free amino acids and dipeptides in portal and peripheral plasma (Adibi, 1971; Adibi *et al.*, 1975; Peters and MacMahon, 1970).

2.2. Absorption of Free Amino Acids

Although dietary proteins are not absorbed, small peptides and free amino acids are absorbed with great efficiency by the small intestine. Enzymatic hydrolysis of dietary proteins releases the free amino acids, which are absorbed from the small-intestinal lumen into the intestinal absorptive cells and then via transcellular transport into the portal circulation (Christensen, 1964; Wilson, 1962). Amino acid absorption is mediated by highly selective processes (Adibi and Gray, 1967; Adibi *et al.*, 1967). Among the 18 dietary amino acids, methionine, leucine, isoleucine, and valine are the most readily absorbed, while glutamic acid and aspartic acid are least readily absorbed. Tryptophan and threonine are the least readily absorbed among the essential amino acids. Essential amino acids, as a group, are better absorbed than nonessential amino acids. D-Amino acids are absorbed solely by passive diffusion, while L-amino acids are actively transported.

There are at least two distinct membrane carrier systems in human intestine, one for basic and one for neutral free amino acids (McCarthy *et al.*, 1964; Shih *et al.*, 1971; Thier *et al.*, 1965). There may also be additional membrane carrier systems for individual free amino acids such as methionine (Hooft *et al.*, 1964), tryptophan (Drummond *et al.*, 1964), and proline (Goodman *et al.*, 1967).

Amino acid absorption is influenced by a variety of factors. The lumen pH affects free amino acid absorption. For example, the absorption rate of leucine is diminished as the pH drops below 2.5 due to its protonation and consequent reduced interaction with its carrier system. The absorption of free glycine is influenced by the presence of galactose (Cook, 1971; Hardcastle and Daniels, 1973) and glucose (Cook, 1971), but absorption of methionine does not appear to be affected by the presence of glucose (Cook, 1972a).

In addition to the absorption of protein as free amino acids, several recent reports have shown that a wide range of dipeptides is absorbed in humans (Adibi, 1971; Adibi and Soleimanpour, 1974; Adibi *et al.*, 1975; Hellier *et al.*, 1972a,b; Meilman *et al.*, 1963; Perry *et al.*, 1967; Silk *et al.*, 1975) and in animals (Matthews, 1975). Practically all tissues contain enzymes capable of hydrolyzing dipeptides (Adibi and Krzysik, 1975; Krzysik *et al.*, 1975), and, consequently, the peripheral venous blood is devoid of dipeptides after intestinal perfusions with large loads of glycylleucine (Adibi, 1971), even though superior mesenteric venous blood has been shown to contain traces of infused dipeptides (Boullin *et al.*, 1973).

Absorption of dipeptides appears to occur in the ileum, duodenum (Adibi, 1971; Silk *et al.*, 1974a), and jejunum (Agar *et al.*, 1953; Newey and Smyth, 1959; Wiggans and Johnston, 1959). Although there are considerable differences between jejunal and ileal absorption rates of free amino acids (Adibi, 1969, 1971; Schedl *et al.*, 1968; Silk *et al.*, 1974b), the absorption rates of dipeptides are very similar in these two segments (Adibi, 1971; Silk *et al.*, 1974b). Additional studies (Asatoor *et al.*, 1970a; Matthews *et al.*, 1968) have established that peptide transport by the small-intestinal epithelium proceeds by mechanisms different from those responsible for free amino acid transport. Carrier-mediated transport systems have been demonstrated for the absorption of dipeptides with fast rates of hydrolysis, glycylglycine and glycyl-L-leucine (Adibi and Soleimanpour, 1974), as well as those with slow rates of hydrolysis, glycylsarcosine and carnosine (Addison *et al.*, 1972; Matthews *et al.*, 1974). The existence of several dipeptide transport systems with rather narrow specificities is proposed by some workers (Addison *et al.*, 1974), while others (Das and Radhakrishnan, 1975) claim that only one dipeptide uptake system with an extremely broad specificity exists.

Although the existence of a tripeptide transport system has yet to be demonstrated, glycylsarcosylsarcosine has been shown to accumulate against an electrochemical gradient (Addison *et al.*, 1974), suggesting the existence of a tripeptide carrier system. It also appears that some tripeptides, such as triglycine (Adibi *et al.*, 1975) and glycylprolylhydroxyproline (Meilman *et al.*, 1963), disappear intact from the intestinal lumen into the mucosal cell without prior hydrolysis. Trileucine (Adibi *et al.*, 1975) and alanylglycylglycine (Silk *et al.*, 1974a) appear to be absorbed either intact or as their dipeptide constituents.

The proportion of dietary protein absorbed as free amino acids, dipeptides, and tripeptides is largely unknown. The importance of dietary protein absorption in peptide form is illustrated by individuals who have hereditary deficiencies of the free, neutral, or basic amino acid carrier systems and who do not appear to suffer from protein malnutrition and absorb peptides normally (Asatoor *et al.*, 1970b, 1972; Hellier *et al.*, 1972b; Navab and Asatoor, 1970; Silk *et al.*, 1975; Tarlow *et al.*, 1972). The absorption of nitrogen in the form of di- and tripeptides from dipeptides (Adibi, 1971; Cook, 1972b; Hellier *et al.*, 1972a; Silk *et al.*, 1973a), tripeptides (Adibi *et al.*, 1975; Silk *et al.*, 1975), and casein hydrolysates (Silk *et al.*, 1973b) is greater than from a corresponding mixture of free amino acids. Because they are less hypertonic than free amino acids and are more efficiently absorbed than free amino acids, small oligopeptides may be very useful in the protein nutrition of patients with limited absorptive capacity.

3. Meal-Related Interorgan Movements of Amino Acids

The concentrations of free amino acids in plasma are well defined and are relatively constant in normal individuals (Scriver and Rosenberg, 1973; Scriver *et al.*, 1971). The steady-state level represents the net balance between release from endogenous protein stores and utilization by various tissues. Muscle contains greater than 50% of the total body pool of free amino acids but lacks the enzymes to dispose of ammonia produced by deamination of amino acids. The liver contains the urea cycle which is essential for nitrogen disposal. Thus, these two organs, along with others, that is, gut and kidney, working in concert play a major role in the homeostasis and turnover of free amino acids.

Muscle is the primary site for the degradation of some amino acids, that is, leucine, isoleucine, valine, alanine, glutamate, and aspartate (Goldberg and Odessey, 1972; Manchester, 1965; Miller, 1962; Odessey and Goldberg, 1972). Other amino acids, such as lysine, serine, proline, threonine, methionine, cysteine, phenylalanine, histidine, tyrosine, and tryptophan, are not degraded to any extent in muscle (Goldberg and Odessey, 1972; Manchester, 1965; Odessey and Goldberg, 1972). The capacity of muscle to degrade leucine, isoleucine, and valine is increased severalfold on fasting (Goldberg and Odessey, 1972). Leucine degradation occurs by transamination followed by decarboxylation to yield isovaleryl coenzyme A (CoA) and finally acetyl-CoA.

Decarboxylation, the rate-limiting step in muscle, is the step which is accelerated during food deprivation (Goldberg and Chang, 1978). Thus, in muscle in the postabsorptive state, there is a net release of glutamine and, to a lesser extent, glycine, lysine, proline, threonine, histidine, leucine, and isoleucine which complements the uptake of alanine, glutamine, and, to a lesser degree, glycine, lysine, threonine, histidine, arginine, phenylalanine, tryosine, and methionine. Alanine and glutamine account for more than 50% of the total α-amino nitrogen released by muscle. These two amino acids are synthesized in muscle *de novo* (Felig, 1975; Goldberg and Odessey, 1974) by a process which utilizes the carbon fragments of pyruvate and lactate, along with the amino groups generated by the degradation of the other muscle amino acids, that is, leucine, isoleucine, valine, and aspartate, which are released from muscle in amounts lower than expected based on their concentrations in muscle tissue. Thus, the branched-chain amino acids in muscle now appear to provide much of the amino groups for the production of alanine (Goldberg and Chang, 1978). The relative amounts of alanine and glutamine produced and released by muscle depend on the ammonia concentration within the tissue. Greater concentrations of ammonia favor the production of glutamine and reduce the production of alanine.

The liver is the major site of uptake of alanine, while the kidney and gut are the major sites of glutamine uptake. The alanine taken up by the liver is a major precursor for glucose and plays an important role in the maintenance of blood glucose. The glutamine taken up by the gut and kidney provides the nitrogen source for alanine and ammonia synthesis in the gut (Matsutaka *et al.*, 1973) and kidney (Cahill and Owen, 1970), respectively. The branched-chain amino acids, particularly valine, are also taken up and metabolized by the brain.

The erythrocytes also play an important role in the interorgan flux of amino acids (Aoki *et al.*, 1972; Elwyn *et al.*, 1968, 1972). It appears that for alanine, serine, threonine, methionine, leucine, isoleucine, and tyrosine, significant tissue exchange occurs via the erythrocytes. Blood cells account for about 30% of the alanine output from muscle and gut and about 20% of the uptake by the splanchnic bed.

4. Protein Synthesis

Protein synthesis requires energy. It occurs on polysomes, which are composed of ribosomes attached to strands of mRNA. Amino acids to be linked together are transported singly on tRNA to the polysomes, where the polypeptide chain is assembled. Polysomes which occur free in the cell sap assemble proteins which are retained within the cell, while polysomes which occur on the cell membranes assemble those proteins destined for export from the cell. The intensity of protein synthesis depends on the supply of amino acids, tRNA, and mRNA, as well as on initiating factors and feedback regulators determined by the amount of protein synthesized. Total-body protein synthesis in the adult male is in the range of 275–300 g/day.

DNA contains the stored information of cellular processes. The cell translates sections of DNA into RNA, from which protein is transcribed. Cells produce an exact DNA replicate for their daughter cells. Only a brief description of how this occurs is presented here. Nuclear DNA is associated with histones and acidic proteins which make up chromosomes, whereas mitochondrial DNA is free and is unassociated with proteins. Nuclear chromosomal proteins regulate, at least in part, which of the encoded information is made available (Stein *et al.*, 1974). Several forms of RNA exist in the nucleus. Ribosomal RNA is made in the nucleolus as a single strand of 45 S RNA and is then cleaved into 28 S and 18 S RNA by an enzymatic process which is regulated by the methylation of ribose residues of specific nucleotides (Greenberg and Penman, 1966). A large RNA is also transcribed from nucleoplasm DNA and gives rise to mRNAs (Perry and Kelley, 1974). The mRNA attaches to polyadenylic acid and finally becomes attached to protein to form an informosome which leaves the nucleus for the cytoplasm (Schumm and Webb, 1972).

DNA-dependent RNA polymerase is essential for the formation of RNA on a DNA template. Three separate RNA polymerases, polymerases I, II, and III, are known to exist (Jacob *et al.*, 1970; Kedinger *et al.*, 1970; Roeder and Rutter, 1969). Polymerase I occurs in the nucleolus and forms 45 S RNA. Polymerase II occurs in the nucleoplasm and forms mRNA. Polymerase III also occurs in the nucleoplasm and forms tRNA and 5 S RNA. The RNA-synthesizing system is sensitive to exogenous stimuli, such as corticosteroid hormone and amino acid supply, and, consequently, is an important mechanism for altering protein synthesis (Godlad and Munro, 1959; Jacob *et al.*, 1969; Sajdel and Jacob, 1971).

The synthesis of peptide chains on ribosomes involves the following three processes: initiation, elongation, and termination of the peptide chain. The initiation process requires ribosome subunits, mRNA, initiating factors, and initiator methionyl-tRNA. The three protein-initiating factors, M_1, M_2, and M_3, were identified and described by Prichard *et al.* (1970, 1971). Recent evidence (Reichman and Penman, 1973) indicates that a new, specific, rapidly-turning-over RNA may be associated with initiation. Elongation involves the successive addition of amino acids to form the peptide chain. The aminoacyl-tRNA species charged with the protein amino acids associate with elongation factor 1 (EF 1) and guanosine triphosphate, which then attaches to the ribosome to insert the specific amino acid dictated by the mRNA codons. The growing peptide chain is then translocated along the ribosome surface by elongation factor 2 (EF2) and guanosine triphosphate. The ribosome also undergoes specific conformational changes during the process of peptide chain elongation (Steinert *et al.*, 1974). Finally, the process of termination, in which the ribosome separates from the mRNA and dissociates into subunits, occurs by a process which requires the binding of a protein-dissociation factor to the ribosome subunits to keep them apart.

The foregoing process results in cell ribosomes which form polyribosomes with a growing peptide chain. The ribosome then detaches, leaving the mRNA, and forms the 60 S and 40 S subunit components which are used for initiation and additional rounds of the cycle. Since chain initiation, chain elongation, and chain termination are continuous and overlapping processes, the proportion of the total ribosomes which occur as polyribosomes, ribosomes, and subunits depends on the balance between initiation, chain elongation, and termination. The rate of the process by which the stored genetic information is translated into a form capable of being read by successive generations of cells (translation) is determined in large part by the rate of peptide chain initiation or peptide chain elongation. A reduced rate of chain initiation leads to the accumulation of ribosomes and subunits. Chain elongation can be reduced by limiting the supply of an essential amino acid. If the supply of an essential amino acid is reduced, chain elongation is retarded only at those points where the specific amino acid has to be inserted into the growing peptide chain, and this produces alterations in the distribution of polysomes on the messenger (Hori *et al.*, 1967).

The interrelationships among amino acid supply, distribution of polysomes on the messenger, and protein synthesis have been extensively studied and reviewed by Munro *et al.* (1975). Each influx of amino acids in a protein-containing meal is associated with a rapid aggregation of ribosomes into polysomes. The aggregation of ribosomes into polysomes is impaired if tryptophan (Fleck *et al.*, 1965; Wunner *et al.*, 1966), threonine, or isoleucine (Ip and Harper, 1973; Pronczuk *et al.*, 1970) are omitted from the dietary amino acid mixtures. Studies on mammalian cells cultured in media devoid of valine, histidine, or methionine have confirmed a rapid disaggregation of polysomes and have shown that polyribosome disaggregation was associated with a marked reduction in the rates of protein synthesis, peptide chain elongation, and even peptide chain initiation through the inhibitory action of accumulated tRNA.

Before many newly formed peptide chains become active cell components, they must undergo further changes. These changes include modification of the constituent amino acids before completion of the peptide chain (i.e., hydroxylation of proline residues of collagen) as well as after completion of the peptide chain (i.e., methylation of histidine residues of actin and myosin and addition of carbohydrates, i.e., glycoproteins). Prothrombin is another protein which undergoes similar changes (Goswami and Munro, 1962; Stenflo and Ganrot, 1972).

5. Protein Breakdown

The process of protein breakdown is not as well understood as that of protein synthesis. The initial step in protein breakdown involves an endoproteolytic cleavage with formation of intermediate peptides and subsequent conversion of these peptides to free amino acids (Goldberg *et al.,* 1976). Sulfhydryl proteases seem to be involved in the degradative process.

Protein breakdown, like protein synthesis, is now believed to be an energy-dependent process, that is, requiring ATP, since inhibition of cellular energy metabolism blocks protein breakdown (Brostrom and Jeffay, 1970; Goldberg, 1972; Schimke, 1974). The blockage is reversed with restoration of energy metabolism (Goldberg *et al.,* 1975). Complete depletion of cellular ATP stores completely blocks protein breakdown. The ATP requirement for protein breakdown appears to be quite low, since protein breakdown is not affected until the cellular ATP pool is reduced to below 10% of its normal size. Energy now seems to be required for multiple steps in the degradation of protein, that is, both for the initial cleavage and for the subsequent disappearance of the intermediate. However, the precise function of ATP and the responsible hydrolytic enzymes and the existence of the aforementioned nonlysosomal system in other cells remain to be established.

Degradation does not appear to depend on concomitant protein synthesis (Goldberg and St. John, 1976) and does not now appear to be a lysosomal process, even though previous hypotheses have focused on a role of ATP in transporting proteins into such organelles or in maintaining the lysosomal interior in the acid pH range. The activity of a highly active proteolytic system, recently isolated from the soluble fraction of reticulocyte lysates, is optimal at pH 7.8 and is without activity in the acid range.

The degradation rates of cell proteins vary widely, and, based on the observation that short-lived protein enzymes are frequently rate limiting, the rate of protein breakdown is thought to regulate the flow of substrates through metabolic pathways. Furthermore, the degradation rate of many key enzymes is reduced when the substrate supply for these enzymes increases (Goldberg *et al.,* 1974). The structural and conformational properties of proteins that affect their inherent sensitivity to proteolytic enzymes also affect their degradation rate (Dice *et al.,* 1973; Goldberg and Dice, 1974; Goldberg and St. John, 1976; Segal *et al.,* 1974). In addition, polypeptides which are larger and have acidic isoelectric points are more readily degraded than are polypeptides which are smaller and have more basic isoelectric points (Goldberg and Chang, 1978).

Abnormal proteins and polypeptides which are incomplete or contain synthetic errors or amino acid analogues are degraded very rapidly. The breakdown of these proteins appears to be a selective process, since the breakdown of normal polypeptides within the same cell is not affected (Goldberg and Chang, 1978). Insulin reduces protein breakdown in liver (Mortimore and Mondon, 1970) and muscle (Goldberg *et al.,* 1974). Exposure of muscle protein to glucose, certain amino acids, stretch, and repeated contraction also reduces the rate of protein breakdown (Goldberg *et al.,* 1974).

Thus, the supply of amino acids not only stimulates protein synthesis but also inhibits protein breakdown in muscle. Among the plasma amino acids, leucine, in particular, promotes protein synthesis and inhibits proteolysis. The mechanism by which this occurs is unknown, but since the stimulatory effect of the branched-chain amino acids on protein synthesis is associated with accumulation of polysomes and depletion of monosomes (Atwell *et al.*, 1977) and since the stimulatory effect is not inhibited by actinomycin D (Goldberg and Chang, 1978), it appears that the action of the branched-chain amino acids is at the translational level and is associated with an enhancing of polypeptide initiation.

It is not known if the leucine effect on protein turnover is due to leucine itself or some degradation product of leucine-tRNA. A deficiency of any single species of tRNA will reduce protein synthesis and increase protein breakdown, and, since muscle tissue rapidly degrades branched-chain amino acids, it is likely that leucyl-tRNA may be the key metabolite of leucine which regulates protein turnover. The ability of leucine to promote protein synthesis and inhibit protein breakdown in muscle may play an important physiological role in regulating muscle mass (Goldberg and Chang, 1978). Also, L-tryptophan appears to stabilize ornithine aminotransferase, an enzyme not involved in the pathway of tryptophan degradation (Chee and Swick, 1976).

Cahill and co-workers (1972) suggested that the rate of skeletal muscle breakdown was 0.5–1.0 g/kg body weight per day. Halliday and McKeran (1975) reported that the rate of skeletal muscle breakdown was 1.8 g/kg body weight per day. Later studies by Bilmazes *et al.* (1978) estimated the rate of skeletal muscle protein breakdown to be 0.7 g/kg body weight per day.

If the breakdown rate of whole-body protein in young men is approximately 3.8 g/kg body weight per day, then skeletal muscle protein breakdown accounts for approximately 20% [(0.7/3.8) × 100] of the total-body protein breakdown in an adult man. Others (Halliday and McKeran, 1975) reported that muscle accounts for 53% of the total-body protein synthesis or breakdown.

6. Interrelationships of Amino Acid Supply and Protein and Nucleic Acid Metabolism

Although the immediate response of mRNA translation to changes in amino acid supply does not necessarily depend upon alterations in RNA synthesis, there is nevertheless evidence that amino acid supply influences ribosome synthesis and transport. The evidence shows that there are amino-acid-dependent changes in RNA polymerase activity (Henderson, 1970), along with changes in the synthesis of ribosomal components (Franze-Fernandez and Pogo, 1971) and in RNA synthesis and maturation (Vaughan, 1972). Limitations in amino acid supply also lead to reduced synthesis of the protein components of the ribosome (Pawlowski and Vaughan, 1972), along with the transfer of the subunits to the cytoplasm (Maden, 1969).

Amino acid supply also influences ribosomal RNA breakdown, tissue purine nucleotide pools, purine biosynthesis, and purine reutilization (Clifford *et al.*, 1972). The interrelationships among these processes are outlined in Fig. 2. When a rat is fed a protein-containing meal, causing an influx of amino

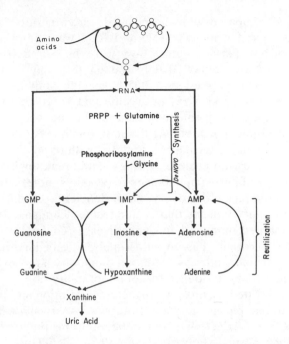

Fig. 2. Schematic outline of the interrelationships between amino acids supply and the synthesis and breakdown of tissue proteins and purines.

acids, there is a transient aggregation of liver monosomes into polysomes, increased protein synthesis, and an accumulation of RNA which is incorporated into monosomes and polysomes. The extent of protein synthesis is determined by the supply of that amino acid which is least abundant in relation to its need. It appears that the effect of amino acid supply on the rate of protein synthesis is determined by the degree of charging of the tRNA which affects both initiation and chain elongation. Feeding also causes an increase in *de novo* purine synthesis as measured by the incorporation of a single pulse of [^{14}C]glycine into RNA and into the purine nucleotide pools.

When the absorptive process is complete or when the rat is fasted or given a protein-deficient diet, the polysomes disaggregate, there is an accumulation of subunits, and the RNA breaks down rather than being incorporated into monosomes and polysomes. The RNA which breaks down results in a slight and transient expansion of the free nucleotide pools, which, in turn, is thought by some investigators (Reem and Friend, 1967, 1969; Wyngaarden and Ashton, 1959) to reduce the rate of *de novo* synthesis of purine nucleotides by feedback inhibition of the first enzyme in the biosynthetic pathway.

In recent times, evidence is beginning to accumulate which shows that tissue purine pools are maintained relatively constant by the relative activities of the *de novo* synthetic and reutilization pathways (Clifford *et al.*, 1972, 1979). The data in Table I show that the intensity of purine synthesis is about four times greater while the intensity of purine reutilization is about four times less in the absorptive compared with the postabsorptive period. In these experiments, the rats were fed at 8:00 P.M. and food was removed at 8:00 A.M. Purine synthesis and purine reutilization were measured from the hepatic incorporation of single pulses of [^{14}C]glycine and [8-^{14}C]hypoxanthine, respectively. The data are expressed as relative specific activities to correct for possible changes in the pool sizes of glycine and hypoxanthine. The data in Table I demonstrate a complementary relationship between purine synthesis and purine reutilization in maintaining tissue purine nucleotide concentrations. Tissue nucleotide pools, in turn, are associated with tissue RNA pools, which, in turn, are associated with protein synthesis and breakdown.

The importance of purine metabolism in protein synthesis has been further emphasized with the recent discovery that a deficiency of one of the enzymes involved in purine metabolism, that is adenosine deaminase, was associated with severe combined immunodeficiency disease (Giblett *et al.*, 1972). The disease is genetic in origin, has an autosomal mode of inheritance, and in it both cell-mediated and antibody-mediated immunity are severely impaired (Meuwissen *et al.*, 1975). Evidence for an abnormality of purine catabolism is provided by the elevated plasma adenosine concentration and increased levels of adenine in urine, plasma and erythrocytes (Agarwal *et al.*, 1976), and lymphocytes (Raivio *et al.*, 1977). ATP concentrations in lymphocytes are ten times greater than normal (Polmar *et al.*, 1976). Deficiency of another key enzyme of purine metabolism, that is, purine nucleoside phosphorylase, also leads to disturbance of T-cell function (Giblett *et al.*, 1975). When the key purine enzymes are deficient, adenosine can increase adenosine 3′,5′-monophosphate concentrations (Wolberg *et al.*, 1975) and can inhibit *S*-adenosylmethionine-dependent transmethylation reactions (Kredich and Martin, 1977). Among the methylation reactions inhibited by *S*-adenosylhomocysteine is the

Table I. Diurnal Meal-Related Rhythms in de Novo Purine Synthesis and Purine Reutilization in Rats[a]

	Time			
	6 P.M.	11 P.M.	3 A.M.	11 A.M.
[^{14}C]*Hypoxanthine uptake*				
Specific activity relative to hypoxanthine				
Adenine	3.83	0.90 [b]	—	0.34
Guanine	1.15	0.48	—	1.06
[^{14}C]*Glycine uptake*				
Specific activity relative to glycine				
Adenine	1.17	3.93[c]	3.14	2.96
Guanine	2.07	8.31	7.61	6.99

[a] The rats had free access to food between 8 P.M. and 8 A.M. *De novo* purine synthesis was measured from the incorporation of a single pulse of [8-^{14}C]glycine into hepatic adenine and guanine pools. Purine reutilization was measured from the incorporation of a single pulse of [^{14}C]hypoxanthine into hepatic adenine and guanine pools.
[b] Four times less than values at 6 P.M.
[c] Four times greater than values at 6 P.M.

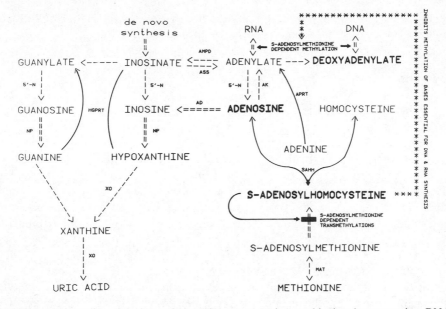

Fig. 3. Schematic outline of the interrelationships among purine, methionine, homocysteine, DNA, and RNA metabolism. Abbreviations for enzymes: AD, adenosine deaminase; AK, adenosine kinase; AMPD, adenylate deaminase; APRT, adenine phosphoribosyltransferase; ASS, adenylo-succinate synthetase; HGPRT, hypoxanthine phosphoriboxyltransferase; MAT, methionine adenosyltransferase; 5'-N, 5'-nucleotidase; NP, nucleoside phosphorylase; SAHH, *S*-adenosylhomocysteine hydrolase; XO, xanthine oxidase.

methylation of purine bases, which is essential for DNA and RNA synthesis. These interrelationships are schematically outlined in Fig. 3.

Adenosine is also implicated in a multiplicity of other biological phenomena, the most dramatic of which is its toxicity to cultured mammalian cells (Green and Chan, 1973; Hershfield *et al.*, 1977; Ishii and Green, 1973; Tattersall *et al.*, 1975; Ullman *et al.*, 1976). Derivatives of purine nucleosides which are 2'-deoxyribonucleosides also appear to be toxic (Carson *et al.*, 1977; Lapi and Cohen, 1977; Scott and Henderson, 1977; Tattersall *et al.*, 1975). Deoxyadenosine, which is also a substrate for adenosine deaminase, is a well-known inhibitor of DNA synthesis after its conversion to deoxyadenosine 5'-triphosphate (Carson *et al.*, 1977; Cohen *et al.*, 1978; Coleman *et al.*, 1978; Klenow, 1959; Moore and Hurlbert, 1966; Simmonds *et al.*, 1978).

It is of interest to correlate the biological effects of adenosine with the well-known growth-retarding effects of low-protein diets which contain a disproportionate amount of one essential amino acid (Harper, 1964; Sullivan *et al.*, 1932). If the growth-depressing effects are equated to toxicity, then on a weight basis, methionine is the most toxic nutritionally important amino acid (Russell *et al.*, 1952; Salmon, 1958; Sauberlich, 1961). Although the basis for the adverse effects of a dietary excess of methionine are at present unknown, it is interesting to speculate that the mechanism of toxicity might involve purine, RNA, and DNA metabolism. Recent studies (Hevia and Clifford, 1978) have shown that dietary protein intake is highly and postively correlated with *de novo* purine synthesis as measured by uric acid output.

7. Overall Total-Body Protein Synthesis and Breakdown

A brief outline of the overall intake, output, and metabolism of protein in the adult human is shown in Fig. 4. The minimum protein intake is the average needed to attain nitrogen equilibrium, that is, 0.45 g/kg body weight. The usual daily intake of protein in the Western diet of an adult male is 90 g (Munro, 1972; Synder *et al.*, 1975). For practical purposes, virtually all of the dietary plus endogenous protein is absorbed and passed to the liver as free amino acids. Thus, the amino acid flux across the gut of a 70-kg man is approximately equivalent to 150 g of amino acids each day, that is, 90 g from diet, plus 70 g endogenous, minus 10 g in feces (Munro, 1972). The usual daily fecal and urinary outputs of nitrogen (as protein) are about 10 and 80 g, respectively, in the adult male (Munro, 1972; Snyder *et al.*, 1975).

Although there are difficulties and limitations with the methodologies for measuring protein synthesis and breakdown (Bilmazes *et al.*, 1978; Waterlow,

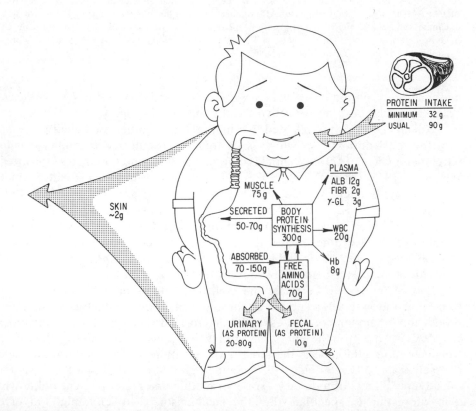

Fig. 4. Total-body protein and amino acid metabolism in the adult. Adapted from Munro (1972).

1969), some estimates of the overall protein metabolism in humans have recently been reported and are summarized in Table II along with appropriate literature values for body potassium, basal metabolic rate, and creatinine clearance.

When the estimates in Table II are combined with literature values (appropriate subjects of comparable age and weight) for body potassium, energy needs (basal metabolic rate), and creatinine clearance, it becomes possible to describe total-body protein synthesis and breakdown relative to body weight, energy expenditure, lean body mass, and age. Results of such calculations, along with regression analyses of amino acid flux, protein synthesis, and protein breakdown, with age as the dependent variable are shown in Figs. 5 and 6.

Amino acid flux, protein synthesis, and protein breakdown per kilogram body weight and per calorie of basal metabolism are presented in Fig. 5. The

Table II. Whole-Body Amino Acid Flux (Q), Protein Synthesis (S), and Protein Breakdown (B) in Human Subjects at Various Ages[a]

Subject	Weight	Age	Q	S	B	K	BMR	CC	Reference
01	74.80	52.00	7.01	3.33	3.46	42.1	22.05	2200	Crane *et al.* (1977)
02	63.90	53.00	6.34	2.77	3.05	41.3	22.12	2180	
03	69.30	58.00	7.27	3.57	3.71	39.5	23.44	2095	
04	58.50	53.00	7.13	3.51	3.45	41.3	22.12	2180	
05	82.90	62.00	7.94	4.39	4.36	38.8	25.25	2035	
06	51.90	20.00	7.99	3.74	3.64	51.8	26.32	2352	
07	64.40	47.00	10.22	5.50	5.44	43.4	22.55	2290	
08	77.40	61.00	8.30	4.39	4.44	38.8	25.25	2050	
09	64.40	53.00	7.46	3.84	3.83	41.3	22.12	2180	
10	64.80	65.00	6.94	3.18	3.35	38.7	25.61	2005	
11	52.50	20.00	8.21	3.88	3.99	51.8	26.32	2352	
PJ	70.00	34.00		5.39		48.6	25.10	2518	James *et al.* (1976)
PS	100.00	31.00		3.87		49.3	25.18	2440	
CH	79.50	56.00		4.63		40.8	22.76	2130	
JW	73.50	58.00		3.86		39.5	23.44	2095	
TE	72.70	64.00		4.18		38.7	25.61	2078	
PP	63.60	46.00		5.64		43.6	22.77	2300	
LG	48.60	25.00		3.68	4.25	51.0	25.35	2430	Long *et al.* (1977)
MS	72.30	25.00		3.70	4.41	51.0	25.35	2430	
WC	2.22	0.06	41.80	25.20	22.70	47.9	54.17	1000	Pencharz *et al.* (1977)
JB	2.00	0.06	35.10	21.10	18.20	47.9	54.17	1000	
LC	1.77	0.01	36.00	21.90	20.00	47.9	54.19	1000	
BM	1.61	0.08	63.80	39.10	37.60	47.9	54.15	1000	
BM	1.81	0.04	35.80	21.30	18.00	47.9	54.18	1000	
SH	1.50	0.08	49.90	30.50	28.00	47.9	554.15	1000	
SH	1.90	0.12	53.40	32.50	29.80	47.9	54.13	1000	

[a] Body weight is given in kilograms and age is in years. Units for Q, S, and B are grams per kilogram body weight per day. Body potassium (K) values represent milliequivalents of potassium per kilogram body weight per day. Basal metabolic rate (BMR) values represent calories per kilogram body weight per day. Creatinine coefficient (CC) values represent creatinine production in urine and are expressed as milligrams of creatinine per kilogram body weight per day. Values for K, BMR, and CC are literature values appropriate for the age and weight of the subjects used for Q, S, and B measurements.

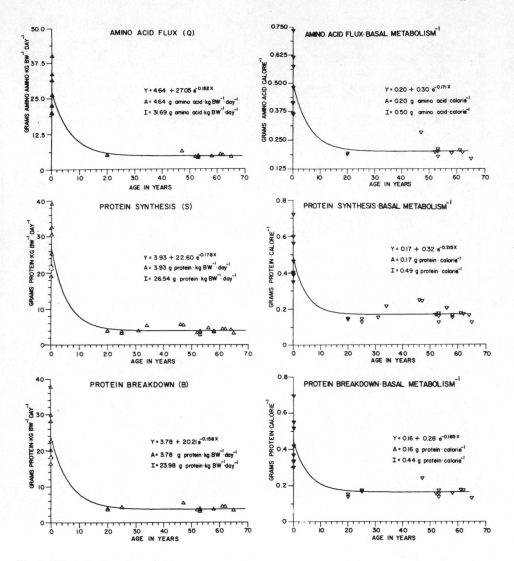

Fig. 5. Whole-body amino acid flux, protein synthesis, and protein breakdown in relation to age, body weight, and energy needs. The data are from Table II, and the regression equation on each panel best describes that relationship.

intercept of the y-axis reflects the values in the newborn, and the asymptote reflects the values in the adult.

Amino acid flux, protein synthesis, and protein breakdown per kilogram of lean body mass and per milligram of urine creatinine are presented in Fig. 6. The y-intercept reflects the values in the newborn, and the asymptote reflects the adult values. The general feature of a reduced intensity of protein metabolism, that is, amino acid flux, protein synthesis, and protein breakdown, with advancing age is clear from the data in Fig. 5 and 6. Although there

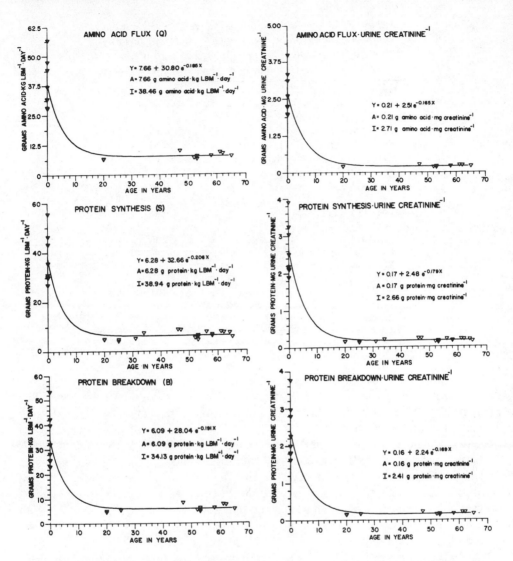

Fig. 6. Whole-body amino acid flux, protein synthesis, and protein breakdown in relation to lean body mass and urine creatinine. Lean body mass was derived from body potassium. The data are from Table II, and the regression equation on each panel best describes that relationship.

appears to be considerable variation in the values for the newborn, the values beyond the first two decades are much better defined. There is a great paucity of values during the first two decades, and the shape of the curves during this period may change substantially as new information becomes available. Despite the limitations on the presently available published data, the amino acid flux and rates of protein synthesis and breakdown are summarized in Table III.

The data in Table III and in Figs. 5 and 6 clearly demonstrate the marked

Table III. *Whole-Body Amino Acid (AA) Flux, Protein*
Synthesis, and Protein Breakdown in Relation to Body
Weight, Basal Metabolism (BMR), Lean Body Mass
(LBM) and Urine Creatinine in Newborns and Adults[a]

	Newborn	Adult
Relative to body weight		
Flux (g AA/kg per day)	31.69	4.64
Synthesis (g protein/kg per day)	26.54	3.93
Breakdown (g protien/kg per day)	23.98	3.78
Relative to BMR		
Flux (g AA/kcal)	0.50	0.30
Synthesis (g protein/kcal)	0.49	0.17
Breakdown (g protein/kcal)	0.44	0.16
Realtive to LBM		
Flux (g AA/ kg per day)	38.46	7.66
Synthesis (g protein/kg LBM per day)	38.94	6.28
Breakdown (g protein/kg LBM per day)	34.13	6.09
Relative to urine creatinine		
Flux (g AA/mg urine creatinine)	2.71	0.21
Synthesis (g protein/mg urine creatinine)	2.66	0.17
Breakdown (g protein/mg urine creatinine)	2.41	0.16

[a] Values are compiled from the data presented in Fig. 3.5 and 6.

reduction (5- to 13-fold) in the intensity of amino acid flux, protein synthesis, and protein breakdown in the adult as compared with newborn humans. The same data also show that the greatest reduction in the intensity of total-body protein metabolism occurs during the first two decades of life and that very little change occurs thereafter. Some investigators (Young *et al.*, 1976, 1977) have reported that the intensity of total-body protein metabolism declines rapidly during the first two decades and then continues to decline, but at a slower rate, with advancing age. The data presented in Figs. 5 and 6 suggest that additional measurements are needed before it can be concluded with confidence that the intensity of protein metabolism continues to decline throughout the life cycle.

The age-related changes in the intensity of total-body protein metabolism are smallest when expressed per calorie of basal metabolism. The synthesis of body protein is well known to require energy, and protein breakdown recently has also been shown to require energy (Brostrom and Jeffay, 1970; Goldberg and St. John, 1976; Schimke, 1974). Whole-body protein synthesis, whole-body energy metabolism, and basal energy metabolism are known to be related, and all decline during growth and development (Munro, 1969; Waterlow and Stephen, 1968). Approximately 2 kcal is needed to support 1 g of total-body protein synthesis in the newborn human, whereas approximately 6 kcal is needed per gram total-body protein in the adult human. The data in Table III clearly show that the differences observed in the intensity of protein metabolism at the various ages are reduced, but not eliminated, when body protein synthesis is related to energy expenditure for the various ages. The 6 kcal of energy needed to support 1 g of total-body protein synthesis in the human is in general agreement with data in animals (sheep) requiring approximately 4 kcal/g, if it is assumed that muscle protein synthesis represents 25% of the total-body protein synthesis (Buttery *et al.,* 1975) in that species.

Dietary protein needs are also well known to decline with age. When total-body protein synthesis is considered in relation to dietary protein needs at various ages, the amount of dietary protein required to support body protein synthesis can be estimated. If the daily protein allowances for the newborn and the adult are 3.2 and 0.5 g/kg body weight, respectively, and if the respective rates of total-body protein synthesis are 26.5 and 3.9 g/kg body weight per day, then total-body protein synthesis per gram dietary protein is 8.3 g (26.5/3.2) for the newborn and 7.9 g (3.9/0.5) for the adult. These calculations indicate that approximately 1 g of dietary protein is required to support the needs of 8 g of total-body protein synthesis. The calculations also show that the amount of dietary protein needed to support a given quantity of total-body protein synthesis remains relatively constant throughout the life cycle. The greater caloric needs to support the synthesis of 1 g of whole-body protein in the adult as compared to the newborn are probably related to the proportionally lesser contribution of total-body protein and the proportionally greater contribution of total-body adipose tissues to overall body metabolism.

The efficiency of total-body protein synthesis can also be evaluated relative to the total flux of amino acids through the free amino acid pool. In Fig. 7, the total amino acid flux, protein synthesis, and protein breakdown in the newborn and in the adult are presented. The breakdown of protein should

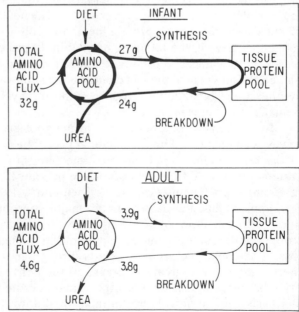

Fig. 7. Schematic outline of the interrelationships among diet, whole-body total amino acid flux, amino acid pool, and tissue protein synthesis and breakdown in the newborn and adult. The values are taken from Table III.

theoretically render all released amino acids available for resynthesis of proteins. If this were quantitative, urea nitrogen excretion on a protein-free diet should equal zero, but this does not occur (Peret and Jacquot, 1972). The proportion of endogenous nitrogen excretion attributable to the incidental breakdown of amino acids as they pass through the amino acid pools during protein turnover is unknown. It is known that hydroxyproline, methylhistidine, and methyllysine are not reutilized for protein synthesis (Munro, 1970; Young *et al.,* 1972). In the newborn, a total flux of 32 g/kg body weight per day will support a net total-body protein synthesis of 3 g/kg body weight per day (27 − 24 = 3). In the adult, a total amino acid flux of 4.6 g/kg body weight per day will support a net total-body protein synthesis of 0.1 g/kg body weight per day (3.9 − 3.8 = 0.1). Expressed in this manner, the efficiency of total-body protein synthesis is 9.4% [(3/32) × 100] for the newborn and only 2.2% [(0.1/4.6) × 100] for the adult. Corresponding values for net muscle protein deposition relative to total-muscle protein synthesis in animals are 3.5% for sheep, 14.7% for cockerels, and 20.9% for rats (Buttery and Boorman, 1976). Although the rate of total-body protein synthesis exceeds the rate of breakdown in the adult, the data in Table III represent regulation of total-tissue protein over the short term.

Since total-body protein mass declines with age, additional long-term regulation of the equilibrium between total-body protein synthesis and breakdown occurs. The total amount of body protein increases during growth and development to reach a maximum during the second and third decades. Thereafter, the total amount of body protein declines. Muscle alone accounts for about 25% of the total-body protein at birth, increases to a maximum of about 45% at the end of the second decade, and thereafter declines. Liver, heart, kidneys, brain, and spleen together account for about 9% of total-body protein in the newborn, decreasing to about 5% in the adult. In view of the importance of muscle in whole-body protein metabolism, it is to be expected that the age-related changes in the intensity of total-body protein metabolism are, in large part, due to changes in total-body muscle protein metabolism. The exact role played by individual organs in changing the amino acid flux and rates of protein synthesis and breakdown throughout the life cycle is, however, unknown in humans at the present time.

In relation to total-body protein metabolism, the metabolic response to injury is characterized by a substantial loss of body protein. The precise source of the lost protein, the biochemical mechanism for the loss, and the identification of improved treatment of injured patients have been extensively studied in recent years (Crain *et al.,* 1977; Kien *et al.,* 1978; O'Keefe *et al.,* 1974). It now appears that muscle protein is a major contributor to the lost protein after injury (Hoover-Plow and Clifford, 1978). Trauma causes a net loss of protein from both the sarcoplasmic and myofibrillar compartments of muscle. The net loss of sarcoplasmic proteins occurs through an increase in their rate of synthesis and an even greater increase in their rate of degradation. Net loss of myofibrillar proteins, on the other hand, occurs through a reduced rate of synthesis without any change in their breakdown rate (Hoover-Plow and Clifford, 1978). It remains to be determined if similar mechanisms account for body protein loss associated with febrile diseases and other types of trauma.

8. General Conclusions

The whole-body metabolism of proteins and amino acids is in a very dynamic state, is intimately associated with purine and energy metabolism, and is markedly influenced by supplies of amino acids from dietary proteins. The total amount of body protein present at a given time represents the balance between protein synthesis and protein breakdown. The intensity of synthesis and breakdown undergoes short-term diurnal meal-related rhythms as well as long-term changes during the life cycle. Protein synthesis is a process which is better understood than protein breakdown. Much of our information on protein breakdown is limited to enzyme proteins and albumin, which may not reflect accurately the catabolism of protein in the whole body. The precise nature of dietary protein and energy needs as essential precursors and regulators of the synthesis and breakdown of body proteins remains a challenging and unsolved problem. Its solution should provide important information to refine our current knowledge of the optimum protein nutriture of humans during the growth, the developmental, and the aging stages of the life cycle, as well as in periods of special needs such as sepsis, surgical trauma, or other trauma.

ACKNOWLEDGMENTS. The unpublished work reported here was supported by the U.S. Public Health Service Grant AM-16726. The author acknowledges Dr. J. L. Koong for statistical presentation of the data in Figs. 5 and 6.

9. References

Addison, J. M., Burston, D., and Matthews, D. M., 1972, Evidence for active transport of the dipeptide glycylsarcosine by hamster jejunum *in vitro, Clin. Sci.* **43:**907.

Addison, J. M., Burston, D., Matthews, D. M., Payne, J. W., and Wilkson, S., 1974, Evidence for active transport of the tripeptide glycylsarcosylsarcosine by hamster jejunum *in vitro, Clin. Sci. Mol. Med.* **46:**30P.

Adibi, S. A., 1969, The influence of molecular structure of neutral amino acids on their absorption kinetics in the jejunum and ileum of human intestine *in vivo, Gastroenterology* **56:**903.

Adibi, S. A., 1971, Intestinal transport of dipeptides in man: Relative importance of hydrolysis and intact absorption, *J. Clin. Invest.* **50:**2266.

Adibi, S. A., and Gray, S. J., 1967, Intestinal absorption of essential amino acids in man, *Gastroenterology* **52:**837.

Adibi, S. A., and Krzysik, B. A., 1975, Transport and hydrolysis of dipeptides by liver, muscle, and kidney, *Clin. Res.* **23:**391A.

Adibi, S. A., and Soleimanpour, M. R., 1974, Functional characterization of dipeptide transport system in human jejunum, *J. Clin. Invest.* **53:**1368.

Adibi, S. A., Gray, S. J., and Menden, E., 1967, The kinetics of amino acid absorption and alteration of plasma composition of free amino acids after intestinal perfusion of amino acid mixtures, *Am. J. Clin. Nutr.* **20:**24.

Adibi, S. A., Morse, E. L., Masilamani, S. S., and Amin, P. M., 1975, Evidence for two different modes of tripeptide disappearance in human intestine: Uptake by peptide carrier system and hydrolysis by peptide hydrolases, *J. Clin. Invest.* **56:**1355.

Agar, W. T., Hird, F. J. R., and Sidhu, G. S., 1953, The active absorption of amino acids by the intestine, *J. Physiol.* **121:**255.

Agarwal, R. P., Crabtree, G. W., Parks, R. E., Jr., Nelson, J. A., Keightley, R., Parkman, R., Rosen, F. S., Stern, R. C., and Polmar, S. H., 1976, Purine nucleoside metabolism in the erythrocytes of patients with adenosine deaminase deficiency and severe combined immunodeficiency, *J. Clin. Invest.* **57:**1025.

Aoki, T. T., Brennan, M. F., Muller, W. A., Moore, F. D., and Cahill, G. F., Jr., 1972, Effect of insulin on muscle glutamate uptake: Whole blood versus plasma glutamate analysis, *J. Clin. Invest.* **51:**2889.

Asatoor, A. M., Bandoh, J. K., Lant, A. F., Milne, M. D., and Navab, F., 1970a, Intestinal absorption of carnosine and its constituent amino acids in man, *Gut* **11:**250.

Asatoor, A. M., Cheng, B., Edwards, K. D. G., Lant, A. F., Matthews, D. M., Milne, M. D., Navab, F., and Richards, A. J., 1970b, Intestinal absorption of two dipeptides in Hartnup disease, *Gut* **11:**380.

Asatoor, A. M., Harrison, B. D. W., Milne, M. D., and Prosser, D. I., 1972, Intestinal absorption of an arginine-containing peptide in cystinuria, *Gut* **11:**373.

Atwell, J. R., Hedden, M. P., Mancusi, V. J., and Buse, M. G., 1977, Branched chain amino acids (BCAA) as regulators of muscle protein synthesis, *Diabetes* **26** (Suppl. 1):373.

Bilmazes, C., Uauy, R., Haverberg, L. N., Munro, H. N., and Young, V. R., 1978, Muscle protein breakdown rates in humans based on N^r-methylhistidine (3-methylhistidine) content of mixed proteins in skeletal muscle and urinary output of N^r-methylhistidine, *Metabolism* **27:**525.

Boullin, D. J., Crampton, R. F., Heading, C. E., and Pelling, D., 1973, Intestinal absorption of dipeptides containing glycine, phenylalanine, proline, β-alanine or histidine in the rat, *Clin. Sci. Mol. Med.* **45:**849.

Brostrom, C. O., and Jeffay, F., 1970, Protein catabolism in rat liver homogenates. A re-evaluation of the energy requirement for protein catabolism, *J. Biol. Chem.* **245:**4001.

Buttery, P. J., and Boorman, K. N., 1976, The energetic efficiency of amino acid metabolism, in: *Protein Metabolism and Nutrition* (D. J. A. Cole, K. N. Boorman, P. J. Buttery, D. Lewis, R. J. Neale, and H. Swan, eds.), pp. 197–206, Butterworths, Boston.

Buttery, P. J., Beckerton, A., Mitchell, R. M., Davies, K., and Annison, E. F., 1975, The turnover rate of muscle and liver protein in sheep, *Proc. Nutr. Soc.* **34:**91A.

Cahill, G. F., Jr., and Owen, O. E., 1970, The role of the kidney in the regulation of protein metabolism, in: *Mammalian Protein Metabolism,* Vol. 4 (H. N. Munro, ed.), pp. 559–581, Academic Press, New York.

Cahill, G. F., Jr., Aoki, T. T., and Marliss, E. B., 1972, Insulin and muscle protein, in: *Handbook of Physiology,* Section 7, *Endocrinology,* Vol. I, *Endocrine Pancreas* (D. F. Steiner and N. Freinkel, eds.), pp. 563–577, American Physiological Society, Washington, D.C.

Carson, D. A, Kaye, J., and Seegmiller, J. E., 1977, Lymphospecific toxicity in adenosine deaminase deficiency and purine nucleoside phosphorylase deficiency: Possible role of nucleoside kinase(s), *Proc. Natl. Acad. Sci. U.S.A.* **74:**5677.

Chee, P. Y., and Swick, R. W., 1976, Effect of dietary protein and tryptophan on the turnover of rat liver ornithine aminotransferase, *J. Biol. Chem.* **251:**1029.

Christensen, H. N., 1964, Free amino acids and peptides in tissues, in: *Mammalian Protein Metabolism,* Vol. 1 (H. N. Munro and J. B. Allison, eds.), pp. 105–124, Academic Press, New York.

Clifford, A. J., Riumallo, J. A., Baliga, B. S., Munro, H. N., and Brown, P. R., 1972, Liver nucleotide metabolism in relation to amino acid supply, *Biochim. Biophys. Acta* **277:**443.

Clifford, A. J., Crane, D. D., Crane, R. T., Smith, N. E., and Ho, C. Y., 1979, Effect of dietary protein on hematologic values and on hepatic adenine and guanine levels in rats, *Fed. Proc. Fed. Am. Soc. Exp. Biol.* **38**:608.

Cohen, A., Hirschhorn, R., Horowitz, S. D., Rubinstein, A., Polmar, S. H., Hong, R., and Martin, D. W., Jr., 1978, Deoxyadenosine triphosphate as a potentially toxic metabolite in adenosine deaminase deficiency, *Proc. Natl. Acad. Sci. U.S.A.* **75**:472.

Coleman, M. S., Donofrio, J., Hutton, J. J., Hahn, L., Daoud, A., Lampkin, B., and Dyminski, J., 1978, Identification and quantitation of adenine deoxynucleotides in erythrocytes of a patient with adenosine deaminase deficiency and severe combined immunodeficiency, *J. Biol. Chem.* **253**:1619.

Cook, G. C., 1971, Impairment of glycine absorption by glucose and galactose in man, *J. Physiol.* **217**:61.

Cook, G. C., 1972a, Intestinal absorption rate of L-methionine in man and the effect of glucose in the perfusing fluid, *J. Physiol.* **221**:707.

Cook, G. C., 1972b, Comparison of intestinal absorption rates of glycine and glycylglycine in man and the effect of glucose in the perfusing fluid, *Clin. Sci.* **43**:443.

Crane, C. W., Picou, D., Smith, R., and Waterlow, J. C., 1977, Protein turnover in patients before and after elective orthopaedic operations, *Br. J. Surg.* **64**:129.

Das, M., and Radhakrishnan, A. N., 1975, Studies on a wide-spectrum intestinal dipeptide uptake system in the monkey and in the human, *Biochem. J.* **146**:133.

Dice, J. F., Dehlinger, P. H., and Schimke, R. T., 1973, Studies on the correlation between size and relative degradation rate of soluble proteins, *J. Biol. Chem.* **248**:4220.

Drummond, K. N., Michael, A. E., Ulstrom, R. A., and Good, R. A., 1964, The blue diaper syndrome: Familial hypercalcemia with nephrocalcinosis and indicanuria, *Am. J. Med.* **37**:928.

Elwyn, D. H., Parikh, H. C., and Shoemaker, W. C., 1968, Amino acid movements between gut, liver, and periphery in unanesthetized dogs, *Am. J. Physiol.* **215**:1260.

Elwyn, D. H., Launder, W. J., Parikh, H. C., and Wise, E. M., Jr., 1972, Roles of plasma and erythrocytes in interorgan transport of amino acids in dogs, *Am. J. Physiol.* **222**:1333.

Felig, P., 1975, Amino acid metabolism in man, *Annu. Rev. Biochem.* **44**:933.

Fleck, A., Shepherd, J., and Munro, H. N., 1965, Protein synthesis in rat liver: Influence of amino acids in diet on microsomes and polysomes, *Science* **150**:628.

Franze-Fernandez, M. T., and Pogo, A. O., 1971, Regulation of the nucleolar DNA-dependent ribonucleic acid polymerase by amino acids in Erlich ascites tumor cells, *Proc. Natl. Acad. Sci. U.S.A.* **68**:3040.

Giblett, E. R., Anderson, J. E., Cohen, F., Pollara, B., and Meuwissen, H. J., 1972, Adenosine-deaminase deficiency in two patients with severely impaired cellular immunity, *Lancet* **2**:1067.

Giblett, E. R., Ammann, A. J., Wara, D. W., Sandman, R., and Diamond, L. K., 1975, Nucleo-side-phosphorylase deficiency in a child with severely defective T-cell immunity and normal B-cell immunity, *Lancet* **1**:1010.

Godlad, G. A. J., and Munro, H. N., 1959, Diet and the action of cortisone on protein metabolism, *Biochem. J.* **73**:343.

Goldberg, A. L., 1972, Degradation of abnormal proteins in *Escherichia coli, Proc. Natl. Acad. Sci. U.S.A.* **69**:422.

Goldberg, A. L., and Chang, T. W., 1978, Regulation and significance of amino acid metabolism in skeletal muscle, *Fed. Proc. Fed. Am. Soc. Exp. Biol.* **37**:2301.

Goldberg, A. L., and Dice, J. F., 1974, Intracellular protein degradation in mammalian and bacterial cells, *Annu. Rev. Biochem.* **43**:835.

Goldberg, A. L., and Odessey, R., 1972, Oxidation of amino acids by diaphragms from fed and fasted rats, *Am. J. Physiol.* **223**:1384.

Goldberg, A. L., and Odessey, R., 1974, Regulation of protein and amino acid degradation in skeletal muscle, in: *Exploratory Concepts in Muscular Dystrophy II* (A. T. Milhorat, ed.), pp. 187–201, American Elsevier, New York.

Goldberg, A. L., and St. John, A. C., 1976, Intracellular protein degradation in mammalian and bacterial cells, *Annu. Rev. Biochem.* **45**:747.

Goldberg, A. L., Howell, E. M., Li, J. B., Martel, S. P., and Prouty, W. F., 1974, Physiological significance of protein degradation in animal and bacterial cells, *Fed. Proc. Fed. Am. Soc. Exp. Biol.* **33**:1112.

Goldberg, A. L., Olden, K., and Prouty, W. F., 1975, Studies on the mechanisms and selectivity of protein degradation in *E. coli,* in: *Intracellular Protein Turnover* (R. T. Schimke and A. Katunuma, eds.), pp. 17–55, Academic Press, New York.

Goldberg, A. L., Kowit, J. D., and Etlinger, J. D., 1976, Studies of the selectivity and mechanisms of intracellular protein degradation, in: *Proteolysis and Physiological Regulation* (D. W. Ribbons and K. Brew, eds.), pp. 313–337, Academic Press, New York.

Goodman, S. I., McIntyre, C. A., Jr., and O'Brien, D., 1967, Impaired intestinal transport of proline in a patient with familial iminoaciduria, *J. Pediatr.* **71**:246.

Goswami, P., and Munro, H. N., 1962, The role of ribonucleic acid in the formation of prothrombin activity by rat liver microsomes, *Biochim. Biophys. Acta* **44**:410.

Gray, G. M., and Santiago, N. A., 1974, Human jejunal surface aminooligopeptidase: Sequential and simultaneous action on a tetrapeptide and its products, *Gastroenterology* **66**:704.

Green, H., and Chan, T. S., 1973, Pyrimidine starvation induced by adenosine in fibroblasts and lymphoid cells: Role of adenosine deaminase, *Science* **182**:836.

Greenberg, H., and Penman, S., 1966, Methylation and processing of ribosomal RNA in HeLa cells, *J. Mol. Biol.* **21**:527.

Halliday, D., and McKeran, R. O., 1975, Measurement of muscle protein synthetic rate from serial muscle biopsies and total body protein turnover in man by continuous intravenous infusion of L-[α-^{15}N]lysine, *Clin. Sci. Mol. Med.* **49**:581.

Hardcastle, P. T., and Daniels, V. G., 1973, Interaction between hexoses and amino acids for transport in rat small intestine, *Comp. Biochem. Physiol.* **45A**:995.

Harper, A. E., 1964, Amino acid toxicities and imbalances, in: *Mammalian Protein Metabolism,* Vol. 2 (H. N. Munro and J. B. Allison, eds.), pp. 87–134, Academic Press, New York.

Heizer, W. D., and Laster, L., 1969, Peptide hydrolase activities of the mucosa of human small intestine, *J. Clin. Invest.* **48**:210.

Heizer, W. D., Kerley, R. L., and Isselbacher, K. J., 1972, Intestinal peptide hydrolases differences between brush border and cytoplasmic enzymes, *Biochim. Biophys. Acta* **264**:450.

Hellier, M. D., Holdsworth, C. D., McColl, I., and Perrett, D., 1972a, Dipeptide absorption in man, *Gut* **13**:965.

Hellier, M. D., Holdsworth, C. D., Perrett, D., and Thirumalai, C., 1972b, Intestinal dipeptide transport in normal and cystinuric subjects, *Clin. Sci.* **43**:659.

Henderson, A. R., 1970, The effect of feeding with a tryptophan-free amino acid mixture on rat liver magnesium ion-activated deoxyribonucleic acid-dependent ribonucleic acid polymerase, *Biochem. J.* **120**:205.

Herschfield, M. S., Snyder, F. F., and Seegmiller, J. E., 1977, Adenine and adenosine are toxic to human lymphoblast mutants defective in purine salvage enzymes, *Science* **197**:1284.

Hevia, P., and Clifford, A. J., 1978, Protein intake, hepatic purine enzyme levels and uric acid production in growing chicks, *J. Nutr.* **108**:46.

Hooft, C., Timmermans, J., Snoeck, J., Antener, I., Oyaert, W., and van den Hende, V., 1964, Methionine malabsorption in a mentally defective child, *Lancet* **2**:20.

Hoover-Plow, J. L., and Clifford, A. J., 1978, The effect of surgical trauma on muscle protein turnover in rats, *Biochem. J.* **176**:137.

Hori, M., Fisher, J. M., and Rabinovitz, M., 1967, Tryptophan deficiency in rabbit reticulocytes: Polyribosomes during interrupted growth of hemoglobin chains, *Science* **155**:83.

Ip, C. C. Y., and Harper, A. E., 1973, Effect of threonine supplementation on hepatic polysome patterns and protein synthesis of rats fed a threonine-deficient diet, *Biochim. Biophys. Acta* **331**:251.

Ishii, K., and Green, H., 1973, Lethality of adenosine for cultured mammalian cells by interference with pyrimidine biosynthesis, *J. Cell Sci.* **13**:429.

Jacob, S. T., Sajdel, E. M., and Munro, H. N., 1969, Regulation of nucleolar RNA metabolism by hydrocortisone, *Eur. J. Biochem.* **7**:449.

Jacob, S. T., Sajdel, E. M., and Munro, H. N., 1970, Different responses of soluble nucleolar RNA polymerase to divalent cations and to inhibition by α-amanitin, *Biochem. Biophys. Res. Commun.* **38**:765.

James, W. P. T., Garlick, P. J., Sender, P. M., and Waterlow, J. C., 1976, Studies of amino acid and protein metabolism in normal man with L-[U-¹⁴C]tyrosine, *Clin. Sci. Mol. Med.* **50**:525.

Kedinger, C., Gniazdowski, M., Mandel, J. L., Jr., Gissinger, F., and Chambon, P., 1970, α-Amanitin: A specific inhibitor of one of two DNA-dependent RNA polymerase activities from calf thymus, *Biochem. Biophys. Res. Commun.* **38**:165.

Kien, C. L., Young, V. R., Rohrbaugh, D. K., and Burke, J. F., 1978, Whole-body protein synthesis and breakdown rates in children before and after reconstructive surgery of the skin, *Metabolism* **27**:27.

Kim, Y. S., Birtwhistle, W., and Kim, Y. W., 1972, Peptide hydrolases in the brush border and soluble fractions of small intestinal mucosa of rat and man, *J. Clin. Invest.* **51**:1419.

Kim, Y. S., Kim, Y. W., and Sleisenger, M. H., 1974, Studies on the properties of peptide hydrolases in the brush-border and soluble fractions of small intestinal mucosa of rat and man, *Biochim. Biophys. Acta* **370**:283.

Klenow, H., 1959, On the effect of some adenine derivatives on the incorporation *in vitro* of isotopically labelled compounds into the nucleic acids of Ehrlich ascites tumor cells, *Biochim. Biophys. Acta* **35**:412.

Kredich, N. M., and Martin, D. W., Jr., 1977, Role of *S*-adenosylhomocysteine in adenosine-mediated toxicity in cultured mouse T-lymphoma cells, *Cell* **12**:931.

Krzysik, B., Peterson, J., and Adibi, S. A., 1975, The potential of blood, liver, muscle, and kidney for dipeptide hydrolysis, *Fed. Proc. Fed. Am. Soc. Exp. Biol.* **34**:466.

Lapi, L., and Cohen, S. S., 1977, Toxicities of adenosine and 2′-deoxyadenosine in L cells treated with inhibitors of adenosine deaminase, *Biochem. Pharmacol.* **26**:71.

Lindberg, T., 1966, Intestinal dipeptidases. Dipeptidase activity in the mucosa of the gastrointestinal tract of the adult human, *Acta Physiol. Scand.* **66**:437.

Lindberg, T., Norden, A., and Josefsson, L., 1968, Intestinal dipeptidases. Dipeptidase activities in small intestinal biopsy specimens from a clinical material, *Scand. J. Gastroenterol.* **3**:177.

Long, C. L., Jeevanandam, M., Kim, B. M., and Kinney, J. M., 1977, Whole body protein synthesis and catabolism in septic man, *Am. J. Clin. Nutr.* **30**:1340.

Maden, B. E. H., 1969, Effects of lysine or valine starvation on ribosome subunit balance in HeLa cells, *Nature* **224**:1203.

Manchester, K. L., 1965, Oxidation of amino acids by isolated rat diaphragm and the influence of insulin, *Biochim. Biophys. Acta* **100**:295.

Matsutaka, H., Aikawa, T., Yamamoto, H., and Ishikawa, E.,·1973, Gluconeogenesis and amino acid metabolism. III. Uptake of glutamine and output of alanine and ammonia by non-hepatic splanchnic organs of fasted rats and their metabolic significance, *J. Biochem.* **74**:1019.

Matthews, D. M., 1975, Absorption of peptides by mammalian intestine, in: *Peptide Transport in Protein Nutrition* (D. M. Matthews and J. W. Payne, eds.), pp. 61–146, American Elsevier, New York.

Matthews, D. M., Craft, I. L., Geddes, D. M., Wise, I. J., and Hyde, C. W., 1968, Absorption of glycine and glycine peptides from the small intestine of the rat, *Clin. Sci.* **35**:415.

Matthews, D. M., Addison, J. M., and Burston, D., 1974, Evidence for active transport of the dipeptide carnosine (β-alanyl-L-histidine) by hamster jejunum *in vitro*, *Clin. Sci. Mol. Med.* **46**:693.

McCarthy, C. F., Borland, J. L., Jr., Lynch, H. J., Jr., Owen, E. E., and Tyor, M. P., 1964, Defective uptake of basic amino acids and L-cystine by intestinal mucosa of patients with cystinuria, *J. Clin. Invest.* **43**:1518.

Meilman, E., Urivetsky, M. M., and Rapoport, C. M., 1963, Urinary hydroxyproline peptides, *J. Clin. Invest.* **42**:40.

Meuwissen, H. J., Pickering, R. J., Pollara, B., and Porter, I. H., 1975, Impairment of adenosine deaminase activity in combined immunological deficiency disease, in: *Combined Immunodeficiency Disease and Adenosine Deaminase Deficiency: A Molecular Defect* (H. J. Meuwissen, ed.), pp. 73–83, Academic Press, New York.

Miller, L. L., 1962. The role of the liver and the non-hepatic tissues in the regulation of free amino acid levels in the blood, in: *Amino Acid Pools* (J. T. Holden, ed.), pp. 708–721, Elsevier, New York.

Moore, E. C., and Hurlbert, R. B., 1966, Regulation of mammalian deoxyribonucleotide biosynthesis by nucleotides as activators and inhibitors, *J. Biol. Chem.* **241:**4802.

Mortimore, G. E., and Mondon, C. E., 1970, Inhibition by insulin of valine turnover in liver. Evidence for a general control of proteolysis, *J. Biol. Chem.* **245:**2375.

Munro, H. N., 1969, Evolution of protein metabolism in mammals, in: *Mammalian Protein Metabolism,* Vol. 3 (H. N. Munro, ed.), pp. 133–182, Academic Press, New York.

Munro, H. N., 1970, Free amino acid pools and their role in regulation, in: *Mammalian Protein Metabolism,* Vol. 4 (H. N. Munro, ed.), pp. 299–386, Academic Press, New York.

Munro, H. N., 1972, Amino acids and protein hydrolysates. Basic concepts for parenteral nutrition, in: *Symposium on Total Parenteral Nutrition,* pp. 7–35, The Food Science Committee, Council on Foods and Nutrition of the American Medical Association, Nashville, Tenn.

Munro, H. N., Hubert, C., and Baliga, B. S., 1975, Regulation of protein synthesis in relation to amino acid supply—A review, in: *Alcohol and Abnormal Protein Synthesis* (M. A. Rothschild, M. Oratz, and S. S. Schreiber, eds.), pp. 33–66, Pergamon Press, New York.

Navab, F., and Asatoor, A. M., 1970, Studies on intestinal absorption of amino acids and a dipeptide in a case of Hartnup disease, *Gut* **11:**373.

Newey, H., and Smyth, D. H., 1959, The intestinal absorption of some dipeptides, *J. Physiol.* **145:**48.

Odessey, R., and Goldberg, A. L., 1972, Oxidation of leucine by rat skeletal muscle, *Am. J. Physiol.* **223:**1376.

O'Keefe, S. J. D., Sender, P. M., and James, W. P. T., 1974, "Catabolic" loss of body nitrogen in response to surgery, *Lancet* **2:**1035.

Pawlowski, P. J., and Vaughan, M. H., 1972, Comparison of the relative synthesis of the proteins of the 50 S ribosomal subunit in growing and valine-deprived HeLa cells, *J. Cell Biol.* **52:**409.

Pencharz, P. B., Steffee, W. P., Cochran, W., Scrimshaw, N. S., Rand, W. M., and Young, V. R., 1977, Protein metabolism in human neonates: Nitrogen-balance studies, estimated obligatory losses of nitrogen and whole-body turnover of nitrogen, *Clin. Sci. Mol. Med.* **52:**485.

Peret, J., and Jacquot, R., 1972, Nitrogen excretion on complete fasting and on a nitrogen-free diet—endogenous nitrogen, in: *Protein and Amino Acid Functions* (E. J. Bigwood, ed.), pp. 73–118, Pergamon Press, New York.

Perry, R. P., and Kelley, D. E., 1974, Existence of methylated messenger RNA in mouse L cells, *Cell* **1:**37.

Perry, T. L., Hansen, S., Tischler, B., Bunting, R., and Berry, K., 1967, Carnosinemia. A new metabolic disorder associated with neurologic disease and mental defect, *N. Engl. J. Med.* **277:**1219.

Peters, T. J., 1970, The subcellular localization of di- and tripeptide hydrolase activity in guinea pig small intestine, *Biochem. J.* **120:**195.

Peters, T. J., 1973, The hydrolysis of glycine oligopeptides by guinea pig intestinal mucosa and by isolated brush borders, *Clin. Sci. Mol. Med.* **45:**803.

Peters, T. J., and MacMahon, M. T., 1970, The absorption of glycine and glycine oligopeptides by the rat, *Clin. Sci.* **39:**811.

Polmar, S. H., Stern, R. C., Schwartz, A. L., Wetzler, E. M., Chase, P. A., and Hirschhorn, R., 1976, Enzyme replacement therapy for adenosine deaminase deficiency and severe combined immunodeficiency, *N. Engl. J. Med.* **295:**1337.

Prichard, P. M., Gilbert, J. M., Shafritz, D. A., and Anderson, W. F., 1970, Factors for the initiation of haemoglobin synthesis by rabbit reticulocyte ribosomes, *Nature* **226:**511.

Prichard, P. M., Picciano, D. J., Laycock, D. G., and Anderson, W. F., 1971, Translation of exogenous messenger RNA for hemoglobin on reticulocyte and liver ribosomes, *Proc. Natl. Acad. Sci. U.S.A.* **68:**2752.

Pronczuk, A., Rogers, Q. R., and Munro, H. N., 1970, Liver polysome patterns of rats fed amino acid imbalanced diets, *J. Nutr.* **100:**1249.

Raivio, K. O., Schwartz, A. L., Stern, R. C., and Polmar, S. H., 1977, Adenine and adenosine metabolism in lymphocytes deficient in adenosine deaminase (ADA) activity, *Adv. Exp. Med. Biol.* **76A:**456.

Reem, G. H., and Friend, C., 1967, Phosphoribosylamidotransferase: Regulation of activity in virus-induced murine leukemia by purine nucleotides, *Science* **157:**1203.

Reem, G. H., and Friend, C., 1969, Properties of 5'-phosphoribosylpyrophosphate amidotransferase in virus induced murine leukemia, *Biochim. Biophys. Acta* **171:**58.

Reichman, M., and Penman, S., 1973, Stimulation of polypeptide initiation *in vitro* after protein synthesis inhibition *in vivo* in HeLa cells, *Proc. Natl. Acad. Sci. U.S.A.* **70:**2678.

Roeder, R. G., and Rutter, W. J., 1969, Multiple forms of DNA-dependent RNA polymerase in eukaryotic organisms, *Nature* **224:**234.

Russell, W. C., Taylor, M. W., and Hogan, J. M., 1952, Effect of essential amino acids on growth of the white rat, *Arch. Biochem. Biophys.* **39:**249.

Sajdel, E. M., and Jacob, S. T., 1971, Mechanism of early effect of hydrocortisone on the transcriptional process: Stimulation of the activities of purified rat liver nucleolar RNA polymerases, *Biochem. Biophys. Res. Commun.* **45:**707.

Salmon, W. D., 1958, The significance of amino acid imbalance in nutrition, *Am. J. Clin. Nutr.* **6:**487.

Sauberlich, H. E., 1961, Studies on the toxicity and antagonism of amino acids for weanling rats, *J. Nutr.* **75:**61.

Schedl, H. P., Pierce, C. E., Rider, A., and Clifton, J. A., 1968, Absorption of L-methionine from the human small intestine, *J. Clin. Invest.* **47:**417.

Schimke, R. T., 1974, Regulation of protein degradation in mammalian tissues, in: *Mammalian Protein Metabolism,* Vol. 4 (H. N. Munro, ed.), pp. 177–228, Academic Press, New York.

Schumm, D. E., and Webb, T. E., 1972, Transport of informosomes from isolated nuclei of regenerating rat liver, *Biochem. Biophys. Res. Commun.* **48:**1259.

Scott, F. W., and Henderson, J. R., 1977, Cytotoxicity of guanine and its naturally occurring derivatives in mammalian cells; reversal by deoxycytidine and adenine, *Can. Fed. Biol. Soc.* **20:**181.

Scriver, C. R., and Rosenberg, L. E., 1973, Distribution of amino acids in body fluids, in: *Amino Acid Metabolism and Its Disorders* (A. L. Schaffer, ed.), pp. 39–60, W. B. Saunders, Philadelphia.

Scriver, C. R., Lamm, P., and Clow, C. L., 1971, Plasma amino acids: Screening, quantitation, and interpretation, *Am. J. Clin. Nutr.* **24:**876.

Segal, H. L., Winkler, J. R., and Miyagi, M. P., 1974, Relationship between degradation rates of proteins *in vivo* and their susceptibility to lysosomal proteases, *J. Biol. Chem.* **249:**6364.

Shih, V. E., Bixby, E. M., Alpers, D. H., Bartsocas, C. S., and Thier, S. O., 1971, Studies of intestinal transport defect in Hartnup disease, *Gastroenterology* **61:**445.

Silk, D. B. A., Perrett, D., and Clark, M. L., 1973a, Intestinal transport of two dipeptides containing the same two neutral amino acids in man, *Clin. Sci. Mol. Med.* **45:**291.

Silk, D. B. A., Marrs, T. C., Addison, J. M., Burston, D., Clark, M. L., and Matthews, D. M., 1973b, Absorption of amino acids from an amino acid mixture simulating casein and a tryptic hydrolysate of casein in man, *Clin. Sci. Mol. Med.* **45:**715.

Silk, D. B. A., Perrett, D., Webb, J. P. W., and Clark, M. L., 1974a, Absorption of two tripeptides by the human small intestine: A study using a perfusion technique, *Clin. Sci. Mol. Med.* **46:**393.

Silk, D. B. A., Webb, J. P. W., Lane, A. E., Clark, M. L., and Dawson, A. M., 1974b, Functional differentiation of human jejunum and ileum: A comparison of the handling of glucose, peptides and amino acids, *Gut* **15:**444.

Silk, D. B. A., Perrett, D., and Clark, M. L., 1975, Jejunal and ileal absorption of dibasic amino acids and an arginine-containing dipeptide in cystinuria, *Gastroenterology* **68:**1426.

Simmonds, H. A., Panayi, G. S., Corrigall, V., 1978, A role for purine metabolism in the immune response: Adenosine-deaminase activity and deoxyadenosine catabolism, *Lancet* **1:**60.

Snyder, W. S., Cook, M. J., Nasset, E. S., Karhausen, L. R., Howells, G. P., and Tipton, I. H., 1975, Physiological data for reference man, in: *Report of the Task Group on Reference Man*, pp. 335–442, International Commission on Radiological Protection, Report No. 23, Pergamon Press, New York.

Stein, G. S., Spelsberg, T. C., and Kleinsmith, L. J., 1974, Nonhistone chromosomal proteins and gene regulation, *Science* **183**:817.

Steinert, P. M., Baliga, B. S., and Munro, H. N., 1974, Available sulphydryl groups of mammalian ribosomes in different functional states, *J. Mol. Biol.* **88**:895.

Stenflo, J., and Ganrot, P. O., 1972, Vitamin K and the biosynthesis of prothrombin. I. Identification and purification of a dicoumarol-induced abnormal prothrombin from bovine plasma, *J. Biol. Chem.* **247**:8160.

Sullivan, M. X., Hess, W. C., and Sebrell, W. H., 1932, Studies on the biochemistry of sulphur. XII. Preliminary studies on amino-acid toxicity and amino acid balance, *Public Health Rep.* **47**:75.

Tarlow, M. J., Seakins, J. W. T., Lloyd, J. K., Matthews, D.M., Cheng, B., and Thomas, A. J., 1972, Absorption of amino acids and peptides in a child with a variant of Hartnup disease and coexistent coeliac disease, *Arch. Dis. Child.* **47**:798.

Tattersall, M. H. N., Ganeshaguru, K., and Hoffbrand, A. V., 1975, The effect of external deoxyribonucleotides on deoxyribonucleoside triphosphate concentrations in human lymphocytes, *Biochem. Pharmacol.* **24**:1495.

Thier, S. O., Segal, S., Fox, M., Blair, A., and Rosenberg, L. E., 1965, Cystinuria: Defective intestinal transport of dibasic amino acids and cystine, *J. Clin. Invest.* **44**:442.

Townes, P. L., 1965, Trypsinogen deficiency disease, *J. Pediatr.* **66**:275.

Ullman, B., Cohen, A., and Martin, D. W., Jr., 1976, Characterization of a cell culture model for the study of adenosine deaminase- and purine nucleoside phosphorylase-deficient immunologic disease, *Cell* **9**:205.

Vaughan, M. H., Jr., 1972, Comparison of regulation of synthesis and of 45 S ribosomal precursor RNA in diploid and heteroploid human cells in response to valine deprivation, *Exp. Cell Res.* **75**:23.

Waterlow, J. C., 1969, The assessment of protein nutrition and metabolism in the whole animal, with special reference to man, in: *Mammalian Protein Metabolism*, Vol. 3 (H. N. Munro, ed.), pp. 325–390, Academic Press, New York.

Waterlow, J. C., and Stephen, J. M. L., 1968, The effect of low protein diets on the turnover rates of serum, liver and muscle proteins in the rat, measured by continuous infusion of L-[U-^{14}C]lysine, *Clin. Sci.* **35**:287.

Wiggans, D. S., and Johnston, J. M., 1959, The absorption of peptides, *Biochim. Biophys. Acta* **32**:69.

Wilson, T. H., 1962, Amino acids, in: *Intestinal Absorption* (T. H. Wilson, ed.), pp. 110–133, W. B. Saunders, Philadelphia.

Wolberg, G., Zimmerman, T. P., Hiemstra, K., Winston, M., and Chu, L. C., 1975, Adenosine inhibition of lymphocyte-mediated cytolysis: Possible role of cyclic adenosine monophosphate, *Science* **187**:957.

Wunner, W. H., Bell, J., and Munro, H. N., 1966, The effect of feeding with a tryptophan-free amino acid mixture on rat liver polysomes and ribosomal ribonucleic acid, *Biochem. J.* **101**:417.

Wyngaarden, J. B., and Ashton, D. M., 1959, The regulation of activity of phosphoribosylpyrophosphate amidotransferase by purine ribonucleotides: A potential feedback control of purine biosynthesis, *J. Biol. Chem.* **234**:1492.

Young, V. R., Alexis, S. D., Baliga, B. S., Munro, H. N., and Muecke, W., 1972, Metabolism of administered nucleic acid charging and quantitative excretion as 3-methylhistidine and its *N*-acetyl derivative, *J. Biol. Chem.* **247**:3592.

Young, V. R., Munro, H. N., and Scrimshaw, N. S., 1976, Muscle and whole body protein metabolism in aging, with special reference to man, in: *Special Review of Experimental Aging Research,* (M. F. Elias, B. E. Eleftherious, and P. K. Elias, eds.), p. 19, EAR, Bar Harbor, Me.

Young, V. R., Scrimshaw, N. S., and Uauy, R., 1977, Changes in protein metabolism with age and protein intake: A brief review, in: *New Developments in Pediatric Research*, Vol. 1 (O. P. Ghai, ed.), pp. 67–80, Interprint, New Delhi, India.

8

Nutrients with Special Functions: Essential Fatty Acids

James F. Mead

1. What Are the Essential Fatty Acids?

The answer to the question in the title of this section goes beyond a simple listing of those acids with essential fatty acid (EFA) activity. Burr and Burr (1930), in their original account of the discovery of these substances, defined them as those fatty acids that could prevent or cure the deficiency symptoms seen in young rats fed a fat-free diet, and they considered three fatty acids to be essential:linoleic, linolenic, and arachidonic. We now know that almost all animals, including humans, can be made to show EFA deficiency symptoms and that a great many polyunsaturated fatty acids (PUFAs) have some EFA activity. Even if we extend the definition to encompass this information, it may still not be adequate. In their review of the subject, Le Breton and Ferret (1960) pointed out that such a definition has two requirements: first, that the substances fulfill some vital function and, second, that they cannot be synthesized from simple precursors. They preferred the term "acides gras indispensables" for these substances. It is now too late to change the terminology, but it may be profitable to consider the meaning of their definition as it applies to our present knowledge of the EFAs.

That the EFAs are required by all animals seems virtually certain. The reasons for this requirement are not entirely clear, but there is ample evidence for vital functions of several types, as will be discussed later in this chapter. The second part of the definition is not nearly so straightforward. First, it has been shown that if any member of a family of PUFAs is supplied, it may serve as an EFA itself or as a precursor from which the true EFA can be synthesized. Thus, dietary linoleic acid may fulfill a necessary function, or it may serve as

James F. Mead • Department of Biological Chemistry, Schools of Medicine and of Public Health and Laboratory of Nuclear Medicine and Radiation Biology, University of California, Los Angeles, California 90024.

a precursor to arachidonic acid, or both. In any event, it is uncertain, at our present state of knowledge, which of the fatty acids are actual dietary requirements and which ultimately carry out the vitamin-like functions ascribed to the class. Linoleic and linolenic acids are synthesized only in plants, while most of the higher PUFAs are derived from them in animals. By this definition, then, linoleic acid and possibly linolenic acid would be the true naturally occurring EFAs. The many desaturation and elongation products derived from them could be considered in the same light as the cofactors derived from vitamins—they may be the true functional EFAs, but since they can be synthesized from certain dietary precursors, they may not quite fulfill the requirements of the definition. Perhaps the best definition is still an extension of the original—any fatty acid that, in the diet, can prevent or cure the symptoms of fat deficiency in appropriate experimental animals.

There is still always the nagging doubt that under certain conditions, some animals can indeed carry out a limited synthesis of linoleic acid. It is very difficult to isolate linoleic acid absolutely tracer-free from an animal given high-activity labeled acetate or other possible precursors. It is generally conceded, however, that such activity will be found exclusively in the carboxyl group (if carboxyl-labeled acetate were given) and that it is the result of a degradation–elongation process in which the carboxyl and α carbons are removed by β-oxidation and replaced by an elongation process, probably involving acetyl coenzyme A (CoA).

2. How Do We Name the Essential Fatty Acids?

In order to discuss any subject, a simple, universal means of communication is required. In the case of the EFAs, it could be a handicap to the discussion to have to refer orally to "all-*cis*-4,7,10,13,16,19-docosahexaenoic acid," and the mere writing of it uses an inordinate amount of space. On the other hand, hardly any scientists in other fields and very few in the lipid area would find "clupanodonic acid" a very familiar term. Actually, as a compromise between these extremes, the trivial names are usually used when they are reasonably short and universally familiar (e.g., linoleic acid). For an unfamiliar fatty acid, in discussions in which the communication of a precise structure is important, the formal descriptive name must still be used. However, in most written and a great deal of oral communication, a shorthand terminology has come into general use. This consists of designating the number of carbon atoms followed by a colon and the number of double bonds. For example, linoleic acid is an 18:2, while clupanodonic acid is a 22:6.

The location of the double bonds (considered to be *cis* unless designated *trans*) is given by two general conventions, depending on which structural features are important to the discussion. Numbering the double bonds from the carboxyl group shows their relationships to that end of the molecule. Thus, 8,11,14–20:3 can be distinguished from 5,11,14–20:3. On the other hand, when the important structural feature of the fatty acid is its biosynthetic family or its precursors, then the double bonds are numbered from the last, or ω, carbon, and it is assumed that all double bonds are in the 1:4 relationship. Thus, 18:2 ω6 or 18:2 (*n*-6) adequately defines linoleic acid, and the relationship of bis(homo)γ-linolenic acid to it is shown by the shorthand 20:3 (*n*-6), another member of the linoleic, or *n*-6, family.*

In this way, it is possible to discuss, either orally or in writing, a wide variety of PUFAs with little chance for misunderstanding and with considerable economy of space and time. Where unforeseen structural features become important, they, too, can be handled, and, at least among those in the field, communication does not present a problem.

3. What Structures Confer Essential Fatty Acid Activity?

Burr and Burr's (1930) original list of three fatty acids—linoleic, linolenic, and arachidonic—has been expanded and modified. All of the lower members of the linoleic, or *n*-6, family—18:2 (*n*-6), 18:3 (*n*-6) (γ-linolenic), and 20:3 (*n*-6) [bis(homo)γ-linolenic]—are very active EFAs, possibly by virtue of their conversion to the most active member, 20:4 (*n*-6) (arachidonic). This may also be the reason for the activity of 22:4 (*n*-6), which can be converted to arachidonic acid by the degradative process mentioned above, termed retroconversion (Schlenk *et al.*, 1969).

The linolenic, or *n*-3, family has never been found to be nutritionally quite the equivalent of the *n*-6 family. In young rats, linolenic acid promotes normal growth but neither prevents nor cures the various skin symptoms associated with EFA deficiency, and experiments in which it is effectively eliminated from a diet containing adequate amounts of linoleic acid have not shown any linolenate requirement for rats (Tinoco *et al.*, 1971). However, information persists that the *n*-3 family has some special function in brain. Certainly, the *n*-3 family is present in greater proportion in the brain phospholipids than in the diet, and 22:6 (*n*-3) is one of the prominent brain fatty acids. There is also some indication that mental processes are in some way dependent on an adequate supply of linolenate (Crawford and Sinclair, 1972) and that it aids in the learning process (Lamptey and Walter, 1976). Beyond its possible function in brain, linolenic acid has also been reported to be the chief EFA in fish (Yu and Sinnhuber, 1972).

The members of the *n*-9, or oleic, family probably have little EFA function, since they are produced in large quantities in EFA deficiency, particularly 20:3 (*n*-9). However, two families of odd-chain fatty acids have been shown by Schlenk and Sand (1967) to have EFA activity. These are the odd-chain *n*-5 and *n*-7 families. Possibly, the structural requirements for full EFA activity are about 20 carbons (19, 20, or 21) and four double bonds in the center of the chain or those fatty acids that are metabolic precursors of these products.

---

* The presently accepted convention for numbering the terminal end of the chain is (*n*-*x*), and it will be used throughout this chapter.

Thus, the most active EFAs of their respective families are 20:4 (*n*-6), 19:4 (*n*-5), and 21:4 (*n*-7). It has also been reported recently that two isomers of arachidonic acid, 20:4 (*n*-6,9,12,16), or 4,8,11,14–20:4, and 20:4 (*n*-2,6,9,12), or 8,11,14,18–20:4, are active in preventing excessive mitochondrial swelling, although not for the usual symptoms of EFA deficiency (Houtsmuller, 1973). Thus, it appears that a completely methylene-interrupted series of double bonds may not be required for all of the functions of the EFAs.

Possible reasons for the importance of these structures are discussed below.

4. How Are the Essential Fatty Acids Obtained?

Actually, it would be difficult for any normal animal on its normal diet to suffer from EFA deficiency. In the case of vegetarians, there is little chance for an inadequate supply of linoleic acid. This acid is formed in the plant by oxidative desaturation of oleic acid and is an integral component of the triglycerides of pasture grasses (about 20%) and many seed oils (about 50% in corn and soy oils). Linolenic acid is often higher in grasses and may be the major fatty acid in some seed oils.

For the carnivore, not only are the vegetable sources directly available from occasional vegetable ingestion, but both precursors 18:2 and 18:3 and their many elongation desaturation products are abundantly present in the phospholipids of most animal tissues. For example, the herbivore, on its vegetable diet, obtains adequate 18:2 and 18:3 and, in turn, deposits these as well as 20:4, 22:6, etc. Thus, the carnivore has a ready dietary supply of both precursors and products (Crawford and Sinclair, 1972) and, in some cases, appears to have lost the ability to convert the precursors to the final products (Rivers *et al.*, 1976).

The fish and other aquatic animals represent a somewhat special case. As a consequence of the usually lower temperature of their environment, most aquatic animals tend to deposit highly unsaturated fatty acids. The resulting membrane phospholipids thus have gel-to-liquid-crystal transition temperatures low enough so that they can remain fluid at the lowest temperature to which the animal is likely to be subjected. The type of unsaturated fatty acids used for this purpose depends on the ultimate food source. For this reason, fresh water fish have a tendency to deposit the more highly unsaturated members of the linoleate, or *n*-6, family, the precursors of which are prominent in freshwater algae, whereas marine fish deposit very highly unsaturated long-chain members of the linolenate, or *n*-3, family after deriving the precursors from the saltwater plankton. An illustration both of the progress of the fatty acids through the food chain and of the effect of temperature is seen in Table I, which gives results of an experiment by Kayama *et al.* (1963), who used the diatom *Chaetoceros* as the sole diet for the brine shrimp, *Artemia salina*, which, in turn, was the sole dietary source for the guppy, *Lebistes reticulatus*. It can be seen that the fatty acids of the diatom were partially deposited

Table I. Fatty Acid Composition of Chaetoceros, Artemia salina, and Lebistes reticulatus Oils[a,b]

Fatty acid	Chaetoceros	Artemia salina	Lebistes reticulatus 17 ± 1°C	24 ± 1.5°C
Shorter chain	Trace	0.4	Trace	
12:0	0.4	Trace	0.2	Trace
13:0	0.7	Trace	Trace	Trace
14:0	13.0	4.8	1.5	0.9
15:0	1.8	1.5	Trace	0.2
14:2	0.6	Trace	0.6	0.5
16:0	18.1	11.6	22.9	36.0
16:1	47.9	44.9	15.9	8.9
16:2	2.7	Trace	0.2	0.2
16:3(?)	4.0	1.7		0.6
16:4(?)	Trace			0.5
18:0	0.5	1.9	8.2	9.8
18:1	8.7	18.4	18.3	15.0
18:2	1.7	0.7	Trace	Trace
18:3	Trace	0.5	1.4	0.8
20:1		0.9		
18:4 and 20:2		0.8	0.3	Trace
20:3			0.2	Trace
20:4		Trace	2.0	2.0
20:5		12.0	4.8	4.6
22:4			1.3	1.0
22:5			6.1	7.3
22:6			16.5	11.5

[a] Weight percentage of total ester.
[b] From Kayama et al. (1963).

unchanged in the crustacean and partially desaturated and elongated before deposition and that the traces of PUFAs in the diatom were increased in the crustacean and again in the guppy, particularly when these fish were raised at a lower temperature (17°C as compared to 25°C).

The experiment serves to emphasize the idea that animals tend to deposit a type of fat that is best suited to their metabolism and environment but that this tendency can be overwhelmed by the fatty acids composing the fats of a high-fat diet, particularly since this diet partially inhibits fatty acid biosynthesis. Nevertheless, the ability of the animal to modify ingested fatty acids remains high and may be the chief factor in the final composition.

5. How Are the Essential Fatty Acids Synthesized and Altered?

Of course, the ultimate source of the precursors of the EFAs in the animal body is their biosynthesis in the green plants. This is accomplished by desaturation of stearylacyl carrier protein (ACP) to oleyl-ACP, since the CoA derivative cannot be desaturated in the plant. An additional double bond can be introduced at the 12 position of oleate (as the CoA derivative or, more often, in a phospholipid) to form linoleate. The introduction of the 15 double bond, however, may not proceed directly, and it has been reported by Jacobson *et al.* (1973) that α-linolenate is formed by elongation of 7,10,13-hexadecatrienoic acid and 5,8,11-tetradecatrienoic acid, which can be formed by direct desaturation of the shorter-chain saturated fatty acids.

In the animal, the precursor fatty acids, both dietary [18:2 (*n*-6) and 18:3 (*n*-3)] and synthetic [18:1 (*n*-9) and 16:1 (*n*-7], are further transformed by a series of elongation and desaturation reactions giving rise to four families of fatty acids that are not interconvertible in the animal body. Figure 1 illustrates the main steps and major products of these pathways (Mead and Fulco, 1976). Using the *n*-6, or linoleate, family as an example, it can be seen that the elongation and desaturation steps alternate with ultimate formation of the major product, arachidonic acid [20:4 (*n*-6)].

It was thought for some time that an alternative pathway exists and, indeed, could be the major one for some species. This was the pathway involving, in the case of linoleate, the formation of 20:2 (*n*-6) and its conversion to 8,11,14-20:3 and thence to arachidonic acid. Evidence for this idea rests largely on the finding of 11,14-20:2 in tissues and in subcellular preparations metabolizing linoleate and the inability to find 18:3 (*n*-6) under most circumstances. However, Ullman and Sprecher (1971) have shown that 11,14-20:2, both *in vivo* and *in vitro*, is actually converted to 5,11,14-20:3, which does not undergo further desaturation to arachidonic acid. Since similar reactions were found for the other PUFA families, it was concluded that there is no $\Delta 8$ desaturase and that a double bond in the 8 position arises from a fatty acid two carbons shorter, with a double bond in the 6 position (see Fig. 1).

With this information, we can begin to understand the factors that control the PUFA composition of animal tissues. The first consideration is the availability of the dietary and synthetic precursors and the affinity of the enzyme systems for them. It was obvious very early that the same enzymes would be involved with different substrates in certain steps of the pathways and that there would be competition among the substrates for the enzymes. This conclusion was illustrated by the experiments of Dhopeshwarkar and Mead (1961), in which it was found that feeding large amounts of oleate to guinea pigs on a diet marginal in linoleate precipitated symptoms of EFA deficiency. Evidently, at these levels, the large amounts of oleate were competing effectively with the linoleate for one of the enzymes in the pathway. This information has been extended for the *in vivo* situation by Holman (1964, Fig. 4) and his collaborators and for the *in vitro* system by Brenner (1971) and his co-workers using a microsomal preparation in the presence of oxygen and reduced pyridine nucleotide. For the $\Delta 6$ desaturase, which may be the rate-controlling step of the series, greatest affinity for the enzyme appears to be conferred by the greatest number of double bonds in the C_{18} substrate. Thus, 18:3 (*n*-3) is desaturated at the highest rate, followed by 18:2 (*n*-6) and 18:1 (*n*-9). In the presence of either of the two dietary acids, little desaturation of oleate occurs.

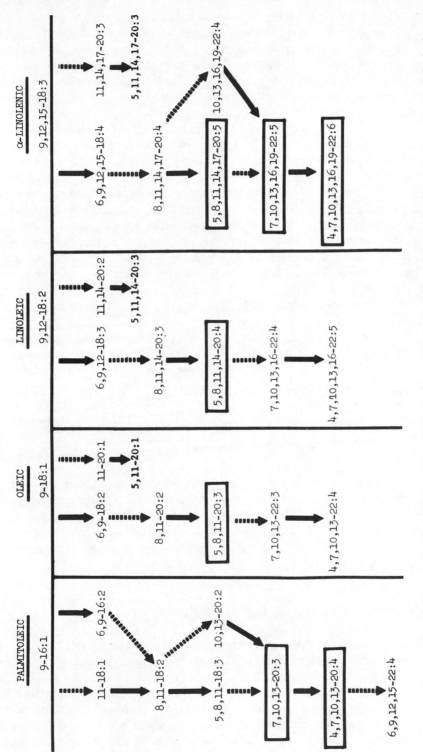

Fig. 1. Pathways of transformation of the major families of PUFAs in higher animals. The solid arrows indicate desaturation steps and the broken arrows, elongation steps. Fatty acids enclosed in boxes represent the major products of each pathway in terms of concentrations in mammalian tissues. From Mead and Fulco (1976).

Linolenate also effectively inhibits the desaturation of linoleate. In the absence of members of the *n*-3 or *n*-6 families, however, oleate is desaturated, and the members of the *n*-9 family, particularly 20:3 (*n*-9), appear in the tissues. This explains the well-documented finding that in EFA deficiency, a trienoic acid [largely 20:3 (*n*-9)] increases dramatically in the tissues and, with reinstatement of a normal diet, as dramatically decreases. There is also competition for the Δ5 desaturase, the chain-elongation enzymes, and for the acyltransferases involved in formation of the phospholipids. Thus, the dietary fatty acids are important regulators of tissue fatty acid composition not only by virtue of their function as precursors but also because of their ability to control the alteration and incorporation of fatty acids of other families. Moreover, it also seems probable that a lower member of a family may be able to compete with some of its products for enzyme sites and limit the extension of its own family.

A further limitation to the indefinite extension of a family comes from the phenomenon termed retroconversion. In this reaction, highly unsaturated long-chain fatty acids, in particular, are shortened by two carbons and, in some cases, hydrogenated. Thus, 7,10,13,16-docosatetraenoate is converted to arachidonate (Ayala *et al.*, 1973), and 4,7,10,13,16,19-22:6 is converted to 7,10,13,16,19-22:5 and, presumably, 22:4 (*n*-6) to arachidonate (Schlenk *et al.*, 1969), both by testicular enzymes. The effect of this reaction is to make some of the steps in the transformation of the longer-chain, more highly unsaturated fatty acids reversible, thus limiting the pathway and increasing the steady-state concentrations of the lower members. In the case of the *n*-6 family, the pathway can be rewritten (Sprecher, 1975):

$$18:2 \ (n\text{-}6) \xrightarrow{\text{Fast}} 18:3 \ (n\text{-}6) \xrightarrow{\text{Very fast}} 20:3 \ (n\text{-}6) \xrightarrow{\text{Fast}} 20:4 \ (n\text{-}6) \rightleftarrows 22:4 \ (n\text{-}6)$$

$$\xrightarrow{\text{Slow}} 20:2 \ (n\text{-}6) \xrightarrow{\text{Slow}} 5,11,14\text{-}20:3$$

We can see from this discussion that the composition of the tissue PUFA depends on the availability of precursor fatty acids, the presence of competitors for the various steps, the occurrence of "dead-end" steps, and the limitation of pathways by retroconversion. When the effects of temperature and hormonal control as well as various dietary effects are included, it is apparent that this is a very complex regulatory process indeed.

6. How Is Essential Fatty Acid Deficiency Established?

It was started above that it is very unlikely that a normal animal consuming its normal diet could suffer from EFA deficiency, and, in this sense, it is largely an experimental disease.

It has been found, through many years of experience, that the earlier in the life of the experimental animal the deficient diet is introduced, the earlier the symptoms will appear and the more severe they will be. Thus, the greatest success in establishment of a deficiency is achieved if the pregnant mothers are introduced to the deficient diet. This cannot be done at too early a stage, however, or the young will be born dead or, in the case of the rat, resorbed.

The composition of the diet is of prime importance, since it must have as little EFA as possible. The only practical way of ensuring that his will be the case is to eliminate all fat from the diet. This is no easy task, since most dietary sources of carbohydrate and protein contain sufficient EFA to prevent or ameliorate the disease. For example, in the original studies of Burr and his colleagues (see Burr and Burr, 1930), it was the substitution of purified sucrose for starch (which contains 0.5% lipid) that permitted him to discover the disease. Once a truly fat-free diet has been formulated, care must be taken to replace those necessary components that are usually associated with fat, such as the fat-soluble vitamins. Other lipid components may than be added, as in the case of saturated long-chain fatty acids (Alfin-Slater *et al.,* 1965; Deuel *et al.,* 1955; Evans and Lepkovsky, 1932) and cholesterol (Holman and Peifer, 1960; Mohrhauer and Holman, 1963) which accelerate and exacerbate the disease. A typical fat-free diet for rats is shown in Table II. Whether or not the mothers are fed the fat-free diet, the earlier the young are weaned to it, the sooner the symptoms appear. This introduces some problems, in that many young animals refuse to eat an artificial diet of this sort, even if weaned at the usual time. One partial solution to this problem is to compact the powdered mixture with the aid of guar gum or a similar substance, thus making it more acceptable to the young animals. Of course, the acceptability of the diet is also dependent on the animal involved, and, in some cases (e.g., the gerbil), it is difficult to compound a diet that in any way resembles the natural food. In any event, as the fat-free diet is introduced at later times in the animal's development, the onset of deficiency symptoms occurs later, until eventually no symptoms are apparent at all. In the case of the mouse, a state of "chronic EFA deficiency" can be produced by starting the diet during the postweaning period near the end of rapid growth. These animals appear normal and remain

Table II. Typical Fat-Free Diet for Rats[a]

Vitamin-free casein (Nutritional Biochemicals)	300 g (or 180 g)
Cellulose (Nutritional Biochemicals)	40 g
Wesson salt mixture	40 g
Sucrose	600 g (or 720 g)
Vitamin mixture in casein[b]	10 g
Choline chloride in casein[c]	10 g
TOTAL	1000 g
Ether solution of vitamins A and E[d]	100 ml/kg

[a] Mohrhauer and Holman (1963).
[b] Vitamin D_2, 200 mg; 2-methylnaphthoquinone, 100 mg; thiamine hydrochloride, 1500 mg; riboflavin, 1500 mg; pyridoxine hydrochloride, 400 mg; calcium pantothenate, 5000 mg; nicotinamide, 5000 mg; inositol, 11,000 mg; *p*-aminobenzoic acid, 3750 mg; folic acid, 50 mg; biotin, 10 mg; vitamin B_{12}, 2.5 mg; casein to make, 500 g.
[c] Choline chloride, 100 g, made up to 1000 g with casein.
[d] Vitamin A acetate, 40 mg; α-tocopherol, 2.80 g; dissolve in 1 liter ether and keep refrigerated.

free of symptoms but if injured or otherwise provoked into some form of cellular proliferation, they immediately show overt signs of EFA deficiency, such as dermatitis and inability to repair wounds (Decker *et al.*, 1950).

If the fat-deficient diet is introduced after cessation of the growth period, no deficiency disease results, and, in general, it can be stated that adult animals will not suffer from this deficiency. Obviously, the stores of EFA laid down during the normal dietary regimen are sufficient to last throughout the lifetime of the animal unless some unusual circumstance initiates rapid growth (however, see below). One means by which adult rats have been made to show deficiency symptoms has been to starve them until they have lost about half their weight and then feed them *ad libitum* with a fat-free diet (Barkie *et al.*, 1947). Of course, this does not constitute an exception to the general rule that an exogenous source of EFA is required only as a result of local or general rapid growth.

An additional point to be considered in establishment of the fat deficiency disease is the sex of the animal. In the case of rats, and possibly also with other experimental animals, the males seem to be much more susceptible to this deficiency and to have a higher requirement of EFA (Greenberg *et al.*, 1950). Possibly, this is, in part, a consequence of the continued growth of the male rat in contrast to that of the female, which reaches a weight plateau, and, in most cases, male experimental animals are routinely used because of the well-known differences in lipid metabolism between males and females of most species.

7. How Do We Recognize Essential Fatty Acid Deficiency in Experimental Animals?

Since most of the critical work on EFA deficiency has been done with the rat, much more is known about the symptoms of the disease in this animal. However, most animals have been shown to be susceptible to the deficiency if it is brought about in the proper manner. This is true for mice (White *et al.*, 1943), guinea pigs (Reid, 1954), hamsters (Christensen and Dam, 1952), rabbits (Holman and Peifer, 1955), dogs (Hansen and Wiese, 1954), pigs (Hill *et al.*, 1957), and even calves (Lambert *et al.*, 1954). The cases for chickens (Bieri *et al.*, 1956) and fish (Castell *et al.*, 1972) are somewhat unusual, but these animals, as well as monkeys (Greenberg and Wheeler, 1966), appear to be no exceptions. The human case will be considered below.

The symptoms originally described by Burr and Burr (1930) for rats have been confirmed many times for these animals and are at least similar enough to those of similar animals to facilitate recognition of the disease when the circumstances of its occurrence are known. These include decreased growth rate, occurrence of a growth plateau considerably below normal (in male rats, about 70% of normal) and a severe dermatitis that appears at about the time of growth cessation, involving particularly the parts of the body usually exposed to some abrasion, such as the soles of the feet and the area around the mouth.

The coat becomes rough, and there is usually some loss of hair in areas affected by dermatitis, such as the interscapular area in the rat. In this animal, also, the tail assumes a scaly appearance and may become necrotic, particularly in very young animals in which the deficiency was established at an early age. These dermal symptoms are not always typical and depend not only on the experimental animal but also on the humidity prevailing in the laboratory in which the experiments are being carried out. One very characteristic symptom is a derangement of water balance, resulting in increased water loss and a very apparent increased consumption. This increased water requirement was found by Basnayake and Sinclair (1956) to be the result of loss through the skin and has been used by Thomasson (1962) as a means of establishing an EFA deficiency in a much shorter time than usual by limiting water consumption. This means has formed the basis of a more rapid bioassay for EFAs.

It is possible to divide the many symptoms of EFA deficiency into relatively few general categories. Thus, the skin symptoms and water loss are both examples of a general derangement of membrane structure and consequent function. Actually, most of the easily observable signs of the deficiency are in this class. There is an increase in capillary permeability and fragility (Kramer and Levine, 1953). The erythrocytes become more fragile and susceptible to osmotic hemolysis (MacMillan and Sinclair, 1958), and mitochondria show abnormal swelling (Hayashida and Portman 1960; Houtsmuller, 1975). It is probable that a loss of selective permeability and deranged structure of the mitochondrial membranes are also responsible for the increased metabolic rate first reported by Wesson and Burr (1931) for fat-deficient rats and the decreased oxidative phosphorylation and respiratory control found by Gerschenson *et al.* (1967) in HeLa cells in a fat-free medium. In all cases, linoleate or arachidonate restored the functions almost to normal.

A second class of deficiency symptoms appears to involve the lipid transport process. It has been demonstrated in many laboratories that fat deficiency leads to a fatty liver (Alfin-Slater *et al.*, 1954) with accumulation of triglyceride and, with cholesterol feeding, cholesteryl esters, particularly in male rats (Nørby, 1965). The mechanism of this accumulation is not entirely clear, but several studies have led to the idea that it is the result of two different, though related, factors. The first has to do with the transport of cholesteryl esters out of the liver via the blood. Since the composition of the cholesteryl esters accumulating in the liver does not seem to have a direct relationship to the dietary fatty acids (Klein, 1958), it seems probable that the delay in removal via the blood is not because of lack of the proper transportable cholesteryl esters (such as the arachidonate). Rather, it may well result from a breakdown in the transport process by failure to form adequate amounts of high-density lipoprotein (HDL), the chief lipoprotein involved in transport of cholesterol out of the tissues (Miller and Miller, 1975). The nascent HDLs, largely cholesterol–phosphatidyl choline bilayers, depend for their formation on the availability of the proper precursor molecules, namely, protein and phosphatidyl choline, which contains the unsaturated acyl group in the 2 position of the glyceride. Thus, a deficiency of PUFAs would result in a decreased rate of synthesis of phosphatidyl choline and, as a result, of HDL, in much the same way as does a deficiency of choline. The explanation may not be exactly the same for the rat, in which the cholesteryl esters are synthesized largely in the liver (Swell and Law, 1966), and for humans, in whom the major controlling step is the transfer of an acyl group from the 2 position of phosphatidyl choline to the 3-hydroxyl of cholesterol in the serum by the enzyme lecithin-cholesterol acyltransferase (LCAT) (Glomset, 1968). However, the evidence that the HDLs are largely involved in this process appears most reasonable.

The second type of information bearing on the removal of cholesterol from the liver involves the catabolism of cholesterol within the liver and the excretion of the resulting bile acids into the intestine via the bile. Again, the HDLs may be involved in accepting cholesterol from other lipoproteins and, either directly or via the low-density lipoproteins (LDLs), from the tissues. However, the most likely mode of action of the PUFAs in the process is in the formation of a cholesterol–phosphatidyl choline–bile acid complex with fairly definite composition and properties (Small *et al.*, 1966), and it is this complex, rather than the bile acids alone, that is excreted in the bile. In fact, so demanding is the formation of the complex during cholesterol excretion that cholesterol feeding causes marginally choline-deficient rats to become acutely deficient (Raulin *et al.*, 1959). Presumably, the phosphatidyl choline involved in this process must contain its usual complement of PUFAs in position 2, and it has been repeatedly reported that cholesterol feeding also exacerbates the symptoms of EFA deficiency (Aaes-Jørgensen and Holman, 1950).

An enormous literature exists documenting the effect of dietary PUFAs on plasma cholesterol. It was first thought that an abnormally high plasma cholesterol in humans might be a symptom of EFA deficiency (Sinclair, 1956), but it soon became evident that the cholesterol-lowering property is shared by the highly unsaturated fatty acids of all the families (Peifer, 1966). The mechanism of this effect may be different in different species, but most of the presently available evidence points to an increase in the biliary excretion of both neutral sterols and bile acids as a result of increased dietary PUFAs (Lewis *et al.*, 1961). Again, the most likely explanation for this effect depends on cholesterol transport to the liver as HDL, but this may not be the only explanation. It has been reported by Maruyama (1965) that the unsaturated cholesteryl esters of the liver are the major precursors of the bile acids, but this, too, could be considered a transport phenomenon.

A third category of deficiency symptoms is not nearly so well defined as the first two and may be discerned only by those who are already aware of its existence. This stems from the function of certain PUFAs as precursors of prostaglandins. The fatty acids involved are, in general, the 20-carbon members of the n-6 family with three and four double bonds and one member of the n-3 family with five double bonds, 5,8,11,14,17-20:5. A few other fatty acids, such as 5,8,11,14-19:4 and 5,8,11,14-21:4, may also fill this function. The 5,8,11 double-bond system appears to be necessary, and it might be recalled that these same fatty acids have been found to have some EFA activity (Beerthuis *et al.*, 1968).

In fact, the structures of fatty acids active as EFAs in membrane formation and those active as prostaglandin precursors may be fortuitously similar. For the proper viscosity characteristics of the membrane bilayer phospholipids (see above), the most effective fatty acid structure appears to involve a chain length of 19, 20, or 21 carbons with four methylene-interrupted double bonds as close to the center as possible. For the prostaglandin precursors, again, a chain length of 19–21 carbons is necessary, and, in addition, regardless of the total number of double bonds, there must be three in the 5, 8, and 11 positions. The conversion of such a structure to the least unsaturated prostaglandins and thromboxanes is shown in Fig. 2.

EFA deficiency in rats has been found to produce an abnormally low tendency of platelet adhesion and aggregation, a symptom undoubtedly connected to the function of EFAs as prostaglandin or thromboxane precursors (Hamberg *et al.,* 1975; Hornstra, 1974). Alteration of the aggregation time by dietary EFAs, however, is not a simple process, since PGE_2 (or thromboxane A_2) derived from arachidonic acid is a potent stimulator of platelet aggregation, whereas PGE_1 (or thromboxane A_1) and prostacyclin, formed in the vascular endothelium, are antagonistic to this action (Marcus, 1978).

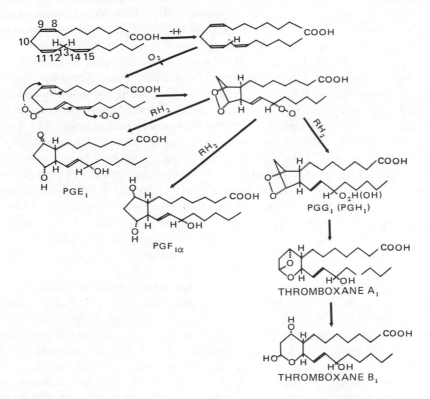

Fig. 2. Mechanism of conversion of 8,11,14-eicosatrienoic acid into PGE, $PGF_{1\alpha}$, PGG_1, PGH_1 and the thromboxanes A_1 and B_1. From Hamberg *et al.* (1975).

Vergroesen and his colleagues (Ten Hoor *et al.,* 1973) have shown that an increase in dietary linoleic acid leads to an increased coronary flow and left-ventricular work in isolated perfused rat hearts. Again, this function appears to stem from the role of linoleic acid as a prostaglandin precursor. Undoubtedly, more of the symptoms of EFA deficiency will be found to have a prostaglandin component, and additional unsuspected symptoms of this type will be found. In general, however, attempts to cure the usual EFA deficiency symptoms with prostaglandins have not been successful (Kupiecki *et al.,* 1968). This may, of course, be due more to the complex nature of prostaglandin interactions and our incomplete understanding of them than to a lack of function of prostaglandins as active metabolic derivatives of the EFAs.

8. What About Deficiency Symptoms in Humans?

Actually, there has never been much doubt among those in the field that human subjects would be as susceptible to EFA deficiency as any other species, but the ability to carry out the necessary experiments has been limited. In the first place, considering the difficulty of producing the deficiency in the adult rat, it was not realistic to suppose that it could be produced in the human adult. This was borne out by an experiment in which W. R. Brown used himself as the experimental subject but, after several months on the low-fat diet, produced no more dramatic symptoms than a slight increase in the triene:tetraene ratio (see below) of his blood lipids (Brown *et al.,* 1938).

Either by design (Hansen *et al.,* 1963) or inadvertently (Cash and Berger, 1969), the symptoms of EFA deficiency have been produced in human infants, and they appear to conform reasonably well to those seen in other animals. The most common feature seen in the deficiency and relieved by proper administration of EFAs is a dermatitis involving dryness, desquamation, and thickening of the skin, often accompanied by unsatisfactory growth.

The human infant may be particularly susceptible to this deficiency, since the tissues of the newborn infant may already be low in EFAs (Wiese *et al.,* 1966), and a further dietary deficiency may be imposed by feeding with cow's-milk formulas, which are often low in linoleate (Holman *et al.,* 1964). If additional stress is applied, such as intestinal surgery or intravenous feeding of a fat-free formula, the deficiency may be precipitated rapidly (Paulsrud *et al.,* 1972).

These conditions are relatively rare and can be avoided or corrected, but in poorly nourished populations of developing countries, a combination of circumstances may commonly exist, leading readily to EFA deficiency. If the mothers are poorly nourished and the infants are weaned to an inadequate diet, growth is poor, but the tissue content of EFAs appears normal. When a high-protein diet (as one based on skim-milk powder) is given, rapid growth resumes, and the symptoms of EFA deficiency are precipitated (Schendel and Hansen, 1959). Perhaps the greatest tragedy of this occurrence is that even if the dietary fat deficiency is corrected, permanent brain damage may have occurred. Thus, the diet used to correct a deficiency of one nutrient must be carefully chosen with respect to other nutrients.

One of the most dramatic findings in recent years has been that it is actually possible to produce an EFA deficiency in the human adult. In several instances, (usually) elderly patients who have had a large portion of their small intestines removed and have consequently been nourished intravenously with glucose–amino acid solutions have developed skin rash that can be cured by intravenous or topically applied unsaturated oils (Collins *et al.*, 1971; Press *et al.*, 1974; Soderhjelm *et al.*, 1971).

In conclusion, it seems evident that, except for certain imposed abnormal conditions, the development of a human EFA deficiency should be a rare occurrence indeed. The requirement for EFAs is small (see below), and normal dietary sources are completely adequate in supplying it.

9. What Actually Is the Function of the Essential Fatty Acids?

This question may, in fact, have been answered in the previous section, in which it was pointed out that many of the symptoms of EFA deficiency could be divided into relatively few groups. Actually, there is some justification in dividing the functions of the PUFAs into only two general types: structural and biosynthetic. In the latter case, the formation of the prostaglandins and thromboxanes has been considered, although attempts to tie EFA deficiency to a lack of these compounds have not been notably successful. In this sense, the EFAs (or the prostaglandin precursors) could be classified as hormones. Their many diverse activities seem to be concerned with potentiation and inhibition, and, in these roles, they act in vanishingly small concentrations of a completely different order of magnitude from that usually associated with EFAs.

The structural role of the PUFAs actually includes both their membrane-related functions and their transport properties. In both cases, the physical properties of the fatty acids are involved as they reflect structure. As discussed above, important structural features are chain length, *cis* unsaturation, and branching. These features are reflected in the melting points of the fatty acids or the viscous properties of the phospholipids in the membrane bilayers and the transport lipoproteins.

Of course, a prerequisite to the structural activity of the fatty acids is that they be incorporated into the correct position of the appropriate lipid. This selective incorporation, whether during *de novo* synthesis or the restructuring process, is also structure dependent. The location of the prostaglandin precursors on the 2 position of the phosphoglycerides assures their ready availability to the prostaglandin-synthesizing enzymes via phospholipase A_2. The incorporation of the proper acids into the proper positions in the membrane phospholipids must be equally important, but here the exact relationships are not nearly so clear. The factors governing these structural specificities have been discussed by Lands (1975).

The importance of the dietary PUFAs is that they are readily converted to more highly unsaturated PUFAs, providing the greatest possible influence on the viscous properties of the total lipid mixture.

What is not well understood at present is why certain of the PUFAs, the EFAs, have all these functions, while others, as the *n*-3 family, are effective for some functions but not for others, and still others, as the *n*-9 family, do not appear to be effective at all. Presumably, the properties of the entire family are at least partially dependent on those of the principal end products.

10. What Is an Adequate Intake of Essential Fatty Acids?

This was originally a very difficult question to answer, since, first, the judging of severity of symptoms is very subjective, and, second, it is not certain that all symptoms respond equally to equal amounts of the same fatty acid. Moreover, as has been pointed out above, different fatty acids may have quantitatively and qualitatively different effects.

Most of the original attempts to quantitate the EFA requirement were based on the cure or prevention of the skin symptoms, or the maintenance of growth, or both. These tests indicated that roughly 50–100 mg/day of linoleic acid were required by male rats and 10–20 mg/day by female rats (Greenberg *et al.*, 1950). This type of assay was brought to its optimal practicability by Thomasson, who used a restricted water supply to hasten the appearance of the symptoms of EFA deficiency (Thomasson, 1962). However, although an assay of this type must obviously be used for the ultimate decision as to the EFA activity of a specific fatty acid, it is still subject to most of the difficulties and uncertainties of the older methods. Nevertheless, it has indicated that for male rats, about 30 mg/day of linoleic acid is optimal and that linoleic acid is about ten times as active in these respects as is linolenic.

The introduction of a simpler, more quantitative method of assay, the triene:tetraene ratio, by Holman and his collaborators (Holman, 1960) was greeted with enthusiasm. The exact meaning of the metabolic changes involved was not clear at the time. Nevertheless, the observation that, accompanying the appearance and exacerbation of other symptoms, there was an increase in triene fatty acids in the lipids of most active tissues and a corresponding decrease in tetraene made the ratio a particularly sensitive indicator of EFA status. The ultimate meaning of the ratio is now known to be the increase in 20:3 (*n*-9) and decrease in 20:4 (*n*-6) as a consequence of competitive inhibition among families of PUFAs for desaturases and possibly acyltransferases, as discussed above. Thus, with the use of gas–liquid chromatography (GLC) in place of the alkaline isomerization analytical method, the assay can be refined to the ratio of 20:3 (*n*-9) to 20:4 (*n*-6). This avoids any problems arising from trienoic or other less usual fatty acids in the diet and at the same time permits rapidity and accuracy. With this method, it has been found that with a number of experimental animals, a plot of [20:3 (*n*-9)]/[20:4 (*n*-6)] versus dietary 18:2 (*n*-6) as percentage of calories describes a hyperbola which can be extrapolated to reveal that for all species tested, optimal dietary linoleate (giving a ratio of less than 0.4) is 1–2% of total calories (Holman, 1960). That this dietary intake is much the same for all species is emphasized by the fact that it is found for human infants, in whom the ratio of $20:3$ (n-9) to $20:4$ (n-6) in serum gives the same type of curve as is the case for rat plasma (Holman *et al.*, 1964) (see Fig. 3).

Of course, it must be kept in mind that the ultimate test for EFA deficiency is not a ratio of fatty acids but a series of symptoms that can be prevented or cured by the proper dietary fatty acids. That the 1–2% of calories may refer only to one set of symptoms has been suggested by Vergroesen *et al.* (1975). While the symptoms attributable to the membrane functions of the EFAs appear to be adequately controlled by a dietary intake of 1–2% of calories, it has been suggested that those depending on prostaglandin formation (see above) may show an optimal response with dietary linoleate as high as 12–14% of calories.

Thus, as is often the case with dietary factors for which the mechanism of action is not entirely clear, an absolute requirement cannot yet be given, and, if the higher figure does not produce any untoward effects, the tendency will be to consider it as optimal in the light of present information.

11. What About Excess Essential Fatty Acids?

Having concluded that the minimum dietary linoleate should be approximately 1–2% of calories (about 5% of total dietary fat) and that the optimal intake may be in the neighborhood of 12–14% of calories (about 40% of total

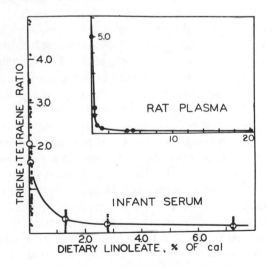

Fig. 3. Triene:tetraene ratios of infant serum and rat plasma related to dietary linoleate. From Holman *et al.* (1964).

dietary fat), it now behooves us to consider whether an upper limit for dietary EFAs should be set.

This question may actually consist of two distinct parts: Is an excess of unsaturated fatty acid or PUFA harmful per se and can PUFA be converted in the tissues to harmful substances?

The first part of the question will be impossible to answer at the present time, particularly since there is no real agreement on the optimal amount of dietary fat of any sort. Perhaps we can merely approach the problem by a consideration of possible areas of concern.

First, it has recently been shown that dietary EFAs act as depressors of some of the enzymes of fatty acid synthesis beyond the well-known feedback inhibition exhibited by all fatty acids—an inhibition dependent largely on the detergent action of the CoA derivatives. At even lower concentrations and in an EFA-replete animal, 18:2, 18:3, and 20:4 depress liver glucose-6-phosphate dehydrogenase and fatty acid synthetase (Clarke *et al.*, 1976). At present, however, it cannot be decided whether this action is beneficial or harmful. One result, of course, is an enhancement of the effect of dietary PUFAs in determining the fatty acid composition of the tissues, since with synthesis inhibited, the deposition of dietary fatty acids will be enhanced.

Second, it has been found by several investigators (Bieri and Evarts, 1975; Witting, 1974) that large amounts of dietary PUFAs increase the requirement for vitamin E and deplete the tissue stores of this vitamin even after the dietary PUFAs have decreased, since the tissue PUFAs have a longer half-life than does tocopherol. Thus, ingestion of a vegetable oil well protected by tocopherol may not confer this protection to the tissues after the oil and vitamin E ingestion have ceased. This subject, however, actually belongs in the second part of the original question.

Third, there is the possibility that, in common with their growth-enhancing effect for all cells, the PUFAs may also enhance the growth of malignant cells. This possibility has been considered by several investigators (J. F. Mead and A. B. Decker, unpublished data; Smedley-Maclean and Hume, 1941), and it seems evident that cancer cells, in common with most normal cells, require a source of EFAs for membrane formation and consequent cell proliferation. Whether this effect is enhanced by increased dietary or tissue PUFAs, however, has not been determined.

Fourth, as has been pointed out above, certain PUFAs, as precursors of prostaglandins and thromboxanes, enhance platelet aggregation and thus thrombus formation, a factor involved in the consequences of atherosclerosis. However, this action of dietary EFAs is not entirely clear, since PGE_2, a product from 20:4 (n-6), enhances aggregation, while PGE_1, from 20:3 (n-6), acts in the opposite sense.

Finally, there is a host of properties of cellular membranes that are regulated in part by the nature of the fatty acid bilayer. These might be transport into and out of the cell, membrane-bound enzyme activity, membrane fragility, osmotic properties, and many others. It cannot be said at present whether an upper limit can be placed on dietary PUFAs to prevent some untoward activity in any of these factors.

In seeking to answer the second half of the original question, we find that an overwhelming amount of information is available but that, in the end, we still cannot come to a definitive conclusion.

The mechanism of PUFA autoxidation has been the subject of much study and many reviews (Bolland and Gee, 1946; Mead, 1976; Uri, 1961), to the point where its details are now very well understood (see Fig. 4), and there seems to be little reason to doubt that, given the proper circumstances, the reaction could occur in living tissues.

That peroxidizing PUFAs could be deleterious in the diet was shown by Barnes *et al.* (1943), and Kaneda and Ishii (1954) as well as Matsuo (1961) showed that highly unsaturated fish oils (poorly protected by antioxidants) could be very toxic when fed at 5% of the weight of the diet to rats. However, it was later shown by Andrews *et al.* (1960) that the hydroperoxides formed as the principal product in such autoxidation (see Fig. 4) are not absorbed from the intestine and that the site of their toxicity must remain in that organ. Thus, the most important consideration is whether a high-PUFA-containing diet leading to a high deposition of PUFAs in the tissues can lead to peroxidation in the body and its consequences.

Of course, all the factors necessary for a peroxidative reaction are present in the tissues, particularly in the region of the cellular membranes, where an

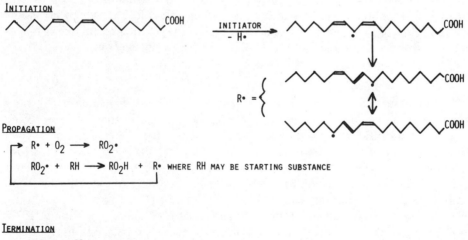

Fig. 4. Simplified outline of the reactions of autoxidation of linoleic acid. From Mead and Fulco (1976).

ordered arrangement of PUFAs in the lipid bilayer can be brought into contact with an iron-containing catalyst, oxygen, and possibly other prooxidants such as ascorbic acid (Barber, 1966). That such a reaction does not overwhelm the organism is largely due to the compartmentalization of these factors. Only in the case of tissue disruption, in which the usually separated components are brought into contact, does peroxidation ensue (Hatefi and Hanstein, 1970).

As a matter of fact, living tissues are well protected against such reactions by a number of devices other than the compartmentalization mentioned above. Chief among these protective agents are the tocopherols, which limit the extent of the radical-chain propagation reaction, particularly in the lipid bilayer of the membranes; the glutathione peroxidase system, which consists of glutathione peroxidase, NADPH glutathine reductase, and glucose-6-phosphate reductase and reduces hydroperoxides to alcohols (Chow and Tappel, 1972), and the superoxide dismutases, which catalyze the conversion of the superoxide radical ion into oxygen and hydrogen peroxide (Fridovich, 1972):

$$2\,O_2^- + 2\,H + \longrightarrow H_2O_2 + O_2$$

There are, however, numbers of prooxidants that can intrude on the normally well-protected tissues and overwhelm the protective systems to the point where a peroxidative reaction can be initiated. Among these are alcohol, which may induce peroxidation of liver mitochondrial membranes (DiLuzio, 1968); nitrogen dioxide (Thomas *et al.*, 1968) and other atmospheric prooxidants, to which we are increasingly subjected; ionizing radiation (Gerschmann *et al.*, 1954), which is inescapable; and carbon tetrachloride, which promotes peroxidation of the liver microsomal lipids after enzymatic homolytic cleavage, with formation of the $\cdot CCl_3$ radical (Slater and Sawyer, 1971):

$$CCl_4 + e^- \longrightarrow \cdot CCl_3 + Cl^-$$

Unfortunately, if we manage to escape oxidative destruction by these many external agents, our own enzyme systems may, in the long run, bring about our downfall. It is apparent that the one-electron reductions of oxygen during normal electron-transfer reactions must result in at least transitory production of free radicals. In particular, it has been shown that several microsomal enzymes, which transfer electrons ultimately to cytochrome P-450 in hydroxylation reactions, can initiate peroxidation of membrane lipids in the absence of the preferred substrate or of antioxidant (Bidlack and Tappel, 1972; McCay *et al.*, 1971). Several mitochondrial enzymes, such as xanthine oxidase, have been shown by Fridovich and his co-workers to produce the superoxide radical ion, which usually reacts with the omnipresent superoxide dismutase to yield hydrogen peroxide, which can be destroyed by catalase, and oxygen (Fridovich, 1972). Although there is no agreement as to whether the superoxide radical can itself initiate radical chains, it may give rise to such potent initiators as singlet oxygen or hydroxyl radical as a consequence of the very reactions that are designed to remove it (Fong *et al.*, 1973; Pederson and Aust, 1973).

$$2 \, O_2^- + 2 \, H^+ \xrightarrow[\text{dismutase}]{\text{Superoxide}} H_2O_2 + O_2$$

$$H_2O_2 + O_2^- \longrightarrow {}^1O_2 + OH^- + \cdot OH$$

Granted that these reactions can occur in the tissues, usually as the result of a breakdown in the normal protective mechanisms, does it necessarily follow that they are promoted by high tissue PUFAs? Certainly, it has been found that in incubation of tissue homogenate, in which peroxides are produced as a result of tissue disruption (Barber, 1966), a high content of PUFAs leads to increased peroxidation, but probably the most definitive evidence comes from the work of Horwitt and his co-workers (Century and Horwitt, 1960; Witting and Horwitt, 1964; Witting *et al.*, 1965) and others (Miller *et al.*, 1964; Witting, 1974) on the production of encephalomalacia and exudative diathesis in chicks. These diseases are increased in severity by increased PUFAs and decreased antioxidant in the diet and, as a result, in the tissues. Indeed, as has been pointed out above, the dietary requirement of tocopherol is related to the amount of dietary PUFAs, and, since the turnover time of tocopherol in the tissues is shorter than that of the PUFAs, even the vegetable oils, which are well protected by antioxidant, may not provide continuing protection once they are discontinued in the diet. Thus, it seems clear that increased dietary PUFAs must be accompanied by increased dietary tocopherol and that the antioxidant must be continued as protection for tissue PUFAs long after dietary PUFAs have been discontinued. In the case of dietary fish oils or other poorly protected fats, this precaution is particularly important.

There remains one important aspect of lipid peroxidation *in vivo* to be considered in this discussion: the formation of aging pigment, lipofuscin, and ceroid in many tissues. These complex fluorescent pigments appear to be formed as a result of the reaction of peroxidized membrane lipids or their products with membrane proteins, denaturing and cross-linking them (Bidlack and Tappel, 1973; Hendley *et al.*, 1961). The pigment lipofuscin increases as a function of age in certain tissues of each species (Hendley *et al.*, 1961) and is, indeed, an index of sorts of chronological age in mammals. However, whether it is harmful to the tissues or not is not known, although in the nematode (Epstein and Gershon, 1972) it appears to be associated with decreased longevity. In any event, it is impossible at the present time to relate tissue PUFA content to longevity. Perhaps the most that can be suggested is that care should be taken that increased dietary PUFAs are accompanied by adequate antioxidant.

12. Conclusions

In this discussion, I have asked, and attempted to answer, a number of questions concerning the EFAs. That the answers may not be satisfactory in every case, particularly the human, is largely a result of the inadequate state of our knowledge. A great deal of study remains to be done before we can consider this subject as closed. Fortunately, the many gifted scientists working in the field will surely produce the answers—some, no doubt, before this chapter is published.

13. References

Aaes-Jørgensen, E., and Holman, R. T., 1950, Essential fatty acid deficiency in mice, *J. Nutr.* **41**:507.

Alfin-Slater, R. B., Aftergood, L., Wells, A. F., and Deuel, H. J., Jr., 1954, Effect of essential fatty-acid deficiency on the distribution of endogenous cholesterol in the plasma and liver, *Arch. Biochem. Biophys.* **52**:180.

Alfin-Slater, R. B., Morris, R. S., Hansen, H., and Proctor, J. F., 1965, Effects of non-essential fatty acids on essential fatty acid deficiency, *J. Nutr.* **87**:168.

Andrews, J. S., Griffith, W. H., Mead, J. F., and Stein, R. A., 1960, Toxicity of air-oxidized soybean oil, *J. Nutr.* **70**:199.

Ayala, S., Gaspar, G., Brenner, R. R., Peluffo, R. O., and Kunau, W., 1973, Fate of linoleic, arachidonic and docosa-7,10,13,16-tetraenoic acids in rat testicles, *J. Lipid Res.* **14**:296.

Barber, A. A., 1966, Lipid peroxidation in rat tissue homogenates; interaction of iron and ascorbic acid as the normal catalytic mechanism, *Lipids* **1**:146.

Barkie, V. H., Nath, H., Hart, E. B., and Elvehjem, C. A., 1947, Essential fatty acid deficiency in the mature rat, *Proc. Soc. Exp. Biol. Med.* **66**:474.

Barnes, R. H., Lundberg, W. O., Hanson, H. T., and Burr, G. O., 1943, The effect of certain dietary ingredients on the keeping quality of body fat, *J. Biol. Chem.* **149**:313.

Basnayake, V., and Sinclair, H. M., 1956, The effect of deficiency of essential fatty acids upon the skin, in: *Biochemical Problems of Lipids* (G. Popjak and E. Le Breton, eds.), pp. 476–484, Butterworths, London.

Beerthuis, R. K., Nugteren, D. H., Pabon, H. J. J., and Van Dorp, D. A., 1968, Biologically active prostaglandins from some new odd-numbered essential fatty acids, *Rec. Trav. Chim.* **87**:461.

Bidlack, W. R., and Tappel, A. L., 1972, A proposed mechanism for the TPNH enzymatic lipid peroxidizing system of rat liver microsomes, *Lipids* **7**:564.

Bidlack, W. R., and Tappel, A. L., 1973, Fluorescent products of phospholipids during lipid peroxidation, *Lipids* **8**:203.

Bieri, J. G., and Evarts, R. R., 1975, Tocopherols and polyunsaturated fatty acids in human tissues, *Am. J. Clin. Nutr.* **28**:717.

Bieri, J. G., Briggs, G. M., Fox, M. R. S., Pollard, C. J., and Ortiz, L. O., 1956, Essential fatty acids in the chick. I. Development of fat deficiency, *Proc. Soc. Exp. Biol. Med.* **93**:237.

Bolland, J. L., and Gee, G. T., 1946, Kinetic studies in the chemistry of rubber and related materials. II. The kinetics of oxidation of unconjugated olefins, *Trans. Faraday Soc.* **42**:236.

Brenner, R. R., 1971, The desaturation step in the animal biosynthesis of polyunsaturated fatty acids, *Lipids* **6**:567.

Brown, W. R., Hansen, A. E., Burr, G. O., and McQuarrie, I., 1938, Effects of prolonged use of extremely low-fat diet on an adult human subject, *J. Nutr.* **16**:511.

Burr, G. O., and Burr, M. M., 1930, On the nature and role of the fatty acids essential in nutrition, *J. Biol. Chem.* **82**:345.

Cash, R., and Berger, C. K., 1969, Acrodermatitis enteropathica: Defective metabolism of unsaturated fatty acids, *J. Pediatr.* **74**:717.

Castell, J. D., Lee, D. J., and Wales, P., 1972, Essential fatty acids in the diet of the rainbow trout (*Salmo gairdneri*): Growth, feed conversion and some gross deficiency symptoms, *J. Nutr.* **102**:77.

Century, B., and Horwitt, M. K., 1960, Role of diet lipids in the appearance of dystrophy and creatinuria in the vitamin-E-deficient rat, *J. Nutr.* **72**:357.

Chow, C. K., and Tappel, A. L., 1972, An enzymatic protective mechanism against lipid peroxide damage to lungs of ozone-exposed rats, *Lipids* **7**:518.

Christensen, F., and Dam, H., 1952, New symptoms of fat deficiency in hamsters: Profuse secretion of cerumen, *Acta Physiol. Scand.* **27**:204.

Clarke, S. D., Romsos, D. R., and Leveille, G. A., 1976, Specific inhibition of hepatic fatty acid synthesis exerted by dietary linoleate and linolenate in essential fatty acid adequate rats, *Lipids* **11**:485.

Collins, F. D., Sinclair, A. J., Royle, J. P., Coats, D. A., Maynard, A. T., and Leonard, R. F., 1971, Plasma lipids in human linoleic acid deficiency, *Nutr. Metab.* **13**:150.

Crawford, M. A., and Sinclair, A. J., 1972, The limitations of whole tissue analysis to define linoleic acid deficiency, *J. Nutr.* **102**:1315.

Decker, A. B., Fillerup, D. L., and Mead, J. F., 1950, Chronic essential fatty acid deficiency in mice, *J. Nutr.* **41**:507.

Deuel, H. J., Jr., Alfin-Slater, R. B., Wells, A. F., Kryder, G. D., and Aftergood, L., 1955, Effect of fat level of the diet on nutrition (XIV). Effect of hydrogenated coconut oil on essential fatty acid deficiency, *J. Nutr.* **55**:337.

Dhopeshwarkar, G. A., and Mead, J. F., 1961, Role of oleic acid in the metabolism of essential fatty acids, *J. Am. Oil Chem. Soc.* **38**:297.

DiLuzio, N. R., 1968, The role of lipid peroxidation and antioxidants in ethanol-induced lipid alterations, *Exp. Mol. Pathol.* **8**:394.

Epstein, J., and Gershon, D., 1972, Studies on aging in nematodes. IV. The effect of antioxidants on cellular damage and life span, *Mech. Ageing Dev.* **1**:257.

Evans, H. M., and Lepkovsky, S., 1932, Vital need of the body for certain unsaturated fatty acids. II. Experiments with high fat diets in which saturated fatty acids furnish the sole source of energy, *J. Biol. Chem.* **96**:157.

Fong, K. L., McCay, P. B., Poyer, J. L., Keele, B. B., and Misra, H., 1973, Evidence that peroxidation of lysosomal membrane is initiated by hydroxyl free radicals produced during flavin enzyme activity, *J. Biol. Chem.* **248**:7792.

Fridovich, I., 1972, Superoxide radical and superoxide dismutase, *Acc. Chem. Res.* **5**:321.

Gerschenson, L. E., Mead, J. F., Harary, I., and Haggerty, D. R., Jr., 1967, Studies on the effects of essential fatty acids on growth rate, fatty acid composition, oxidative phosphorylation and respiratory control of HeLa cells in culture, *Biochim. Biophys. Acta* **131**:40.

Gerschmann, R., Gilbert, D. L., Nye, S. W., Dwyer, P., and Fenn, W. O., 1954, Oxygen poisoning and x-irradiation: A mechanism in common, *Science* **119**:623.

Glomset, J. A., 1968, The plasma lecithin:cholesterol acyltransferase reaction, *J. Lipid Res.* **9**:155.

Greenberg, L. D., and Wheeler, P., 1966, Comparison of fatty acid composition of erythrocytes and buffy coat in essential fatty acid deficient monkeys, *Fed. Proc. Fed. Am. Soc. Exp. Biol.* **25**:766.

Greenberg, S. M., Calbert, C. E., Savage, E. E., and Deuel, H. J., Jr., 1950, Effect of fat level of the diet on nutrition. VI. Interrelation of linoleate and linolenate on supplying essential fatty acid requirement, *J. Nutr.* **41**:473.

Hamberg, M., Svensson, J., and Samuelsson, B., 1975, Thromboxanes: A new group of biologically active compounds derived from prostaglandin endoperoxides, *Proc. Natl. Acad. Sci. U.S.A.* **72**:2994.

Hansen, A. E., and Wiese, H. F., 1954, Studies with dogs maintained on diets low in fat, *Proc. Soc. Exp. Biol. Med.* **52**:205.

Hansen, A. E., Wiese, H. F., Boelsche, A. N., Haggard, M. E., Adam, D. J. D., and Davis, H., 1963, Relation of linolenic acid to infant feeding; A review, *Pediatrics* **31** (Suppl. 1, Pt. 2):171.

Hatefi, Y., and Hanstein, W. G., 1970, Lipid oxidation in biological membranes. I. Lipid oxidation in submitochondrial particles and microsomes induced by chaotropic agents, *Arch. Biochem. Biophys.* **138**:73.

Hayashida, T., and Portman, O. W., 1960, Swelling of liver mitochondria from rats fed diets deficient in essential fatty acids, *Proc. Soc. Exp. Biol. Med.* **103**:656.

Hendley, D. D., Strehler, B. L., Reporter, M. C., and Gee, M. V., 1961, Further studies on human cardiac age pigment, *Fed. Proc. Fed. Am. Soc. Exp. Biol.* **20**:298.

Hill, E. G., Warmanen, E. L., Hayes, H., and Holman, R. T., 1957, Effects of essential fatty acid deficiency in young swine, *Proc. Soc. Exp. Biol. Med.* **95**:274.

Holman, R. T., 1960, The ratio of trienoic:tetraenoic acids in tissue lipids as a measure of essential fatty acid requirement, *J. Nutr.* **70**:405.

Holman, R. T., 1964, Nutritional and metabolic interrelationships between fatty acids, *Fed. Proc. Fed. Am. Soc. Exp. Biol.* **23**:1062.

Holman, R. T., and Peifer, J. J., 1955, EFA deficiency in the rabbit, *Hormel Inst. Univ. Minn. Annu. Rep.*, p. 41.

Holman, R. T., and Peifer, J. J., 1960, Acceleration of essential fatty acid deficiency by dietary cholesterol, *J. Nutr.* **70**:411.

Holman, R. T., Caster, W. O. and Wiese, H. F., 1964, The essential fatty acid requirement of infants and the assessment of their dietary intake of linoleate by serum fatty acid analysis, *Am. J. Clin. Nutr.* **14**:70.

Hornstra, G., 1974, Dietary fats and arterial thrombosis, *Haemostasis* **2**:21.

Houtsmuller, U. M. T., 1973, Differentiation in the biological activity of polyunsaturated fatty acids, in: *Dietary Lipids and Postnatal Development* (C. Galli, G. Jacini, and A. Pecile, eds.), pp. 145–156, Raven Press, New York.

Houtsmuller, U. M. T., 1975, Specificity of the biological effects of polyunsaturated fatty acids, in: *The Role of Fats in Human Nutrition* (A. J. Vergroesen, ed.), pp. 331–351, Academic Press, New York.

Jacobson, B. S., Kannangara, C. G., and Stumpf, P. K., 1973, Biosynthesis of α-linolenic acid by disrupted spinach chloroplasts, *Biochem. Biophys. Res. Commun.* **34**:646.

Kaneda, T., and Ishii, S., 1954, Nutritive value or toxicity of highly unsaturated fatty acids, *J. Biochem. (Tokyo)* **41**:327.

Kayama, M., Tsuchiya, Y., and Mead, J. F., 1963, A model experiment of aquatic food chain with special significance in fatty acid conversion, *Bull. Jpn. Soc. Sci. Fish.* **49**:452.

Klein, P. D., 1958, Linoleic acid and cholesterol metabolism in the rat. I. The effect of dietary fat and linoleic acid levels on the content and composition of cholesterol esters in liver and plasma, *Arch. Biochem. Biophys.* **76**:56.

Kramer, J., and Levine, V. E., 1953, Influence of fats and fatty acids on the capillaries, *J. Nutr.* **50**:149.

Kupiecki, F. P., Sekhar, N. C., and Weeks, J. R., 1968, Effects of infusion of some prostaglandins in essential fatty acid-deficient and normal rats, *J. Lipid Res.* **9**:602.

Lambert, M. R., Jacobson, N. L., Allen, R. L., and Zalatel, J. H., 1954, Lipide deficiency in the calf, *J. Nutr.* **52**:259.

Lamptey, M. S., and Walter, B. L., 1976, A possible essential role for dietary linolenic acid in the development of the young rat, *J. Nutr.* **106**:86.

Lands, W. E. M., 1975, Selectivity of microsomal acyl-transferases, in: *The Essential Fatty Acids* (W. W. Hawkins, ed.), pp. 15–26, The University of Manitoba, Winnipeg.

Le Breton, E., and Ferret, S., 1960, Quelques aspects actuels du problème des acides gras indispensables et essentials. La notion des familles d'AGP et ses conséquences, *Expo. Annu. Biochim. Méd.* **18**:155.

Lewis, B., Pilkington, T. R. E., and Hodd, K. A., 1961, A mechanism for the action of unsaturated fat in reducing serum cholesterol, *Clin. Sci.* **20**:249.

MacMillan, A. L., and Sinclair, H. M., 1958, The structure and function of essential fatty acids, in: *Proceedings of the International Conference on Biochemical Problems of Lipids* (H. M. Sinclair, ed.), p. 208, Butterworths, London.

Marcus, A. J., 1978, The role of lipids in platelet function: With particular reference to the arachidonic acid pathway, *J. Lipid Res.* **19**:793.

Maruyama, I., 1965, Effect of essential fatty acids and pyridoxine on the formation of gallstones, especially cholesterol stones, *Arch. Jpn. Chir. (Nihon Geka Hukan)* **34**:19.

Matsuo, N., 1961, Studies on the toxicity of fish oil. III. Toxicity of autoxidized highly unsaturated fatty acid ethyl ester applied to rat, *Tokushima J. Exp. Med.* **8**:90.

McCay, P. B., Poyer, J. L., Pfeifer, P. M., May, H. E., and Gilliam, J. M., 1971, A function for α-tocopherol: Stabilization of the microsomal membrane from radical attack during TPNH-dependent oxidations, *Lipids* 6:297.

Mead, J. F., 1976, Free radical mechanisms of lipid damage and consequences for cellular membranes, in: *Free Radicals in Biology,* Vol. I (W. A. Pryor, ed.), pp. 51–68, Academic Press, New York.

Mead, J. F., and Fulco, A. J., 1976, *The Unsaturated and Polyunsaturated Fatty Acids in Health and Disease,* pp. 60–61, Charles C. Thomas, Springfield, Ill.

Miller, D., Leong, K. C., Knold, G. M., Jr., and Gruger, E., Jr., 1964, Exudative diathesis and muscular dystrophy induced in the chick by esters of polyunsaturated fatty acids, *Proc. Soc. Exp. Biol. Med.* 116:1147.

Miller, G. J., and Miller, H., 1975, Plasma high-density lipoprotein concentration and development of ischemic heart disease, *Lancet* 1:16.

Mohrhauer, H., and Holman, R. T., 1963, The effect of dose level of essential fatty acids upon fatty acid composition of the rat liver, *J. Lipid Res.* 4:151.

Nørby, J. G., 1965, Effects of giving a fat-free diet for up to 10 weeks on the male weanling rat, *Br. J. Nutr.* 19:209.

Paulsrud, J. R., Pensler, L., Whitten, C. F., Stewart, S., and Holman, R. T., 1972, Essential fatty acid deficiency in infants induced by fat-free intravenous feeding, *Am. J. Clin. Nutr.* 25:897.

Pederson, T. C., and Aust, S. D., 1973, The role of superoxide and singlet oxygen in lipid peroxide promoted by xanthine oxidase, *Biochem. Biophys. Res. Commun.* 52:1071.

Peifer, J. J., 1966, Hypercholesterolemic effects induced in the rat by specific types of fatty acid unsaturation, *J. Nutr.* 88:351.

Press, M., Hartop, P. J. and Prottey, C., 1974, Correction of essential fatty acid deficiency in man by the cutaneous application of sunflower-seed oil, *Lancet* 1:597.

Raulin, J., Clément, J., and Blum, J.-C., 1959, Étude de la toxicité de l'association alimentaire "Acides gras libres et cholesterol," modalités de l'action de l'extra choline et role de la sécrétion biliare, *Arch. Sci. Physiol.* 13:79.

Reid, M. E., 1954, Production and counteraction of a fatty acid deficiency in the guinea pig, *Proc. Soc. Exp. Biol. Med.* 86:708.

Rivers, J. P. W., Hassam, A. G., Crawford, M. A. and Brambell, M. R., 1976, The inability of the lion, *Panthera leo,* to desaturate linoleic acid, FEBS Lett. 67:269.

Schendel, H. E., and Hansen, J. D. L., 1959, Studies of serum polyenic fatty acids in infants with kwashiorkor, *S. Afr. Med. J.* 33:1005.

Schlenk, H., and Sand, D. M., 1967, A new group of essential fatty acids and their comparison with other polyenoic fatty acids, *Biochim. Biophys. Acta* 144:305.

Schlenk, H., Sand, D. M., and Gellerman, J. L., 1969, Retroconversion of docosahexaenoic acid in the rat, *Biochim. Biophys. Acta* 187:201.

Sinclair, H. M., 1956, Deficiency of essential fatty acids and atherosclerosis, etcetera, *Lancet* 1:381.

Slater, T. F., and Sawyer, B. C., 1971, The stimulatory effects of carbon tetrachloride and other halogenoalkanes on peroxidative reactions in rat liver fractions *in vitro, Biochem. J.* 123:805.

Small, D. M., Bourges, M., and Dervichian, D. G., 1966, Ternary and quaternary aqueous systems containing bile salt, lecithin and cholesterol, *Nature* 211:816.

Smedley-Maclean, I., and Hume, E. M., 1941, Fat-deficiency disease of rats. The influence of tumour growth on the storage of fat and of polyunsaturated acids in the fat-starved rat, *Biochem. J.* 35:996.

Soderhjelm, L., Wiese, H. F., and Holman, R. T., 1971, The role of polyunsaturated fatty acids in human nutrition and metabolism, in: *Progress in the Chemistry of Fats and Other Lipids* Vol. IX (R. T. Holman, ed.), pp. 555–586, Pergamon Press, Oxford.

Sprecher, H., 1975, in: *The Essential Fatty Acids* (W. W. Hawkins, ed.), pp. 29–43, The University of Manitoba, Winnipeg.

Swell, L., and Law, M. D., 1966, Labeling of liver and serum cholesterol esters after the injection of cholesterol-4-C^{14} and cholesterol-4-C^{14} esters, *Arch. Biochem. Biophys.* 113:143.

Ten Hoor, F., Van de Graaf, H. M., and Vergroesen, A. J., 1973, Effects of dietary erucic and linoleic acid on myocardial function in rats, in: *Recent Advances in Studies on Cardiac Structure and Metabolism,* Vol. 3 (N. S. Dhalla, ed.), pp. 31–57, Elsevier, Amsterdam.

Thomas, H. V., Mueller, P. K., and Lyman, R. L., 1968, Lipoperoxidation of lung lipids in rats exposed to nitrogen dioxide, Science **159**:532.

Thomasson, H. J., 1962, Les acides gras essentials, *Rev. Fr. Corps Gras (Journ. Inf. Corps Gras Aliment.*

Tinoco, J., Williams, M. A., Hincenbergs, I., and Lyman, R. L., 1971, Evidence for nonessentiality of linolenic acid in the diet of the rat, *J. Nutr.* **101**:937.

Ullman, D., and Sprecher, H., 1971, An *in vitro* and *in vivo* study of the conversion of eicosa-11,14-dienoic acid to eicosa-5,11,14-trienoic acid and of the conversion of eicosa-11-enoic acid to eicosa-5,11-dienoic acid in the rat, *Biochim. Biophys. Acta* **248**:186.

Uri, N., 1961, Physico-chemical aspects of autoxidation, in: *Autoxidation and Antioxidants,* Vol. I (W. O. Lundberg, ed.), pp. 55–106, Interscience, New York.

Vergroesen, A. J., De Deckere, E. A. M., Ten Hoor, F., Hornstra, G., and Houtsmuller, U. M. T., 1975, Physiological functions of essential fatty acids, in: *The Essential Fatty Acids* (W. W. Hawkins, ed.), pp. 5–14, The University of Manitoba, Winnipeg.

Wesson, L. G., and Burr, G. O., 1931, Metabolic rate and respiratory quotients of rats on a fat-deficient diet, *J. Biol. Chem.* **91**:525.

White, E. A., Foy, J. R., and Cerecedo, L. R., 1943, Essential fatty acid deficiency in the mouse, *Proc. Soc. Exp. Biol. Med.* **54**:301.

Wiese, H. F., Bennet, M. J., Braun, I. H. G., Yamanaka, W., and Coon, E., 1966, Blood serum lipid patterns during infancy, *Am. J. Clin. Nutr.* **18**:155.

Witting, L. A., 1974, Vitamin E–polyunsaturated lipid relationship in diet and tissues, *Am. J. Clin. Nutr.* **27**:952.

Witting, L. A., and Horwitt, M. K., 1964, Effect of degree of fatty acid unsaturation in tocopherol deficiency, *J. Nutr.* **82**:19.

Witting, L. A., Harmon, E. M., and Horwitt, M. K., 1965, Extent of tocopherol depletion versus onset of creatinuria in rats fed saturated and unsaturated fat, *Proc. Soc. Exp. Biol. Med.* **120**:718.

Yu, T. C., and Sinnhuber, R. O., 1972, Effect of dietary linolenic acid and docosahexaenoic acid on growth and fatty acid composition of rainbow trout (*Salmo gairdneri*), *Lipids* **7**:450.

9

Nutrients with Special Functions: Cholesterol

David Kritchevsky and Susanne K. Czarnecki

Cholesterol is one of the most ubiquitous compounds in the animal kingdom. Its chemistry, biosynthesis, and metabolism have been studied for over two centuries. The relationship between cholesterol and certain human diseases, such as atherosclerosis and cholelithiasis, has stimulated extensive research on this compound. Numerous reviews covering cholesterol chemistry and biology have been published (Bloch, 1965; Dempsey, 1974; Dietschy and Wilson, 1970a–c), as have several books (Cook, 1958; Kritchevsky, 1958; Nes and McKean, 1977; Sabine, 1977).

The early history (as summarized by Kritchevsky, 1958) of cholesterol began with the recognition by Vallisnieri in 1733 that gallstones were soluble in alcohol or turpentine. Thirty years later, Poulletier de la Salle showed that the principal constituent of gallstones could be crystallized from alcohol as glistening white leaflets. Chevreul, in 1815, showed that this substance was an unsaponifiable lipid and named it *cholésterine* (Greek *cholē*, bile + *stereos*, solid). Soon after, cholesterol was found to be present in blood, brain, tumors, and egg yolk. Gobley demonstrated in 1846 that bile, gallstone, and egg-yolk cholesterol were identical.

The chemistry of cholesterol was elucidated between 1850 and 1940. The empirical formula was determined by Reinitzer in 1888 based upon the analysis of cholesteryl acetate dibromide. Most of the investigation on the structure of cholesterol was conducted by Wieland and Windaus in the 1920s and has been reviewed by Kritchevsky (1958) and Fieser and Fieser (1959).

The discovery of the mechanisms of cholesterol biosynthesis is one of the most interesting and important achievements in biochemistry. As a result of

David Kritchevsky and Susanne K. Czarnecki • The Wistar Institute of Anatomy and Biology, Philadelphia, Pennsylvania 19104.

the brilliant research of Schoenheimer, Rittenberg, Bloch, Lynen, Popjak, and Cornforth and their many colleagues, most of the enzymatic steps in the biosynthesis of cholesterol are known. Their studies of this complex pathway revealed many new metabolic intermediates and also shed light on the biogenesis of other related compounds (Bloch, 1965). The early balance studies by Dezani (cf. Kritchevsky, 1958) showed that mice on a diet of extracted corn-meal and casein produced and excreted cholesterol. Many other investigators confirmed this finding in various animal species (Channon, 1925; Schoenheimer and Breusch, 1933), including humans (Gamble and Blackfan, 1920). Schoenheimer and Breusch (1933), who studied cholesterol synthesis and metabolism in a quantitative manner by balance experiments in mice, demonstrated that the rates of cholesterol biosynthesis and metabolism were affected by the presence of dietary cholesterol and its principal degradation products, the bile acids. Although these early balance studies demonstrated that cholesterol can be synthesized and metabolized, they yielded no information on the nature of the precursors. The availability of isotopes made possible more thorough studies on the mechanisms of cholesterol biogenesis (Rittenberg and Schoenheimer, 1935; Schoenheimer and Rittenberg, 1935a,b). Rittenberg and Schoenheimer (1937) found that cholesterol isolated from mice maintained on heavy water was progressively enriched in deuterium and calculated that 50% of the isotope in the sterol was derived from labeled body water. The authors concluded that cholesterol arises from the coupling of many small molecules. That same year, Sonderhoff and Thomas (1937) found that when yeast was incubated with trideuteroacetate, the unsaponifiable fraction contained about twice the deuterium content of fatty acids. This demonstrated the apparent direct utilization of acetate for sterol synthesis and supported the earlier suggestion of Rittenberg and Schoenheimer. In the early 1940s, Bloch and Rittenberg (1942a,b) were the first to demonstrate *in vivo* that rats and mice could convert labeled acetate to cholesterol. These authors found that in rats, 13% of the deuterium in acetate was incorporated into cholesterol in both the steroid nucleus and the isooctyl side chain. Later experiments on rat liver slices with methyl- and carboxyl-labeled acetate revealed that both carbon atoms were converted into cholesterol in equal amounts and that all carbon atoms of cholesterol were derived from acetate (Little and Bloch, 1950). Systematic degradation of biologically labeled cholesterol revealed the origin of each carbon, and this labeling pattern became the guide for the investigation of the pathway of cholesterol biosynthesis (Bloch, 1965). On the basis of earlier observations, Langdon and Bloch discovered that squalene, an open-chain isoprenoid, was an obligatory intermediate late in cholesterol biosynthesis (Langdon and Bloch, 1953a,b). Later research revealed the pathway of the assembly of isoprene units from acetate to squalene (Nes and McKean, 1977). It was shown that mevalonic acid was a necessary intermediate, and eventually the key role of 3-methyl-3-hydroxyglutaryl coenzyme A (CoA) reductase was demonstrated. This enzyme converts hydroxymethylglutaryl-CoA, a direct product of acetate condensation, to mevalonic acid, the rate-determining step in cholesterol biosynthesis. The regulation of this enzyme has been reviewed (Rodwell *et al.*, 1976).

Metabolically, cholesterol is the key intermediate in the biosynthesis of many biologically important steroids, such as adrenocortical hormones (Hechter, 1958), sex hormones (Dorfman and Ungar, 1965), and bile acids (Nair and Kritchevsky, 1971, 1973). Structurally, it is an indispensable constituent of all cell membranes. Studies on cholesterol distribution in humans suggest that the total body content of cholesterol is about 140–145 g (0.2%), most of which is in the brain, nervous system, connective tissue, and muscle (Masoro, 1968).

The blood, which contains about 8% of total-body cholesterol, has been the most widely studied source of tissue cholesterol (Masoro, 1968). Blood cholesterol represents the sum of the metabolic flux between tissue compartments and plasma (Goodman and Noble, 1968; Goodman *et al.*, 1973). Plasma cholesterol is present in both free (30–40%) and esterified forms (60–70%). Cholesterol and nearly all the other plasma lipids are transported as a part of large complexes, the lipoproteins. The lipoproteins are polydisperse macromolecules varying in size, hydrated density, and chemical composition (Jackson *et al.*, 1976). The circular diagrams in Figs. 1–4 illustrate the four main classes of lipoproteins in normal human sera and summarize their characteristic molecular properties and apoprotein content. Lipoproteins are classified by analytic ultracentrifugation in terms of S_f value, the rate of flotation in a dense salt solution (de Lalla and Gofman, 1954; Lindgren *et al.*, 1972). The S_f rates are a function of lipoprotein size, shape, and density. Preparative ultracentrifugation is the most common method used for lipoprotein isolation. By cumulative flotation, the lipoproteins are sequentially separated according to their buoyant densities by stepwise adjustment of serum density with concentrated salt solutions or a solid salt (Havel *et al.*, 1955; Radding and Steinberg, 1960). Lipoprotein fractionation can also be achieved by column chromatography (Rudel *et al.*, 1974), electrophoresis (Hatch and Lees, 1968), or precipitation by polyanions and metals (Burstein *et al.*, 1970).

Although lipoprotein classification is partially operational, it is physiologically justified, since most lipoprotein classes play distinctly different roles in lipid metabolism (Osborne and Brewer, 1977). Table I presents some chemical and metabolic features of the apoproteins. Although all apoproteins function as lipid-transport proteins, they differ in lipid-binding properties (Morrisett *et al.*, 1977) and in ability to act as cofactors for lipolytic enzymes responsible for lipoprotein metabolism (Schaefer *et al.*, 1978).

The lipoproteins form a dynamic system in which both exchange and net transfer of protein and lipid occur during metabolism. The catabolism of certain lipoproteins produces other lipoproteins altered in molecular properties and chemical composition. The structural relationship among lipoproteins during their metabolism is illustrated by the conversion of very-low-density lipoprotein (VLDL) to low-density lipoprotein (LDL) (Lewis, 1976). VLDL primarily transports endogeneous triglycerides and carries about 10% of the total plasma cholesterol at a free-to-esterified-cholesterol ratio (FC:EC) of 1. VLDL catab-

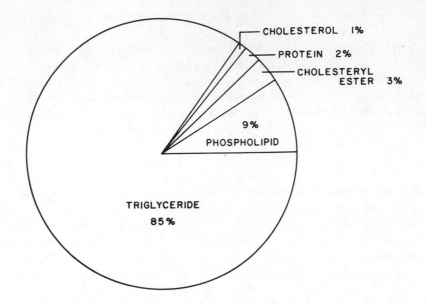

Fig. 1. Chylomicron. Source: small intestine; apoproteins: AI, AII, B, E; diameter: 75–1000 nm; mol. wt.: $> 0.4 \times 10^9$; S_f (1.063 g/ml): 400–10^5; density: < 0.95 g/ml; electrophoretic mobility: origin.

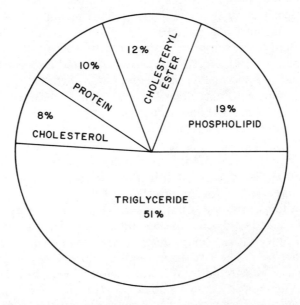

Fig. 2. Very-low-density lipoprotein (VLDL). Source: liver, small intestine; apoproteins: AI, AII, B, CI–III, E; diameter: 30–80 nm; mol. wt.: 5–10 $\times$ 10^6; S_f (1.063 g/ml): 20–400; density: 0.95–1.006 g/ml; electrophoretic mobility: pre-β.

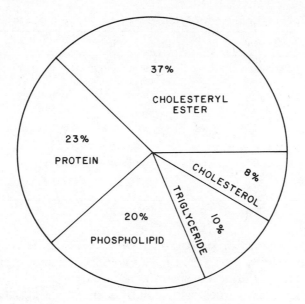

Fig. 3. Low-density lipoprotein (LDL). Source: VLDL, liver; apoproteins: B, CI–III; diameter: 19–25 nm; mol. wt.: 2.7–4.8 × 10⁶; S_f(1.063 g/ml): 0–20; density: 1.006–1.063 g/ml; electrophoretic mobility: β.

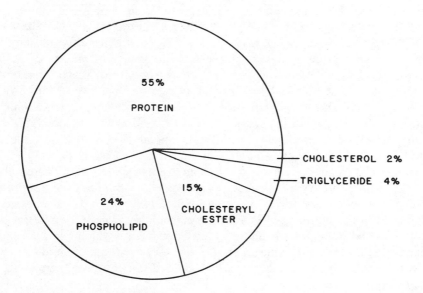

Fig. 4. High-density lipoprotein (HDL). Source: liver, small intenstine; apoproteins: AI, AII, B, CI–III, D, E–AII; diameter: 4–10 nm; mol. wt.: 1.8–3.9 × 10⁵; S_f(1.20 g/ml): 0–9; density: 1.063–1.210 g/ml; electrophoretic mobility: α.

Table I. Characteristics of Human Plasma Apoproteins[a]

Apoprotein	Mol. wt.	Site of synthesis	Function
AI	28,331	Liver, small intestine	LCAT activation
AII	17,000	Liver, small intestine	?
B	250,000	Liver, small intestine	Triglyceride transport
CI	6,331	Liver	LCAT, LPL activation
CII	8,837	Liver	LPL activation
CIII$_{0-2}$	8,764	Liver	LPL inhibition
D	19,000–35,000	?	Cholesterol ester transfer?
E	37,000	Liver, small intestine	?
E–AII	46,000	?	?
F	29,000	?	?

[a] After Schaefer et al. (1978).

olism occurs in a series of steps during which triglyceride is progressively removed by lipoprotein lipase (LPL), and apo-C, FC, and phospholipid are transferred to HDL. During the lipoprotein interconversion process, apo-B remains with the parent particle and comprises an increasingly larger portion of the total protein mass as VLDL is metabolized to intermediate-density lipoprotein (IDL) and ultimately to LDL. The apo-C proteins, a group of low-molecular-weight apoproteins, are very important in the metabolism of triglyceride-rich lipoproteins. Apo-CII is a specific protein cofactor for LPL, although other apoproteins may also affect LPL activity (Eisenberg, 1976; Eisenberg and Levy, 1975). LDL is the most abundant lipoprotein in human serum and carries 60–70% of total serum cholesterol at a EC:FC ratio of 2. Both LDL and HDL are quasispherical particles in which the neutral lipid, primarily cholesteryl ester, is sequestered in a central core (Atkinson et al., 1977, 1978; Deckelbaum et al., 1977). The largest lipoproteins, the chylomicrons, are synthesized by the small intestine and transport dietary cholesterol and triglyceride. Chylomicron metabolism is similar to VLDL metabolism (Fielding 1978; Zilversmit, 1978). Both the liver and small intestine produce disc-shaped nascent HDL rich in FC and lecithin, and a serum enzyme, lecithin-cholesterol acyltransferase (LCAT), transform this nascent HDL into circulating spherical particles of a different chemical composition (Glickman and Green, 1977; Green et al., 1978; Hamilton, 1978; Havel, 1978).

Many disturbances of lipid metabolism are associated with characteristic changes in lipoprotein profiles and, in some cases, with abnormal lipoproteins (Fredrickson and Levy, 1978). Lees and Hatch (1963) found that the addition of albumin to the buffer used in serum electrophoresis permitted better separation of α- and β-lipoprotein and revealed a pre-β band which migrated slightly ahead of the β band. Fredrickson et al. (1967a–e) used this method to identify various types of lipemias in a way which helped to determine their origin and rationalize their treatment.

In 1951, Barr and co-workers reviewed the available literature on α- and β-lipoprotein cholesterol and found that young women (18–35 years of age), who show a lower mortality rate from heart disease than do men of the same age, carried considerably more of their serum cholesterol in the α-lipoprotein fraction. They also found that elevations in the cholesterol-rich β-lipoprotein fraction were correlated with diseases such as diabetes or nephritis, both of which are associated with increased risk of coronary disease. Olson (1959) showed that susceptibility to atherosclerosis in animals was a function of their plasma β-lipoprotein concentrations. Recently, Miller and Miller (1975) linked susceptibility to heart disease with plasma HDL–cholesterol concentrations. Their observations were confirmed by those of Rhoads *et al.* (1976). Slack *et al.* (1977) have found that HDL–cholesterol levels generally decrease with age but are usually higher in women than in men. Glueck *et al.* (1977) have reported on relatively elevated HDL levels in octogenarian kindreds. In 1968, Rothblat *et al.* showed that a medium whose cholesterol:protein:phospholipid ratio resembled that of β-lipoprotein facilitated entry of cholesterol into tissue culture cells and that another medium whose ratios of these three components resembled that of α-lipoprotein enhanced removal of cellular cholesterol. Recent work suggests that LDL and HDL facilitate entry and egress of cellular cholesterol, respectively (Carew *et al.* 1976; Goldstein and Brown, 1977).

The absorption of cholesterol was shown by Mueller (1915, 1916) to require bile and secretions of the pancreas and to proceed from the intestine via the lymph. Intestinal cholesterol is derived largely from diet and bile, with a small amount coming from intestinal secretion or sloughed cells. FC forms a micelle with bile salts, monoglycerides, and lysolecithin, and this micelle is transported into the intestinal mucosa. In the mucosa, cholesterol of endogeneous and exogeneous origin is esterified and assembled with triglyceride and protein into chylomicron particles, which are released into the lymph (Hofman, 1976).

Early studies (Frohlicher and Sullman, 1934; Mueller, 1915, 1916) suggested that dietary fat was indispensable for cholesterol absorption, but it has since been shown that rabbits (Popjak, 1946; Turner, 1933), rats (Bollman and Flock, 1951), and chickens (Stamler *et al.*, 1948) can absorb cholesterol on a fat-free diet.

Not all ingested cholesterol is absorbed. The capacity for absorption varies among species. For example, when fed 50 mg cholesterol, a rat absorbs 65% of the dose, but when fed 288 mg, only 32% of the dose is absorbed (Lin *et al.*, 1955). A 1.5-kg rabbit absorbs about 500 mg cholesterol/day, and a 70-kg man absorbs 200–400 mg daily (Dietschy and Wilson, 1970a–c).

Dietary cholesterol also affects cholesterol biosynthesis. About 85% of circulating cholesterol in a rat is of dietary origin, whereas only 40–60% in a man is accounted for by diet (Dietschy and Wilson, 1970a–c).

The principal products of cholesterol metabolism are the bile acids. The liver converts cholesterol to either cholic or chenodeoxycholic acid (primary bile acids), and the intestinal flora remove the 7α-hydroxyl group to give the secondary bile acids, deoxycholic and lithocholic (Nair and Kritchevsky, 1971, 1973).

An important reason for the intense interest in cholesterol metabolism is the association between serum cholesterol levels and risk of coronary disease.

Stamler (1979) has summarized a number of epidemiologic studies which relate hypercholesterolemia and coronary risk. However, impressions that serum cholesterol levels may not be a reliable indicator of coronary risk at the individual level, as well as the findings in clinical trials (Olson, 1979) that reduced risk may not correlate with reductions in serum cholesterol levels, lead to continuing controversy.

Epidemiologic studies indicate that heart disease is a life-style disease and that the three major risk factors—hypercholesterolemia, elevated blood pressure, and excessive cigarette smoking (Kagan et al., 1962)—are only part of a broad spectrum of risk factors (Strasser, 1972). The extent to which each factor contributes to risk in a given individual is not clear.

Since a good deal of circulating cholesterol is of dietary origin and since cholesterol is a member of the lipid class of compounds, it is not surprising that most studies of serum cholesterol levels have centered on the effects of dietary fats.

Various analyses of dietary variables have shown that cardiovascular mortality (Stamler, 1979; Yerushalmey and Hilleboe, 1957; Yudkin, 1963) is best correlated with total protein and animal protein. Ingestion of animal protein generally implies ingestion of cholesterol.

It is well documented that vegetarians have lower cholesterol levels than does the general population (Hardinge and Stare, 1954a,b; Sacks et al., 1975). Still, although 34% less cholesterol is available in Finland than in the United States, the Finns have a higher rate of coronary disease (Connor and Connor, 1972). The Framingham study (Kannel and Gordon, 1970) showed that cholesterol levels are not necessarily dependent on cholesterol intake. In several recent studies (Porter et al., 1977; Slater et al., 1976), the addition of eggs to subjects' diets did not affect serum cholesterol levels. Nichols et al. (1976) completed a study of over 4000 subjects and found no correlation between dietary habits and serum lipid levels. Obviously, the diet–cholesterol question is a very complicated one.

The American diet has undergone considerable alteration since 1909 (Page and Friend, 1978). In the years 1909–1913, calories were derived 56% from carbohydrate, 32% from fat, and 12% from protein. In 1975, the breakdown was carbohydrate, 46%; fat, 42%; and protein, 12%. There have been sharp changes within these categories as well.

Table II presents a summary of the nutrient content of the American diet in selected years between 1909 and 1975. Although there has been no change in total available protein, the ratio of animal to vegetable protein found in the diet has doubled, and the ratio of animal to vegetable fat has decreased dramatically. In 1909, the principal sources of nutrient fat were butter (about 37% of all fat) and lard (about 32% of all fat). The main sources of fat today are salad and cooking oils (38%), vegetable shortenings (28%), and vegetable margarine (16%). The amount of edible beef fat available has risen from 3 to 5% of all nutrient fat.

The total amount of meat, poultry, and fish ingested in the United States has risen from 172 to 222 lb/person per year (Table III). Fish consumption has

Table II. Nutrient Content of the American Diet[a,b]

Year	Calories	Protein (g)	Fat (g)	Carbohydrate (g)
1909–1913	3480	102	125	492
1925–1929	3460	95	134	476
1935–1939	3260	90	132	436
1947–1949	3230	95	140	403
1957–1959	3140	95	143	375
1965	3150	96	144	372
1970	3300	100	156	380
1975	3220	90	152	370

[a] Per person per day.
[b] After Page and Friend (1978).

remained constant, but meat and poultry consumption has risen 17 and 194%, respectively. Intake of eggs has not changed, but consumption of dairy products, excluding butter, has risen 25%. In the years 1909–1913, the total consumption of fats and oils, of which 44% was butter, amounted to 41 lb/person per year; in 1975, fat and oil consumption, of which only 7% was butter, had risen to 57 pounds. The ratio of complex to simple carbohydrates has fallen from 2.15 to 0.89, and simple calculation shows that Americans eat less total carbohydrate and more simple sugar today than 60 years ago.

It is not easy to demonstrate a dose response of serum cholesterol to dietary cholesterol, suggesting a wide individual variation based, in part, on variations in the total diet. Quintao *et al.* (1971) fed a cholesterol-containing formula diet for several months to eight patients whose previous diets had not contained this sterol. The amount absorbed varied from 17 to 50 g, and the one patient who absorbed the most cholesterol showed the least decrease in synthesis. The amount of cholesterol in the plasma compartment increased in some subjects but not in others, suggesting that sterol balance may be a function of individual metabolic patterns.

Table III. Changes in the American Diet[a,b]

Year	Meat	Poultry	Fish	Eggs	Dairy[c]	Butter	Other fats and oils
1909–1913	141	18	13	37	177	18	41
1925–1929	129	17	14	40	191	18	48
1935–1939	120	16	13	36	202	17	49
1947–1949	141	22	13	47	236	11	46
1957–1959	144	34	13	45	239	8	49
1965	148	41	14	40	235	6	51
1970	165	49	15	40	226	5	56
1975	158	49	15	35	216	5	57

[a] Pounds per person per year.
[b] After Page and Friend (1978).
[c] Excluding butter; given as milk equivalent, that is, calcium content.

The saturation of dietary fat consumed by humans affects serum cholesterol levels. In a study comparing a number of fats of different levels of saturation, Ahrens (1957) found an inverse correlation between iodine value and hypercholesterolemic effect. The most cholesterolemic fat was coconut oil. Saturated fats are more atherogenic for rabbits than are unsaturated fats (Kritchevsky, 1970; Kritchevsky *et al.*, 1954, 1956). Although serum cholesterol levels may respond to diet in most subjects, there are large variations in total sterol balance (Quintao *et al.*, 1971). Schreibman (1975) described a study of human subjects in which corn oil was substituted isocalorically for butter. Although serum cholesterol levels fell by 35% (370 to 240 mg/dl), fecal sterol excretion was unchanged.

Ahrens (1957) found that peanut oil had no effect relative to corn oil when fed to human subjects. However, peanut oil fed as part of a cholesterol-rich diet is inordinately atherogenic for rats (Gresham and Howard, 1960; Scott *et al.*, 1964), rabbits (Kritchevsky *et al.*, 1971), and monkeys (Vesselinovitch *et al.*, 1974). Peanut oil differs from most fats in that it contains 4–6% long-chain, saturated fatty acids (arachidic, behenic, lignoceric) which appear exclusively in the 3 position of the component triglycerides (Myher *et al.*, 1977). When peanut oil is subjected to autointeresterification (randomization), the triglyceride fatty acids are redistributed within the fat so that each component fatty acid is uniformly distributed among the three positions of the triglycerides. Randomized peanut oil is no more atherogenic for rabbits than is corn oil (Kritchevsky *et al.*, 1973a), demonstrating that, under certain conditions, the structure of a fat is as important in determining atherogenicity as is its fatty acid spectrum. The feeding of peanut oil results in an aortic lesion that is more fibrous than that produced by the feeding of other fats, such as corn oil or coconut oil. This is true whether the diet contains cholesterol or not (Kritchevsky *et al.*, 1976).

Kummerow (1974) has suggested that the use of hydrogenated shortenings contributes a relatively large amount of *trans* fatty acids to the American diet and that these acids may exert an untoward atherogenic effect. A comparison of the effects of oleic (*cis*) and elaidic (*trans*) acids on the development of atherosclerosis in rabbits has demonstrated few differences (McMillan *et al.*, 1963; Weigensberg and McMillan, 1964a,b).

Cholesteryl esters of *cis* fatty acids are synthesized and hydrolyzed more readily than are esters of *trans* fatty acids (Kritchevsky and Baldino, 1978; Sgoutas 1968, 1970). Pancreatic lipase does not discriminate between triglycerides containing *cis* and those containing *trans* fatty acids (Jensen *et al.*, 1964). There also appears to be no difference in absorption or oxidation between *cis* and *trans* fatty acids (Anderson, 1968; Ono and Fredrickson, 1964).

There has been increased interest in the effects of animal and vegetable protein on cholesterol metabolism and atherosclerosis. In 1909, Ignatowski postulated that animal protein might have an atherogenic effect on humans. He showed that diets containing animal protein caused atherosclerosis in rabbits, but Anitschkow's demonstration (Antischkow and Chalatow, 1913) that cholesterol per se was atherogenic drew attention to it and away from animal protein.

Newburgh and his collaborators (Newburgh and Clarkson, 1923; Newburgh and Squier, 1920) showed that rabbits fed powdered beef developed atherosclerosis. The development of lesions was a function of the amount of protein in the diet and of the duration of feeding. Meeker and Kesten (1940, 1941) showed that casein was more atherogenic for rabbits than was soy protein in both the presence and absence of cholesterol. Nath *et al.* (1958) found that casein (40%) was significantly more cholesterolemic for rats than was wheat gluten.

Lofland and his associates (1961, 1966) demonstrated that the type of protein fed could affect serum cholesterol levels in monkeys and pigeons. The effect could be mediated by the level of protein fed, 25% being more atherogenic than 8%, as well as by the type of fat present in the diet. These investigators varied the amount and type of dietary protein, fat, and cholesterol. Statistical analysis of their data revealed no independent variable (Lofland and Clarkson, 1968).

Strong and McGill (1967) fed baboons diets low (8% of calories) or high (20% of calories) in protein and low (0.01%) or high (0.5%) in cholesterol and containing 40% of calories as unsaturated (iodine value, 109) or saturated (iodine value, 53) fat. In the high-cholesterol, saturated fat group, the low-protein diet led to increased aortic lipid deposition. In the other groups, the high-protein diet was always more cholesterolemic and atherogenic.

Carroll and Hamilton (1975) fed rabbits a diet containing 30% protein and 1% fat. As can be seen from Table IV, animal protein was generally more cholesterolemic than was vegetable protein. It should be noted that when the

Table IV. Influence of Dietary Protein on Plasma Cholesterol Levels in Rabbits[a,b]

Protein	n	Plasma cholesterol[c] (mg/dl ± SEM)
Casein	20	200 ± 22 (64.0)
Whole egg	4	235 ± 89 (44.7)
Skim milk	6	230 ± 40 (70.9)
Lactalbumin	5	215 ± 69 (57.2)
Beef	5	160 ± 60 (62.5)
Pork	6	110 ± 17 (68.2)
Egg white	6	105 ± 28 (66.7)
Wheat gluten	6	80 ± 21 (68.8)
Peanut meal	4	75 ± 27 (66.7)
Soy concentrate	6	25 ± 5 (56.0)
Soy isolate	6	15 ± 5 (33.3)
Control (commercial ration)	22	70 ± 5 (62.9)

[a] Diets fed 28 days.
[b] After Hamilton and Carroll (1976).
[c] Values in parentheses represent percentage ester.

diet contained glucose, the average cholesterol level of the casein-fed group was over 200 mg/dl and that of the soy-fed group, 70 mg/dl. When glucose was replaced by raw potato starch, cholesterol levels in both groups of rabbits were normal (45 ± 5 mg/dl).

These results illustrate the interdependence of dietary components. It is important that this point be recognized. For instance, Howard *et al.* (1965) demonstrated that rabbits fed semipurified diets containing 25% casein exhibited higher cholesterol levels and more severe atherosclerosis than did rabbits fed diets containing soy protein. Kritchevsky *et al.* (1977) confirmed these observations and showed them to be true when the dietary fiber is cellulose; when the fiber is alfalfa, the differences in protein effect are negligible.

It has recently been reported (Kritchevsky, 1979) that a diet containing 25% beef protein and 14% beef tallow is considerably more atherogenic and cholesterolemic for rabbits than one containing textured vegetable protein (TVP). When the protein is beef–TVP (1:1), lesions and cholesterolemia are no more severe than those seen in rabbits fed TVP alone. This is another demonstration of positive interaction among dietary components.

In humans, Sirtori *et al.* (1977, 1979) have shown that substitution of vegetable protein for animal protein in the diet usually prescribed for hyper-lipemic patients enhances the hypolipemic effect of the diet. Thus, low-lipid diet reduced cholesterol and triglyceride levels by 5 and 11%, respectively. When a soy diet was fed, serum cholesterol levels fell by 19% and triglyceride levels by 17%. It has also been demonstrated that vegetable protein is effective in lowering serum cholesterol and triglyceride levels even when the diet contains practically no unsaturated fat.

Carbohydrates can also effect lipemia (Macdonald, 1973). Plasma triglyceride levels become elevated in subjects fed diets high in carbohydrate (Bierman and Porte, 1968; Knittle and Ahrens, 1964). Fructose or sucrose can be triglyceridemic when substituted for starch (Kuo, 1965; Macdonald and Braithwaite, 1964; Nikkila and Pelkonen, 1966). Grande (1974) summarized results from a number of studies which show that isocaloric substitution of starch (such as bread, fruit, legumes, potatoes) for sucrose lowers cholesterol levels, often significantly. The effect of the fiber (which was added when complex carbohydrate was added) cannot be discounted. When rabbits are fed a semipurified cholesterol-free diet containing 40% carbohydrate, fructose and sucrose are more cholesterolemic than is either glucose or lactose (Kritchevsky *et al.*, 1968, 1973b).

In the past few years, much has been said and written about the hypocholesterolemic and antiatherogenic potential of dietary fiber. The interest in this area has been spurred by the writings of Burkitt *et al.* (1974) and Trowell (1972), who have suggested that the black African is free from heart disease because of the fiber content of his diet. Mendeloff (1976) has pointed out the unlikelihood that the course of a disease of multiple etiology could be altered by a change in one dietary component.

''Dietary fiber'' is a generic term for plant substances resistant to hydrolysis by intestinal enzymes. There is no agreement on a suitable definition of fiber, and there is no single recognized method that allows analysis of all types of fiber. The physiological effects of specific types of fiber appear to depend on structure. The four major types of fiber—cellulose, hemicellulose, pectin, and lignin—differ in composition and in metabolic effect (Kritchevsky, 1977). One factor contributing to the great public interest in fiber may be the fact that it is easier to add a substance to the diet than to delete one.

Several reviews of fiber and lipid metabolism have recently been published (Kay and Strasberg, 1978; Kritchevsky, 1978; Story and Kritchevsky, 1976). In rats fed semipurified, fiber-free diets containing cholesterol (0.5–10%), pectin and vegetable gums lower cholesterol levels in serum and liver, but cellulose, agar, and alginic acid do not (Kiriyama *et al.,* 1969; Tsai *et al.,* 1976). Pectin reduces cholesterol levels in rats (Wells and Ershoff, 1961), rabbits (Ershoff, 1963), and chickens (Fisher *et al.,* 1966). In a series of experiments in which rats were fed a high-fat, high-cholesterol diet containing native starches, Vijayagopal and Kurup (1970) showed that a hypocholesterolemic effect could be correlated with the fiber content of the starch. In humans, bran has virtually no effect on serum lipids (Truswell and Kay, 1976), but pectin and guar gum exert a distinct cholesterol-lowering effect (Jenkins *et al.,* 1975; Kay and Truswell, 1977).

One of the most interesting findings has been that diets high in fiber lower cholesterol and triglyceride levels (Stone and Connor, 1963), lower blood glucose levels, and improve glucose tolerance in diabetics (Kiehm *et al.,* 1976; Miranda and Horwitz, 1978). Such diets also decrease dependence on insulin or oral hypoglycemic agents.

The mechanism by which fiber affects lipid metabolism is unclear. Some types of fiber increase fecal excretion of steroids (Portman and Murphy, 1958), and the possibility that fiber binds bile salts and thus reduces cholesterol absorption has been suggested (Kritchevsky *et al.,* 1974, 1975; Kyd and Bouchier, 1972). Much more research will be needed before specific effects of fiber can be delineated and attributed to particular aspects of structure. Hellendoorn (1978) has suggested that many effects of fiber are due to the volatile fatty acids formed when fiber is degraded by intestinal flora.

The foregoing shows that many aspects of the diet can affect cholesterolemia. Interaction among dietary components may vitiate the effects of substances which, under other conditions, exert a hypercholesterolemic effect.

A case in point is milk, which, although indicted because of its cholesterol content (Segall, 1977), has been shown to be hypocholesterolemic for humans (Howard and Marks, 1977) and rats (Kritchevsky *et al.,* 1979; Malinow and McLaughlin, 1975; Nair and Mann, 1977). Since both skim and whole milk exert the same effect, the active principle must reside in the aqueous portion.

In view of the correlation between serum cholesterol levels and the risk of coronary disease, and in the absence of any more accurate indicator of susceptibility to coronary disease, it is wise to be prudent; but, in the face of the gaps in our knowledge, it is foolish to become hysterical. The information which is being continually published on diet and cholesterolemia should, in most cases, be looked upon as hypothesis-generating data (as Stavraky, 1976, suggested in a different context). Perhaps the best current dietary advice is a reduction in total calories.

ACKNOWLEDGMENT. This work was supported, in part, by grants (HL-03299; HL-05209; CA-09171) and a Research Career Award (HL-0734) from the National Institutes of Health.

References

Ahrens, E. M., Jr., 1957, Nutritional factors and serum lipid levels, *Am. J. Med.* **23**:928.

Anderson, R. L., 1968, Oxidation of the geometric isomers of Δ9,12-octadecadienoic acid by rat liver mitochondria, *Biochim. Biophys. Acta* **152**:531.

Anitschkow, N., and Chalatow, S., 1913, Über experimentelle cholesterinst eatose und ihre bedeutung fur die entstehung einige pathologische prozesse, *Zentralbl. Allg. Pathol. Pathol. Anat.* **24**:1.

Atkinson, D., Deckelbaum, R. J., Small, D. M., and Shipley, G. G., 1977, Structure of human plasma low density lipoproteins: Molecular organization of the central core, *Proc. Natl. Acad. Sci. U.S.A.* **77**:1042.

Atkinson, D., Tall, A. R., Small, D. M., and Mahley, R. W., 1978, Structural organization of the lipoprotein HDL_c from atherosclerotic swine. Structural features relating the particle surface and core, *Biochemistry* **17**:3930.

Barr, D. P., Russ, E. M., and Eder, H. A., 1951, Protein lipid relationship in human plasma II. In atherosclerosis and related conditions, *Am. J. Med.* **11**:480.

Bierman, E. L., and Porte, D., Jr., 1968, Carbohydrate intolerance and lipemia, *Ann. Intern. Med.* **68**:926.

Bloch, K., 1965, The biological synthesis of cholesterol, *Science* **150**:19.

Bloch, K., and Rittenberg, D., 1942a, The biological formation of cholesterol from acetate, *J. Biol. Chem.* **143**:297.

Bloch, K., and Rittenberg, D., 1942b, On the utilization of acetate for cholesterol formation, *J. Biol. Chem.* **145**:625.

Bollman, J. L., and Flock, E. V., 1951, Cholesterol in intestinal and hepatic lymph in the rat, *Am. J. Physiol.* **164**:480.

Burkitt, D. P., Walker, A. R. P., and Painter, N. S., 1974, Dietary fiber and disease, *J. Am. Med. Assoc.* **229**:1068.

Burstein, M., Scholnick, M. R., and Morjin, R., 1970, Rapid method for the isolation of lipoproteins from human serum by precipitation with polyanions, *J. Lipid Res.* **11**:583.

Carew, T. E., Koschinson, T., Mayos, S. B., and Steinberg, D., 1976, Mechanism by which high density lipoproteins may slow atherogenic process, *Lancet* **1**:1315.

Carroll, K. K., and Hamilton, R. M. G., 1975, Effects of dietary protein and carbohydrate on plasma cholesterol levels in relation to atherosclerosis, *J. Food Sci.* **40**:18.

Channon, M. J., 1925, Cholesterol synthesis in the animal body, *Biochem. J.* **19**:424.

Connor, W. E., and Connor, S. L., 1972, The key role of nutritional factors in the prevention of coronary heart disease, *Prev. Med.* **1**:49.

Cook, R. P. (ed.), 1958, *Cholesterol: Chemistry, Biochemistry and Pathway*, Academic Press, New York.

Deckelbaum, R. J., Shipley, G. G., and Small, D. M., 1977, Structure and interactions of lipids in human plasma low density lipoproteins, *J. Biol. Chem.* **252**:744.

de Lalla, O. F., and Gofman, J. W., 1954, Ultracentrifuge analysis of serum lipoproteins, *Methods Biochem. Anal.* **1**:459.

Dempsey, M. E., 1974, Regulation of steroid biosynthesis, *Annu. Rev. Biochem.* **43**:967.

Dietschy, J. M., and Wilson, J. D., 1970a–c, Regulation of cholesterol metabolism, *N. Engl. J. Med.* **282**:1179; 1241; 1128.

Dorfman, R. I., and Ungar, F., 1965, *Metabolism of Sex Hormones,* Academic Press, New York.

Eisenberg, S., 1976, Mechanisms of formation of low density lipoproteins metabolic pathways and their regulation, in: *Low Density Lipoproteins* (E. C. Day and R. S. Levy, eds.), pp. 73–92, Plenum Press, New York.

Eisenberg, S., and Levy R. I., 1975, Lipoprotein metabolism, *Adv. Lipid Res.* **13**:1.

Ershoff, B. M., 1963, Effects of pectin N. F. and other complex carbohydrates on hypercholesterolemia and atherosclerosis, *Exp. Med. Surg.* **21**:108.

Fielding, C. J., 1978, Origin and properties of remnant lipoproteins, in: *Disturbances in Lipid and Lipoprotein Metabolism* (J. M. Dietschy, A. M. Gotto, and J. A. Ontko, eds.), pp. 83–98, Williams and Wilkins, Baltimore.

Fieser, L. F., and Fieser, M., 1959, *Steroids,* Reinhold, New York.

Fisher, M., Soller, W. G., and Griminger, P., 1966, The retardation by pectin of cholesterol induced atherosclerosis in fowl, *J. Atheroscler. Res.* **6**:292.

Fredrickson, D. S., and Levy, R. I., 1978, Familial hyperlipoproteinemia, in: *The Metabolic Basis of Inherited Disease,* 3rd ed. (J. B. Stanbury, J. B. Wyngaarden, and D. S. Fredrickson, eds.), pp. 545–614, McGraw-Hill, New York.

Fredrickson, D. S., Levy, R. I., and Lees, R. S., 1967a–e, Fat transport in lipoproteins–an integrated approach to mechanisms and disorders, *N. Engl. J. Med.* **276**:34; 94; 148; 215; 273.

Frolicher, E., and Sullman, H., 1934, Die veresterung von cholesterin bei die resorption aus dem darm, *Biochem. Z.* **274**:21.

Gamble, J. L., and Blackfan, K. D., 1920, Evidence indicating a synthesis of cholesterol by infants, *J. Biol. Chem.* **42**:401.

Glickman, R. M., and Green, P. M. R., 1977, The intestine as a source of apolipoprotein A1, *Proc. Natl. Acad. Sci. U.S.A.* **74**:2569.

Glueck, C. J., Gartside, P. S., Steiner, P. M., Miller, M., Todhunter, T., Hoaf, J., Pucke, M., Terrana, M., Fallat, R. W., and Kashyap, M. L., 1977, Hyperalpha and hypobeta lipoproteinemia in octogenarian kindreds, *Atherosclerosis* **27**:387.

Goldstein, J. L., and Brown, M. S., 1977, The low density lipoprotein pathway and its relation to atherosclerosis, *Annu. Rev. Bicohem.* **46**:897.

Goodman, D. S., and Noble, R. Pl, 1968, Turnover of plasma cholesterol in man, *J. Clin. Invest.* **47**:231.

Goodman, D. S., Noble, R. P., and Dell, R. B., 1973, Three-pool model of the long term turnover of plasma cholesterol in man, *J. Lipid Res.* **14**:178.

Grande, F., 1974, Sugars in cardiovascular disease, in: *Sugars in Nutrition* (H. I. Sipple and K. W. McNutt, eds.), pp. 401–437, Academic Press, New York.

Green, P. M. R., Tall, A. R., and Glickman, R. M., 1978, Rat intestine secretes discoid high density lipoprotein, *J. Clin. Invest.* **61**:528.

Gresham, G. A., and Howard, A. N., 1960, The independent production of atherosclerosis and thrombosis in the rat, *Br. J. Exp. Pathol.* **41**:395.

Hamilton, R. L., 1978, Hepatic secretion and metabolism of high density lipoproteins, in: *Disturbances in Lipid and Lipoprotein Metabolism* (J. M. Dietschy, A. M. Gotto, and J. A. Ontko, eds.), pp. 155–172, Williams and Wilkins, Baltimore.

Hamilton, R. M. G., and Carroll, K. K., 1976, Plasma cholesterol levels in rabbits fed low fat, low cholesterol diets: Effects of dietary proteins, carbohydrates and fibre from different sources, *Atherosclerosis* **24**:47.

Hardinge, M. G., and Stare, F. J., 1954a, Nutritional studies of vegetarians I. Nutritional, physical and laboratory studies, *Am. J. Clin. Nutr.* **2**:73.

Hardinge, M. G., and Stare, F. J., 1954b, Nutritional studies of vegetarians II. Dietary and serum levels of cholesterol, *Am. J. Clin. Nutr.* **2**:83.

Hatch, F. T., and Lees, R. S., 1968, Practical methods for plasma lipoprotein analysis, *Adv. Lipid Res.* **6**:1.

Havel, R. J., 1978, Origin of HDL, in: *High Density Lipoproteins and Atherosclerosis* (A. M. Gotto, N. E. Miller, and M. F. Oliver, eds.), pp. 21–36, Elsevier/North Holland Biomedical Press, Amsterdam.

Havel, R. J., Eder, H. A., and Bragdon, J. M., 1955, The distribution and chemical composition of ultracentrifugally separated lipoproteins in human serum, *J. Clin. Invest.* **34:**1345.

Hechter, D., 1958, Conversion of cholesterol to steroid hormones, in: *Cholesterol* (R. P. Cook, ed.), pp. 309–348, Academic Press, New York.

Hellendoorn, E. W., 1978, Some critical observations in relation to "dietary fibre," the methods for its determination and the current hypotheses for the explanation of its physiological action, *Voeding* **39:**230.

Hofman, A. F., 1976, Fat digestion: The interaction of lipid digestion products with micellar bile acid solutions, in: *Lipid Absorption: Biochemical and Clinical Aspects* (K. Rommel, H. Goebell, and R. Bohmer, eds.), pp. 3–22, MTP Press, Lancaster, England.

Howard, A. N., and Marks, J., 1977, Hypocholesterolaemic effect of milk, *Lancet* **2:**255.

Howard, A. N., Gresham, G. A., Jones, D., and Jennings, I. W., 1965, The prevention of rabbit atherosclerosis by soya bean meal, *J. Atheroscler. Res.* **5:**330.

Ignatowski, A., 1909, Über die wirkung des tierischem eiweisses auf die aorta und die parenchymatösen organe der kaninchen, *Virchows Arch. Pathol. Anat. Physiol.* **198:**248.

Jackson, R. L., Morrisett, J. D., and Gotto, A. M., 1976, Lipoprotein structure and metabolism, *Physiol. Rev.* **56:**2509.

Jenkins, D. J. A., Leeds, A. R., Newton, C., and Cummings, J. M., 1975, Effect of pectin, guar gum and wheat fibre on serum cholesterol, *Lancet* **1:**1116.

Jensen, R. G., Sampugna, J., and Pereira, R. L., 1964, Pancreatic lipase lipolysis of synthetic triglycerides containing a *trans* fatty acid, *Biochim. Biophys. Acta* **84:**481.

Kagan, A., Kannel, W. B., Dawber, T. R., and Revotskie, N., 1962, The coronary profile, *Ann. N.Y. Acad. Sci.* **97:**883.

Kannel, W. B., and Gordon, T., 1970, *The Framingham Study: Section 24: The Framingham Diet Study: Diet and the Regulation of Serum Cholesterol,* U.S. Public Health Serv., Washington, D.C.

Kay, R. M., and Strasberg, S. M., 1978, Origin, chemistry, physiological effects and clinical importance of dietary fibre, *Clin. Invest. Med.* **1:**9.

Kay, R. M., and Truswell, A. S., 1977, Effect of citrus pectin on blood lipids and fecal steroid excretion in man, *Am. J. Clin. Nutr.* **30:**171.

Kiehm, T. G., Anderson, J. W., and Ward, R., 1976, Beneficial effects of a high carbohydrate high fiber diet on hyperglycemic diabetic man, *Am. J. Clin. Nutr.* **29:**895.

Kiriyama, S., Okozaki, Y., and Yoshida, A., 1969, Hypocholesterolemic effect of polysaccharides and polysaccharide-rich foodstuff in cholesterol-fed rats, *J. Nutr.* **97:**382.

Knittle, J. L., and Ahrens, E. H., 1964, Carbohydrate metabolism in two forms of hyperglyceridemia, *J. Clin. Invest.* **43:**485.

Kritchevsky, D., 1958, *Cholesterol,* Wiley, New York.

Kritchevsky, D., 1970, Role of cholesterol vehicle in experimental atherosclerosis, *Am. J. Clin. Nutr.* **23:**1105.

Kritchevsky, D., 1977, Dietary fiber: What it is and what it does, *Ann. N.Y. Acad. Sci.* **300:**283.

Kritchevsky, D., 1978, Fiber, lipids and atherosclerosis, *Am. J. Clin. Nutr.* **31:**565.

Kritchevsky, D., 1979, Vegetable protein and atherosclerosis, *J. Am. Oil Chem. Soc.* **56:**135.

Kritchevsky, D., and Baldino, A. R., 1978, Pancreatic cholesteryl ester synthetase: Effects of *trans* unsaturated and long chain saturated fatty acids, *Artery* **4:**480.

Kritchevsky, D., Moyer, A. W., Tesar, W. C., Logan, J. B., Brown, R. A., Davies, M. C., and Cox, H. R., 1954, Effect of cholesterol vehicle in experimental atherosclerosis, *Am. J. Physiol.* **178:**30.

Kritchevsky, D., Moyer, A. W., Tesar, W. C., McCandless, R. F. J., Logan, J. B., Brown, R. A., and Englert, M. E., 1956, Cholesterol vehicle in experimental atherosclerosis. II. Influence of unsaturation, *Am. J. Physiol.* **185:**279.

Kritchevsky, D., Sallata, P., and Tepper, S. A., 1968, Experimental atherosclerosis in rabbits fed cholesterol-free diets. II. Influence of various carbohydrates, *J. Atheroscler. Res.* **8:**697.

Kritchevsky, D., Tepper, S. A., Vesselinovitch, D., and Wissler, R. W., 1971, Cholesterol vehicle in experimental atherosclerosis. XI. Peanut oil, *Atherosclerosis* **14:**53.

Kritchevsky, D., Tepper, S. A., Vesselinovitch, D., and Wissler, R. W., 1973a, Cholesterol vehicle in experimental atherosclerosis. XIII. Randomized peanut oil, *Atherosclerosis* **17**:225.

Kritchevsky, D., Tepper, S. A., and Kitagawa, M., 1973b, Experimental atherosclerosis in rabbits fed cholesterol-free diets. III. Comparison of fructose and lactose with other carbohydrates, *Nutr. Rep. Int.* **7**:193.

Kritchevsky, D., Davidson, L. M., Shapiro, I. L., Kim, H. K., Kitagawa, M., Malhotra, S., Nair, P. P., Clarkson, T. B., Bersohn, I., and Winter, P. A. D., 1974, Lipid metabolism and experimental atherosclerosis in baboons: Influence of cholesterol-free semi-synthetic diets, *Am. J. Clin. Nutr.* **27**:29.

Kritchevsky, D., Tepper, S. A., Kim, H. K., Moses, D. E., and Story, J. A., 1975, Experimental atherosclerosis in rabbits fed cholesterol-free diets. IV. Investigations into the source of cholesteremia, *Exp. Mol. Pathol.* **22**:11.

Kritchevsky, D., Tepper, S. A., Kim, H. K., Story, J. A., Vesselinovitch, D., and Wissler, R. W., 1976, Experimental atherosclerosis in rabbits fed cholesterol-free diets. V. Comparison on peanut, corn, butter and coconut oils, *Exp. Mol. Pathol.* **24**:375.

Kritchevsky, D., Tepper, S. A., Williams, D. W., and Story, J. A., 1977, Experimental atherosclerosis in rabbits fed cholesterol-free diets, Part 7. Interaction of animal or vegetable protein with fiber, *Atherosclerosis* **26**:397.

Kritchevsky, D., Tepper, S. A., Morrissey, R. B., Czarnecki, S. K., and Klurfeld, D. M., 1979, Influence of whole or skim milk on cholesterol metabolism in rats, *Am. J. Clin. Nutr.* **32**:597.

Kummerow, F. A., 1974, Current studies on relation of fat to health, *J. Am. Oil Chem. Soc.* **51**:255.

Kuo, P. T., 1965, Dietary sugar in the production of hypertriglyceridemia in patients with hyperlipemia and atherosclerosis, *Trans. Assoc. Am. Physicians* **78**:97.

Kyd, P. A., and Bouchier, I. A. D., 1972, Cholesterol metabolism in rabbits with oleic acid induced cholelithiasis, *Proc. Soc. Exp. Biol. Med.* **141**:846.

Langdon, R. G., and Bloch, K., 1953a, The biosynthesis of squalene, *J. Biol. Chem.* **200**:129.

Langdon, R. G., and Bloch, K., 1953b, The utilization of squalene in the biosynthesis of cholesterol, *J. Biol. Chem.* **200**:135.

Lees, R. S., and Hatch, F. T., 1963, Sharper separation of lipoprotein species by paper electrophoresis in albumin containing buffer, *J. Lab. Clin. Med.* **61**:518.

Lewis, B., 1976, *The Hyperlipidaemias: Clinical and Laboratory Practice,* Blackwell Scientific Publications, London.

Lin, T. M., Karvinen, E., and Ivy, A. C., 1955, Capacity of the rat intestine to absorb cholesterol, *Proc. Soc. Exp. Biol. Med.* **89**:422.

Lindgren, F. T., Jensen, L. C., and Hatch, F. T., 1972, The isolation and quantitative analysis of serum lipoproteins, in: *Blood Lipids and Lipoproteins: Quantitation, Composition and Metabolism* (G. J. Nelson, ed.), pp. 181–274, Wiley, New York.

Little, H. N., and Bloch, K., 1950, Studies on the utilization of acetate for biological synthesis of cholesterol, *J. Biol. Chem.* **183**:33.

Lofland, H. B., and Clarkson, T. B., 1968, Interrelated effects of nutritional factors on serum lipids and atherosclerosis, in: *Proceedings of Symposia on Dairy Lipids and Lipid Metabolism* (M. F. Brink and D. Kritchevsky, eds.), pp. 135–148, AVI, Westport, Conn.

Lofland, H. B., Clarkson, T. B., and Goodman, H. O., 1961, Interactions among dietary fat, protein and cholesterol in atherosclerosis-susceptible pigeons; effects on serum cholesterol and aortic atherosclerosis, *Circ. Res.* **9**:919.

Lofland, H. B., Clarkson, T. B., Rhyne, L., and Goodman, H. O., 1966, Interrelated effects of dietary fats and proteins on atherosclerosis in the pigeon, *J. Atheroscler. Res.* **6**:395.

Macdonald, I. 1973, Effects of dietary carbohydrates on serum lipids, in: *Effects of Carbohydrates on Lipid Metabolism* (I. Macdonald, ed.), pp. 216–243, S. Karger, Basel.

Macdonald, I., and Braithwaite, D. M., 1964, The influence of dietary carbohydrates on the lipid pattern in serum and in adipose tissue, *Clin. Sci.* **27**:23.

Malinow, M. R., and McLaughlin, P., 1975, The effect of skim milk on plasma cholesterol in rats, *Experientia* **31**:1012.

Masoro, E. J., 1968, *Physiological Chemistry of Lipids in Mammals,* p. 117, W. B. Saunders, Philadelphia.

McMillan, G. C., Silver, M. D., and Weigensberg, B. I., 1963, Elaidinized olive oil and cholesterol atherosclerosis, *Arch. Pathol.* **76:**106.

Meeker, D. R., and Kesten, H. D., 1940, Experimental atherosclerosis and high protein diets, *Proc. Soc. Exp. Biol. Med.* **45:**543.

Meeker, D. R., and Kesten, H. D., 1941, Effect of high protein diets on experimental atherosclerosis in rabbits, *Arch. Pathol.* **31:**147.

Mendeloff, A. I., 1976, A critique of "fiber deficiency," *Am. J. Dig. Dis.* **21:**109.

Miller, G. J., and Miller, N. E., 1975, Plasma HDL concentration and development of ischaemic heart disease, *Lancet* **1:**16.

Miranda, P. M., and Horwitz, D. L., 1978, High fiber diets in the treatment of diabetes mellitus, *Ann. Intern. Med.* **88:**482.

Morrisett, J. D., Jackson, R. L., and Gotto, A. M., 1977, Lipid–protein interactions in the plasma lipoproteins, *Biochim, Biophys. Acta* **472:**93.

Mueller, J. M., 1915, The assimilation of cholesterol and its esters, *J. Biol. Chem.* **22:**1.

Mueller, J. M., 1916, The mechanism of cholesterol absorption, *J. Biol. Chem.* **27:**463.

Myher, J. J., Marai, L., Kuksis, A., and Kritchevsky, D., 1977, Acylglycerol structure of peanut oils of different atherogenic potential, *Lipids* **12:**775.

Nair, C. R., and Mann, G. V., 1977, A factor in milk which influences cholesteremia in rats, *Atherosclerosis* **26:**363.

Nair, P. P., and Kritchevsky, D. (eds.), 1971, *The Bile Acids: Chemistry, Physiology, and Metabolism, Vol. 1, Chemistry,* Plenum Press, New York.

Nair, P. P., and Kritchevsky, D. (eds.), 1973, *The Bile Acids: Chemistry, Physiology, and Metabolism, Vol. 2, Physiology and Metabolism,* Plenum Press, New York.

Nath, N., Harper, A. E., and Elvehjem, C. A., 1958, Dietary protein and serum cholesterol, *Arch. Biochem. Biophys.* **77:**234.

Nes, W. R., and McKean, M. L., 1977, *Biochemistry of Steroids and Other Isopentenoids,* University Park Press, Baltimore.

Newburgh, L. M., and Clarkson, S., 1923, The production of atherosclerosis in rabbits by feeding diets rich in meat, *Arch. Intern. Med.* **31:**653.

Newburgh, L. M., and Squier, T. L., 1920, High protein diets and atherosclerosis in rabbits: A preliminary report, *Arch. Intern. Med.* **26:**38.

Nichols, A. B., Ravenscroft, C., Lamphiear, D. E., and Ostrander, L. D. Jr., 1976, Independence of serum lipid levels and dietary habits. The Tecumseh study, *J. Am. Med. Assoc.* **236:**1948.

Nikkila, E. A., and Pelkonen, R., 1966, Enhancement of alimentary hypertriglyceridemia by fructose and glycerol in man, *Proc. Soc. Exp. Biol. Med.* **123:**91.

Olson, R. E., 1959, Prevention and control of chronic disease. I., Cardiovascular disease—with particular attention to atherosclerosis, *Am. J. Public Health* **49:**1120.

Olson, R., 1979, Is there an optimum diet for the prevention of coronary heart disease? in: *Nutrition, Lipids and Coronary Heart Disease: A Global View of Nutrition in Health and Disease,* Vol. I (R. I. Levy, B. M. Rifkind, B. H. Dennis, and N. Ernst, eds.), pp. 349–364, Raven Press, New York.

Ono, K., and Fredickson, D. S., 1964, The metabolism of ^{14}C-labeled *cis* and *trans* isomers of octadecaenoic and octadecadienoic acids, *J. Biol. Chem.* **239:**2482.

Osborne, J. C., and Brewer, H. B., 1977, The plasma lipoproteins, *Adv. Protein Chem.* **31:**253.

Page, L., and Friend, B., 1978, The changing United States diet, *Bioscience* **28:**192.

Popjak, G., 1946, The effect of feeding cholesterol with or without fat on the plasma lipids of the rabbit: The role of cholesterol in fat metabolism, *Biochem. J.* **40:**608.

Porter, M. W., Yamanaka, W., Carlson, S. D., and Flynn, M. A., 1977, Effect of dietary egg on serum cholesterol and triglycerides in human males, *Am. J. Clin. Nutr.* **30:**490.

Portman, O. W., and Murphy, P., 1958, Excretion of bile acids and β-hydroxysterols by rats, *Arch. Biochem. Biophys.* **76:**367.

Quintao, E., Grundy, S. M., and Ahrens, E. H., 1971, Effects of dietary cholesterol on the regulation of total body cholesterol in man, *J. Lipid Res.* **12:**233.

Radding, C. M., and Steinberg, D., 1960, Studies on the synthesis and secretion of serum lipoproteins by rat liver slices, *J. Clin. Invest.* **39:**1560.

Rhoads, G. G., Gulbrandsen, C. L., and Kagan, A., 1976, Serum lipoprotein and coronary heart disease in a population study of Hawaii-Japanese men, *N. Eng. J. Med.* **294**:293.

Rittenberg, D., and Schoenheimer, R., 1935, Deuterium as an indicator in the study of intermediary metabolism. II. Methods, *J. Biol. Chem.* **111**:169.

Rittenberg, D., and Schoenheimer, R., 1937, Deuterium as an indicator in the study of intermediary metabolism. XI. Further studies on biological uptake of deuterium into organic substance of special reference to fat and cholesterol formation, *J. Biol. Chem.* **121**:235.

Rodwell, V. W., Nordstrom, J. L, and Mitschelen, J. J., 1976, Regulation of HMG CoA reductase, *Adv. Lipid Res.* **14**:1.

Rothblat, G. M., Buchko, M. K., and Kritchevsky, D., 1968, Cholesterol uptake by L5178Y tissue culture cells, *Biochim. Biophys. Acta* **164**:327.

Rudel, L. L., Lee, J. A., Morris, M. D., and Felts, J. M., 1974, Characterization of plasma lipoproteins separated and purified by agarose-column chromatography, *Biochem. J.* **139**:89.

Russ, E. M., Eder, H. A., and Barr, D. P., 1951, Protein lipid relationships in human plasma. I. In normal individuals, *Am. J. Med.* **11**:468.

Sabine, J. R., 1977, *Cholesterol,* Marcel Dekker, New York.

Sacks, F. M., Castelli, W. P., Donner, A., and Maas, E. M., 1975, Plasma lipids and lipoproteins in vegetarians and controls, *N. Eng. J. Med.* **292**:1148.

Schaefer, E. J., Eisenberg, S., and Levy, R. I., 1978, Lipoprotein–apoprotein metabolism, *J. Lipid Res.* **19**:667.

Schoenheimer, R., and Breusch, F., 1933, Synthesis and destruction of cholesterol in the organism, *J. Biol. Chem.* **103**:439.

Schoenheimer, R., and Rittenberg, D., 1935a, Deuterium as an indicator in the study of intermediary metabolism, *J. Biol. Chem.* **111**:163.

Schoenheimer, R., and Rittenberg, D., 1935b, Deuterium as an indicator in the study of intermediary metabolism. III. The role of the fat tissues, *J. Biol. Chem.* **111**:175.

Schreibman, P. M., 1975, Diet and plasma lipids, *Adv. Exp. Med. Biol.* **60**:159.

Scott, R. F., Morrison, E. S., Thomas, W. A., Jones, R., and Nam, S. C., 1964, Short term feeding of unsaturated vs saturated fat in the production of atherosclerosis in the rat, *Exp. Mol. Pathol.* **3**:421.

Segall, J. J., 1977, Is milk a coronary health hazard? *Br. J. Prev. Soc. Med.* **31**:81.

Sgoutas, D. S., 1968, Hydrolysis of synthetic cholesterol esters containing *trans* fatty acids, *Biochim. Biophys. Acta* **164**:317.

Sgoutas, D. S., 1970, Effects of geometry and position of ethylenic bond upon acyl coenzyme A-cholesterol-*O*-acyltransferase, *Biochemistry* **9**:1826.

Sirtori, C. R., Agradi, E., Conti, F., Mantera, O., and Gatti, E., 1977, Soybean-protein diet in treatment of type II hyperlipoproteinaemia, *Lancet* **1**:275.

Sirtori, C. R., Conti, F., Sirtori, M., Girantranceschi, G., Zucchi, C., Zoppi, S., Agradi, E., Tavazzi, L., Mantero, O., Gatti, E., and Kritchevsky, D., 1979, Clinical experience with the soybean protein diet in the treatment of hypercholesterolemia, *Am. J. Clin. Nutr.* **32**:1645.

Slack, J., Noble, N., Meade, T. W., and North, W. R. S., 1977, Lipid and lipoprotein concentrations in 1604 men and women in working populations in North West London, *Br. Med. J.* **2**:353.

Slater, G., Mead, J., Dhopeshwarkar, G., Robinson, S., and Alfin-Slater, R. B., 1976, Plasma cholesterol and triglycerides in men with added eggs in the diet, *Nutr. Rep. Int.* **14**:249.

Sonderhoff, R., and Thomas, H., 1937, Die enzymatische dehydrieung der trideutero-essigsaure, *Ann. Chem.* **530**:195.

Stamler, J., 1979, Population studies, in: *Nutrition, Lipids and Coronary Heart Disease: A Global View of Nutrition in Health and Disease,* Vol. I (R. I. Levy, B. M. Rifkind, B. H. Dennis, and N. Ernst, eds.), pp. 25–88, Raven Press, New York.

Stamler, J., Bolene, C., Levinson, E., Dudley, M., and Katz, L. N., 1948, Blood and tissue lipids in the chick fed cholesterol in various forms, *Am. J. Physiol.* **155**:470.

Stavraky, K. M., 1976, The role of ecologic analysis in studies of the etiology of disease: A discussion with reference to large bowel cancer, *J. Chronic Dis.* **29**:435.

Stone, D. B., and Connor, W. E., 1963, The prolonged effects of a low cholesterol, high carbo-
    hydrate diet upon the serum lipids in diabetic patients, *Diabetes* **12**:127.
Story, J. A., and Kritchevsky, D., 1976, Dietary fiber and lipid metabolism, in: *Fiber in Human
    Nutrition* (G. A. Spiller and R. J. Amen, eds.), pp. 171–184, Plenum Press, New York.
Strasser, T., 1972, Atherosclerosis and coronary heart disease: The contribution of epidemiology,
    W.H.O. Chron. **26**:7.
Strong, J. P., and McGill, H. C. Jr., 1967, Diet and experimental atherosclerosis in baboons, *Am.
    J. Pathol.* **50**:669.
Trowell, H. T., 1972, Ischemic heart disease and dietary fiber, *Am. J. Clin. Nutr.* **25**:926.
Truswell, A. S., and Kay, R. M., 1976, Bran and blood lipids, *Lancet* **1**:367.
Tsai, A. C., Elias, J., Kelley, J. J., Lin, R. S. C., and Robson, J. R. K., 1976, Influence of certain
    dietary fibers on serum and tissue cholesterol levels in rats, *J. Nutr.* **106**:118.
Turner, K. B., 1933, Studies on the prevention of cholesterol atherosclerosis in rabbits. I. The
    effects of whole thyroid and of potassuim iodide, *J. Exp. Med.* **58**:115.
Vesselinovitch, D., Getz, G. S., Hughes, R. M., and Wissler, R. W., 1974, Atherosclerosis in the
    rhesus monkey fed three food fats, *Atherosclerosis* **20**:303.
Vijayagopal, P., and Kurup, P. A., 1970, Effect of dietary starches on the serum aorta and hepatic
    lipid levels in cholesterol-fed rats, *Atherosclerosis* **11**:257.
Weigensberg, B. I., and McMillan, G. C., 1964a, Lipids in rabbits fed elaidinized olive oil and
    cholesterol, *Exp. Mol. Pathol.* **3**:201.
Weigensberg, B. I., and McMillan, G. C., 1964b, Serum and aortic lipids in rabbits fed cholesterol
    and linoleic acid stereoisomers, *J. Nutr.* **83**:314.
Wells, A. F., and Ershoff, B. M., 1961, Beneficial effects of pectin in prevention of hypercholes-
    terolemia and increase in liver cholesterol in cholesterol fed rats, *J. Nutr.* **74**:87.
Yerushalmey, J., and Hilleboe, M. E., 1957, Fat in the diet and mortality from heart disease: A
    methodologic note, *N.Y. State Med. J.* **57**:2343.
Yudkin, J., 1963, Nutrition and palatability with special reference to obesity, myocardial infarction
    and other diseases of civilization, *Lancet* **1**:1335.
Zilversmit, D. B., 1978, Assembly of chylomicrons in the intestinal cell, in: *Disturbances in Lipid
    and Lipoprotein Metabolism* (J. M. Dietschy, A. M. Gotto, and J. A. Ontko, eds.), pp. 69–
    82, Williams and Wilkins, Baltimore.

10

Nutrients with Special Functions: Dietary Fiber

Jon A. Story and David Kritchevsky

1. Introduction

The study of dietary fiber has recently attracted a great deal of interest. Most of the current discussion originated with a group of epidemiological studies comparing Western nations with undeveloped societies of rural Africa.

The most publicized of these studies was by Burkitt *et al.* (1974), who reported that ischemic heart disease, appendicitis, diverticular disease, gallstones, varicose veins, hiatus hernia, hemorrhoids, and colon cancer were very rarely seen in rural Africa. This they attributed, in part, to the consumption of a diet high in fiber. Several authors have presented data that tend to support this "fiber theory." Cleave (1956, 1974) preceded that current fiber hypothesis when he postulated that many of the diseases common to the Western world were related to the consumption of refined carbohydrate. Trowell has presented evidence supporting the relationship of increased fiber intake with a decreased incidence of heart disease (1972a,b, 1975a) and has expanded the theory to include diabetes mellitus (1973) and obesity (1975b). These papers provided a nucleus of information around which the current interest in fiber has revolved.

It must be remembered, however, that the study of dietary fiber is not completely new. Work from the laboratory of A. R. P. Walker in the 1960s (Walker and Arvidson, 1954; Walker and Walker, 1969; Walker *et al.*, 1961) pointed out the effects of fiber and discussed theories concerning fiber's relationship to bowel motility, heart disease, and cancer. Walker (1976) has recently reviewed these studies in detail.

The correlations made in the reports cited above do not constitute cause-and-effect relationships. Many differences other than the level of dietary fiber

Jon A. Story and David Kritchevsky ● The Wistar Institute of Anatomy and Biology, Philadelphia, Pennsylvania 19104.

exist between the populations studied. Life expectancy in rural Africa, for example, is lower than that in the Western world. This could prejudice findings, since many of the diseases in question occur late in life. Other differences in diet and life-style also raise serious doubts about a strictly causal relationship between fiber and disease. In relating disease conditions to lack of fiber in the American diet (Burkitt *et al.*, 1974), it should be pointed out that despite the reduced levels of fiber in the American diet (Scala, 1975), the incidence of appendicitis has dropped by 30% since 1940, and the incidence of colon cancer has not changed between 1940 and 1975.

With the foregoing precautions in mind, we propose to review the status of fiber research because, in spite of the shortcomings of the epidemiological data, some of the reported effects of fiber have been borne out by experimentation. There have been two scientific books published over the past few years that review the effects of dietary fiber (Burkitt and Trowell, 1975; Spiller and Amen, 1976). In this chapter, we will review and update the information available concerning the influence of fiber in nutrition.

2. A Definition of Fiber

One of the major problems inherent in the study of the effects of fiber has been the failure of workers to agree on a definition of fiber. The evolution of a definition has been hampered by a methodology that is not sufficiently sophisticated to determine the components of fiber. The result has been that, until recently, several terms describing fiber were used interchangeably, and many typical fiber sources that were not completely and accurately analyzed were employed in experiments.

Until recently, all data concerning the fiber content of foods were expressed in terms of crude fiber content. Methods for determining crude fiber were developed in the 19th century (Horwitz, 1970; Mangold, 1934). "Crude fiber" is defined as the residue remaining after sequential treatment of food with solvents, dilute acid, dilute base, and ashing. Unfortunately, newer, more exact methods have shown that this process recovers variable quantities of the various components of fiber and thus gives erroneous and inconsistent results (Southgate, 1969b; Van Soest, 1973; Van Soest and McQueen, 1973).

The recent interest in fiber has resulted in a resurgence in work on methods of estimating the composition of fiber. Two analytical methods have emerged as the most accurate. Van Soest (1963; Goering and Van Soest, 1970) has developed a method utilizing detergents to fractionate and quantitate all the constituents of the cell wall. Southgate (1969a,b) has developed a method of systematic extraction and quantitation of the various components of fiber. Both techniques are quite accurate and have been shown to give similar results for both cellulose and lignin content (McConnell and Eastwood, 1974). The consensus of opinion indicates that the Van Soest method is best suited for forages available in large quantities and for foods that do not contain starch or

Table I. Dietary Fiber Composition of Vegetables (g/100 g)[a]

Food	Total dietary fiber	Noncellulosic polysaccharides	Cellulose	Lignin
Pea, frozen	7.8	5.5	2.1	0.2
Bean, baked, canned	7.3	5.7	1.4	0.2
Corn, canned	5.7	5.0	0.6	0.1
Parsnip, raw	4.9	3.8	1.1	Trace
Broccoli, boiled	4.1	2.9	0.9	0.1
Carrot, boiled	3.7	2.2	1.5	Trace
Potato, raw	3.5	2.5	1.0	Trace
String bean, boiled	3.4	1.9	1.3	0.2
Brussels sprout, boiled	2.9	2.0	0.8	0.1
Cabbage, boiled	2.8	1.8	0.7	0.4
Turnip, raw	2.2	1.5	0.7	Trace
Onion, raw	2.1	1.6	0.5	Trace
Cauliflower, boiled	1.8	0.7	1.1	Trace
Lettuce, raw	1.5	0.5	1.0	Trace
Tomato, raw	1.4	0.7	0.4	0.3
Pepper, cooked	0.9	0.6	0.2	Trace

[a] After Southgate *et al.* (1976).

significant amounts of lipids. The Southgate method is more time-consuming but is very reliable for human diets and has the added advantage of providing estimates of noncellulosic polysaccharides (Cummings, 1976). The term "crude fiber" should be retired, and data from either one or both of these methods should be adopted in the near future. Southgate (1976a) has reviewed, in detail, the methods of fiber analysis and has determined the fiber content of many human foods (Southgate *et al.*, 1976) (Tables I and II).

Table II. Dietary Fiber Composition of Breads and Cereals (g/100 g)[a]

Food	Total dietary fiber	Noncellulosic polysaccharides	Cellulose	Lignin
Breads				
Whole-meal	8.5	6.0	1.3	1.2
Brown	5.1	3.6	1.3	0.2
White	2.7	2.0	0.7	Trace
Cereals				
Bran	44.0	32.7	8.1	3.2
All-Bran	26.7	17.8	6.0	2.9
Puffed Wheat	15.4	10.4	2.6	2.4
Shredded Wheat	12.3	8.8	2.6	0.8
Corn Flakes	11.0	7.3	2.4	1.3
Grape-Nuts	7.0	5.1	1.3	0.6
Sugar Puffs	6.1	4.0	1.0	1.1
Special K	5.5	3.7	0.7	1.1
Rice Krispies	4.5	3.5	0.8	0.2

[a] After Southgate *et al.* (1976).

3. The Composition of Fiber

Fiber consists of the structural components of plants, which are, for the most part, indigestible by humans. The major components are cellulose, hemicellulose, lignin, pectins, and gums.

Cellulose is the most abundant and best-known component of fiber. It consists of unbranched polymers of 1-4β-D-glucose with an average molecular weight of approximately 6×10^5 (3000 carbohydrate units).

Hemicellulose is a much more complex polysaccharide. It contains a mixture of pentoses and hexoses, many of which are branched, and uronic acids, which are sugars that contain a terminal carboxyl group. Each hemicellulose molecule contains an average of 150–200 carbohydrate units.

Of the major components of fiber, lignin is the only one of noncarbohydrate composition. Lignin is composed of polymers of substituted phenylpropanes. The molecular weight and polymer size of lignin vary a great deal depending on the source. Estimates of molecular weight range from 1000 to 8000.

Pectins are usually classified as dietary fiber. However, their solubility in water prompts some investigators to exclude them from this category. Pectins are present in small amounts in plant cell walls and intercellular spaces. The basic structure consists of a polymer of 1-4β-D-galacturonic acid with several other sugars; the average molecular weight is 6–9 $\times$ 10^4.

Two closely related fiber components are gums and mucilages. They are highly branched chains of galacturonic or glucuronic acids with a few other sugars.

There are also several fiber-associated substances that may have some nutritional importance. Chief among these are phytic acid, silica, cutin, and protein.

Cummings (1976) and Southgate (1976b) have reviewed extensively the chemistry of dietary fiber.

4. Intestinal Function and Disease

4.1. Stool Weight and Transit Time

The effects of fiber on laxation have been well known since the time of Hippocrates (McCance and Widdowson, 1955). Cowgill and his collaborators (Cowgill and Anderson, 1932; Cowgill and Sullivan, 1933) described the laxative effects of bran. Williams and Olmstead (1936) did a thorough study of the laxative properties of foods such as cabbage and cereals. They analyzed the foods for fiber content and tried to correlate laxative properties with composition but met with little success. Their work presaged the complexity of the problem. Burkitt (1971, 1973) reported rapid (25–40 hr) transit time in native African school children eating a high-residue diet; in a similar-aged-group of English boarding-school boys consuming a low-residue diet, transit time was much slower (70+ hr). The weight of feces excreted was also much different:

It was four times greater in the African children. Other workers have reported similar findings of decreased transit time and increased fecal bulk in relation to cellulose in the diet in other African populations (Walker, 1961; Walker *et al.*, 1970).

The experimental addition of 16 g of bran/day to human diets for three weeks resulted in a twofold increase in fecal weight (Eastwood *et al.*, 1973). Both wet weight and dry weight increased as a result of the increased intake of indigestible carbohydrate and the increased water-holding capacity of the bran.

McConnell *et al.* (1974) have tabulated the water-holding capacity of a large number of fruits and vegetables in order to allow for a wider selection of foods to increase fecal excretion.

Similarly, Wyman *et al.* (1976) found that raw bran increased wet and dry weight of feces excreted. However, they reported that cooked bran did not cause a similar response. Coarse raw bran was effective in increasing speed of transit at 16 and 20 g/day, but cooked bran was not effective even at 22 g/day. Bran has been shown to decrease transit time and increase fecal weight in many other experiments (Connell and Smith, 1974; Findlay *et al.*, 1974; Kirwan *et al.*, 1974; Parks, 1974; Payler *et al.*, 1975).

An interesting facet of bran's effect on fecal transit is its ability to "normalize" transit time. Harvey *et al.* (1973) reported on eight subjects whose transit times ranged from one to seven days (mean, 2.6). After at least four weeks on 30 g bran/day, the range was from one to four days (mean, 2.1). This convergence on two days was described as a normalization, with the understanding that any definition of normal is arbitrary.

4.2. Diverticular Disease

Diverticular disease is a condition of the large intestine in which the mucosa protrudes through the muscle layers that have become weakened as a result of increased pressures inside the colon (Price, 1975). These increased pressures are thought to result from the long-term consumption of a low-fiber diet. In developed countries, 30–40% of the population will develop diverticula (Parks, 1974), and about 10% of these cases will require surgery (Painter, 1969). Painter and Burkitt (1971) suggested the use of an unrefined diet in treatment of diverticular disease. Subsequent experiments with patients have shown this treatment to be useful in relief of symptoms (Findlay *et al.*, 1974; Painter *et al.*, 1972; Plumley and Francis, 1973). Painter *et al.* (1972) fed from 3 to 45 g of bran/day (mean, 13 g) and observed very encouraging relief of symptoms (89%) and a relatively low rate of dropouts (8%).

Bran functions by increasing the mass of feces in the colon, thus decreasing the intraluminal pressure. Pressure measurements before and after bran treatment in patients with diverticular disease have confirmed this decrease in pressure (Findlay *et al.*, 1974).

Tests concerning the beneficial effects of other high-fiber foods have not been carried out. However, it is likely that these foods will be as useful as bran in treatment of diverticular disease. Water-holding capacity may be a factor in their efficacy.

The treatment of diverticular disease with a high-fiber diet is becoming more widespread, thus reversing the standard treatment of the past 50 years (Spriggs and Marper, 1927). This use of fiber, regardless of how beneficial, still does not imply an etiological link, as has been suggested by the epidemiological evidence.

4.3. Cancer of the Colon

Dietary components have long been implicated in colon cancer (Gregor *et al.*, 1969). There are currently two major hypotheses concerning the development of colon cancer. Hill (1976) has reviewed the correlation between levels of dihydroxy bile acids and the incidence of colon cancer. Underlying his hypothesis is the suggestion that the dihydroxy bile acids are converted to carcinogenic polynuclear hydrocarbons by bacterial action. The other hypothesis states that high-fiber diets, because of their effect on transit time, reduce *in vivo* residence time of carcinogenic or cocarcinogenic materials and thus reduce risk of colon cancer. At this writing, neither hypothesis has been subjected to rigorous experimental testing.

Epidemiological evidence has indicated that a high-fiber diet may protect against the development of colon cancer (Burkitt *et al.*, 1974). Burkitt (1969; Burkitt *et al.*, 1972) has suggested that the low-residue diet of people in the Western world results in increased exposure of the lining of the colon to the feces and the carcinogens they may contain. In addition, the slower transit time may result in more bacterial transformation of dietary or secretory materials, the products of which may be carcinogenic.

A comparison of populations with low and high risk of developing colon cancer does, in fact, show differences in bacterial flora. Aries *et al.* (1969) found that low-risk populations had a lower ratio of anaerobic to aerobic bacteria, which supports the theory of slow-moving fecal contents. Moore *et al.* (1969) have shown that a complete change to a vegetarian diet did not alter the types of bacteria present in feces.

A comparison of the fecal flora of vegetarians with that of people eating a mixed diet (Aries *et al.*, 1971) showed no differences in type or number of organisms, but cultures of the bacteria did demonstrate a functional difference vis-à-vis steroid metabolism. Bacteria from the vegetarians had a much lower level of bile acid 7-dehydroxylase, the enzyme responsible for conversion of primary to secondary bile acids. Correspondingly, lower levels of secondary bile acids were found in the feces of the vegetarian group. Others have also reported differences in excretion of coprostanol and coprostanone, the bacterial degradation products of cholesterol, in the feces of vegetarians (Reddy and Wynder, 1973).

Fiber apparently does have an effect on bile acid excretion. As can be seen in Table III, Eastwood *et al.* (1973) found that 16 g of bran/day increased both the concentration and the total excretion of bile acids. Jenkins *et al.*

Table III. *Effects of Supplementary Dietary Fiber on Bile Acid Excretion in Humans*

Reference	Supplementary fiber (type and amount)	Duration of feeding	Fecal neutral steroid		Fecal acidic steroid	
			mg/g dry feces (% change)	mg/day (% change)	mg/g dry feces (% change)	mg/day (% change)
Eastwood et al. (1973)	Bran (16 g/day)	3 wk	N.R.[a]	N.R.	+20	+32
Cummings et al. (1976)	Bran (28 g/day)	3 wk	N.R.	N.R.	-35	+40
Jenkins et al. (1975a)	Bran (36 g/day)	3 wk	-55	+36	-57	+40
Baird et al. (1977)	Bran (39 g/day)	3 wk	N.C.[b]	N.C.	-41	-2
	Bagasse (10.5 g/day)	12 wk	-41	-10	N.C.	+50

[a] N.R., not reported.
[b] N.C., no change.

(1975a) found that 36 g of bran/day increased total excretion of both acidic and neutral steroids but greatly reduced the concentration of steroids in the feces. Baird *et al.* (1977) fed 39 g of bran/day and observed no change in neutral steroid excretion but a great decrease in acidic steroid per gram of feces and a small decrease in total excretion. Sugar-cane fiber (bagasse), however, decreased neutral steroid excretion both per gram of feces and per day and increased total acidic steroid excretion without changing the concentration. Thus, the effects of fiber on steroid excretion cannot be easily summarized. Total acidic excretion seems to be increased, but, due to the great increase in fecal bulk, the concentration may, in fact, be decreased.

If steroids, other materials present in the intestine, or their metabolites act as carcinogens or cocarcinogens in the colon, as has been suggested (Hill, 1976), and if fiber does, in fact, decrease the incidence of colon cancer, fiber could have four possible mechanisms of action. It might (1) dilute the potential carcinogen or cocarcinogen, thus reducing its effect; (2) decrease the length of time the colon is exposed to these materials by increasing fecal transit time through the colon; (3) decrease production of these carcinogens by altering bacterial flora or their functional abilities; or (4) adsorb these materials, thus reducing their availability to the lining of the colon. (This aspect of fiber will be discussed in detail in the following section.) The most likely explanation of the role of fiber in colon cancer is some combination of these effects, a combination that may vary with the composition and amount of fiber being fed as well as with all the other components of the diet.

The incidence of colon cancer in the United States has not changed in the last three or four decades (Enstrom, 1975), despite the apparent reduction in dietary fiber during that time. Within the United States, bowel cancer mortality is highest in the northeastern states and lowest in the southern states (Blot *et al.,* 1976). The pattern of fiber intake in these areas has not been studied. The possible connection between dietary fiber and colon cancer remains a hypothesis that must be put to a rigorous test.

5. Lipid Metabolism and Atherosclerosis

5.1. Animals

As has been mentioned, in the early 1970s, epidemiological data indicated that high-fiber diets may protect against heart disease. The data suggested that fiber played a role in lipid metabolism and atherosclerosis in experimental animals.

Kritchevsky (1964) reviewed a large number of experiments involving the feeding of types of fat to rabbits and examination of their effects on serum lipids and aortic atherosclerosis. He found apparent disagreement concerning the effects of dietary saturated fat on lipid metabolism. None of the fats was hyperlipemic or atherogenic when fed with commercial laboratory rations;

however, when fed as part of a semipurified diet, saturated fat was definitely atherogenic and cholesterolemic (Lambert *et al.*, 1958; Malmros and Wigand, 1959). Since the two types of diet (commercial and semipurified) had only saturated fat in common, it was concluded that some other dietary component, most likely the fiber, was responsible for the disparate results (Kritchevsky, 1964). Subsequently, Kritchevsky and Tepper (1965, 1968) found that saturated fat fed with ether-extracted commercial ration was not atherogenic. This was further indication that some other dietary component was protecting the rabbits against the effects of saturated fat.

Moore (1967) had made a similar observation. He pointed out that when wheat straw or peat was included in a high-fat, semipurified atherogenic diet, the degree of hyperlipemia and atherogenesis was significantly lower than when the diet contained cellulose or celluloid.

Alfalfa has also been shown to be hypocholesterolemic. Cookson *et al.* (1967) found that alfalfa fed to rabbits in sufficient quantities (90%) prevented the cholesterolemia and atherosclerotic lesions that otherwise resulted from cholesterol feeding (600 mg/day). This hypocholesterolemia was accompanied by a great increase in excretion of neutral steroids, which indicated that alfalfa was interfering with cholesterol absorption (Horlick *et al*, 1967).

Kritchevsky *et al.* (1973; 1974b) have also reported a decrease in cholesterol absorption with alfalfa. Compared with cellulose in isocaloric, semipurified diets, alfalfa resulted in greater excretion of both neutral and acidic steroids.

Specific components of fiber have different effects on lipid metabolism. Cellulose seems to have little effect on cholesterol levels in either cholesterol-containing or cholesterol-free diets. Several workers have shown that cellulose is ineffective in reducing the accumulation of cholesterol in the serum and liver of rats fed cholesterol (Kiriyama *et al.*, 1969; Tsai *et al.*, 1976; Wells and Ershoff, 1961). Large quantities of cellulose in semipurified, cholesterol-free diets result in increased cholesterol levels in rabbits (Hamilton and Carroll, 1976). Kiriyama and Tsai and their associates have also shown that cellulose and agar are cholesterolemic for rats fed semipurified, cholesterol-containing diets.

Pectin and lignin, on the other hand, have been shown to reduce serum and tissue cholesterol levels in rats. Lin *et al.* (1957) fed 500 mg of pectin and 50 mg of cholesterol daily and found a large increase in excretion of saponifiable lipids in the feces. Leveille and Sauberlich (1966) reported a large increase in bile acid but not in neutral sterol excretion in rats fed pectin with 1% cholesterol. Pectin is also hypocholesterolemic in rabbits fed cholesterol-containing (Berenson *et al.*, 1975) and/or cholesterol-free (Hamilton and Carroll, 1976) diets. Lignin, fed to rats at a level of 30 g/day, results in a significant lowering of serum lipids (Judd *et al.*, 1976).

Recently, Story *et al.* (1977) have compared the effects of 5% each of cellulose, lignin, and pectin in rats fed 0.5% cholesterol in a semipurified diet (Table IV). Addition of cholesterol resulted in a 540% increase in liver cholesterol over the group fed a cholesterol-free diet. Liver cholesterol levels

Table IV. Lipid Levels in Rats Fed 0.5% Cholesterol with Various Components of Fiber[a]

	B[b]	BC	BC + 5% cellulose	BC + 5% lignin	BC + 5% pectin
Survival	5/6	6/6	6/6	6/6	6/6
Weight gain (g)	240 ± 11	259 ± 14	215 ± 9	212 ± 13	229 ± 20
Liver weight (g)	10.5 ± 0.2	13.1 ± 0.7	10.2 ± 0.5	11.7 ± 0.5	10.3 ± 0.8
Liver, % body weight	4.42 ± 0.25	3.53 ± 0.13	3.11 ± 0.10	3.61 ± 0.27	2.99 ± 0.08
Cholesterol					
Serum (mg/dl)	97.8 ± 12.7	79.9 ± 8.7	80.4 ± 10.9	77.3 ± 6.9	82.4 ± 8.7
Serum (% ester)	81	86	88	83	85
Liver (mg/g)	2.8 ± 0.2	17.9 ± 3.4	12.6 ± 2.1	6.1 ± 2.0	4.5 ± 0.7
Liver (% ester)	31	84	82	61	54
Triglycerides					
Serum (mg/dl)	71.2 ± 11.6	60.3 ± 14.2	49.3 ± 6.9	79.3 ± 8.7	60.8 ± 9.9
Liver (mg/g)	26.1 ± 8.7	49.6 ± 10.3	31.6 ± 4.5	40.7 ± 7.2	20.8 ± 3.5

[a] Abbreviations: B, basal diet; BC, basal diet + 0.5% cholesterol.
[b] 60% sucrose, 24% casein, 10% corn oil, 5% salt mix, and 1% vitamin mix; cholesterol and fiber added at expense of sucrose.

were lowered by 30% with the addition of cellulose to the diet, 66% with the addition of lignin, and 75% with pectin. Liver triglycerides were also lowered by all three components of fiber. Serum cholesterol levels were not changed.

The effects of fiber on atherosclerosis have been studied in several animal models. In addition to the aforementioned rabbit experiments of Kritchevsky and Tepper (1965, 1968), Cookson *et al.* (1967), and Moore (1967), Fisher *et al.* (1966) found that 3% pectin reduced diet-induced atherogenesis in chickens. Baboons and vervet monkeys develop more severe aortic sudanophilia when fed a semipurified diet than when fed a control diet of bread, fruits, and vegetables (Kritchevsky *et al.*, 1974a, 1977a).

Fiber seems to affect the atherogenicity of other dietary variables. Kritchevsky *et al.* (1977b) fed rabbits either casein or soya protein in a semipurified diet with alfalfa, wheat straw, or cellulose as fiber. Casein is more atherogenic and cholesterolemic than soya protein (Hamilton and Carroll, 1976; Meeker and Kesten, 1940). As can be seen in Table V, casein was more cholesterolemic with each type of fiber. Within protein groups, cholesterolemia was always cellulose > wheat straw > alfalfa. When one looks at the atherosclerosis data, an interesting picture emerges. With cellulose as fiber, casein is more atherogenic than soya protein. When wheat straw is used, casein is still slightly more atherogenic. However, when alfalfa is the fiber, the incidence of atheroma is similar. Thus, both alfalfa and, to a lesser extent, wheat straw seem to have the ability to moderate the effects of casein.

Portman (1960) reviewed a large body of data examining the role of diet in bile acid metabolism. Rats fed commercial laboratory ration excreted more cholic acid and β-hydroxysterols, and the half-life and pool size of their cholic acid decreased in comparison to rats fed no fiber. A component of the grain

Table V. Interaction of Fiber and Protein in Rabbit
Atherosclerosis[a]

Protein[b]	Fiber[b]	Serum cholesterol (mg/dl)	Atherosclerosis[c]
Casein	Cellulose	402	1.50
	Wheat straw	375	1.03
	Alfalfa	193	0.63
Soya	Cellulose	248	1.25
	Wheat straw	254	0.91
	Alfalfa	159	0.73

[a] After Kritchevsky *et al.* (1977b).
[b] Diet contained 40% sucrose, 25% protein, 15% fiber, 14% hydrogenated coconut oil, 5% salt mix, and 1% vitamin fortification.
[c] Average of arch and thoracic portion of aorta graded 0–4 after Sudan IV staining.

present in commercial diets was found to be responsible for the observed differences (Table VI). This component was not fat.

Eastwood and Boyd (1967) examined the distribution of bile salts along the intestine and between solid and liquid portions of the contents. Substantial quantities of bile salts were found bound to the solid material in the intestine, apparently unavailable for reabsoprtion. Eastwood and Hamilton (1968) further examined this phenomenon by binding bile salts to a dry grain preparation *in vitro*. They found that this high-fiber grain did, in fact, bind bile salts.

The binding of bile acids and bile salts to fiber has since been thoroughly examined. Kritchevsky and Story (1974) found that alfalfa and other types of fiber normally found in animal diets bound substantial quantities of bile salts. Cellulose bound no bile salts. Others have also found that many materials in the diets of both humans and experimental animals bind appreciable quantities of bile salts (Balmer and Zilversmit, 1974; Birkner and Kern, 1974; Kritchevsky

Table VI. Influence of the Components of a Commercial Diet on Cholic
Acid Metabolism in Rats[a]

Diet	Cholic acid half-life (days)	Cholic acid pool (mg/kg)	Cholic acid excretion (mg/kg per day)
Commercial diet (C)[b]	2.0	100	35
Extracted C[c]	2.2	97	31
Semipurified (SP)[d]	4.2	62	10
SP + lipid from C	2.8	50	12
SP + nonsaponifiables from C	3.7	64	12

[a] After Portman (1960).
[b] Purina Laboratory Chow.
[c] Ethanol-extracted chow with vitamins and corn oil added to levels present in SP.
[d] 67.6% sucrose, 20% casein, 8% corn oil, 4% salts, 0.1% inositol, 0.2% choline, and 0.1% *p*-aminobenzoic acid and vitamin fortification.

et al., 1976; Story and Kritchevsky, 1975). Recently, Eastwood *et al.* (1976) reported a method by which the adsorbed bile salts could be distinguished from those dissolved in the solvent adhering to the fiber. Extensive rinsing with buffer after an initial adsorption did not dislodge the bile salts from the fiber. This method should be very useful for comparing the adsorption of types of fiber with widely different water-holding capacities.

It is important to note that the adsorption affinities of various types of fiber for the various bile salts are very different (Kritchevsky and Story, 1975; Story and Kritchevsky, 1976). Measurement of the binding capacity of a certain type of fiber for one bile salt cannot be assumed to indicate its capacity for all the bile salts.

This information about bile salt binding offers a possible explanation for the hypercholesterolemic effects of fiber. If fiber binds bile salts in the small intestine, these bile salts are unavailable for interaction with cholesterol and other lipids essential for micelle formation, which is, in turn, necessary for fat absorption. In addition, if these bile salts remained bound to fiber and were excreted, they would have to be replaced by further synthesis from cholesterol. The result would be a twofold drain on the body cholesterol: Excretion of cholesterol not absorbed due to the unavailability of bile salts would increase, and bile salts would be lost in the feces. These losses of both cholesterol and bile salts have been observed in experimental animals. If the losses are great enough, cholesterol levels decrease and atherosclerosis is inhibited, as has also been observed in experimental animals. The real question that remains is whether this theory is viable in humans.

5.2. Humans

As was the case with animal experimentation, data concerning the effects of fiber on atherosclerosis in humans were available before the rebirth of interest in fiber in the early 1970s. Hardinge and colleagues (Hardinge and Stare, 1954; Hardinge *et al.,* 1958) had observed that vegetarians consumed 50% more fiber (crude fiber) and had cholesterol levels which were 28% below those of the general population. Walker and Arvidson (1954) had noticed the absence of heart disease among the South African Bantu and associated it with increased fiber intake.

The fiber source most thoroughly tested with regard to its influence on serum lipids in humans has been bran. It was chosen because it had proven efficacious in diverticular disease, and it was a familiar source of fiber. Unfortunately, it appears to be completely ineffective in lowering serum lipids. Truswell and Kay (1976) have recently summarized the data concerning the effect of bran on serum lipids. Ten studies were cited in which an average of 14 subjects was fed 14–100 g of bran for 3–19 weeks, and no significant effects were seen on serum cholesterol or triglyceride levels.

Bile acid metabolism is, however, affected by bran, as was noted earlier (Table III). These patterns of excretion are inconsistent and apparently not of sufficient magnitude to influence serum lipids.

Biliary bile acid dynamics, however, seem to be influenced by bran. Pomare and Heaton (1973) found that 30 g of bran/day fed for four weeks resulted in reduced deoxycholate and increased chenodeoxycholate levels in aspirated duodenal contents (bile). More recently, Pomare *et al.* (1976) have shown that bran (20–108 g/day for four to six weeks) increased the cheno-deoxycholate pool in the bile by 27% and reduced the deoxycholate pool by 33%. The ability of bran to alter biliary bile acids may be important in treating cholelithiasis, since the increase in chenodeoxycholate levels has been shown to result in a dissolution of gallstones.

To date, the most promising fiber-associated substance for lowering serum lipid levels in humans has been pectin. As early as 1961 (Keys *et al.*, 1961), pectin was shown to be hypocholesterolemic. Given in doses of 6–36 g/day for 14–28 days, pectin generally lowered serum cholesterol levels (Table VII). Fahrenbach *et al.* (1965), however, found that pectin had no effect on serum cholesterol. Kay and Truswell (1977) reported that 15 g of pectin/day resulted in a 15% drop in serum cholesterol, a 16% increase in fecal excretion of neutral steroids, and a 40% increase in excretion of bile acids.

Guar gum has also been reported to be hypocholesterolemic in humans. Fahrenbach *et al.* (1965) fed 6 g/day to 23 subjects for 66 days and observed a 5% reduction in serum cholesterol. In two different studies, guar gum fed for 45 days resulted in an 11% lowering. Jenkins *et al.* (1975b) fed 36 g/day for two weeks and lowered serum cholesterol by 16%.

Fiber in the diet appears to be related to lipid metabolism and atherosclerosis. Both animal experiments and data from human subjects provide convincing evidence that certain types and specific components of fiber can lower lipid levels and retard atherosclerosis. However, as has been widely shown with bran in humans, it is impossible to make a responsible universal statement about the effectiveness of fiber in lowering lipids or influencing the atherogenic process.

Table VII. The Effect of Pectin on Serum Cholesterol Levels in Humans

Reference	Pectin (g/day)	Duration (days)	Serum cholesterol (% change)
Keys *et al.* (1961)	15	20	−5
Fahrenbach *et al.* (1965)	6	66	0
Fahrenbach *et al.* (1965)	12	52	0
Palmer and Dixon (1966)	10	28	−5
Palmer and Dixon (1966)	6	28	−4
Jenkins *et al.* (1975b)	36	14	−12
Durrington *et al.* (1976)	12	21	−9
Kay and Truswell (1977)	15	21	−15

6. Diabetes Mellitus

Cleave and Campbell (1966) and Trowell (1973, 1974) have suggested that the increased consumption of fiber-depleted carbohydrate foods has played an etiological role in the increased incidence of diabetes. Trowell argues that the incidence of diabetes in England and Wales can be correlated with the fiber content of the flour available for baking bread. Thus, in the years 1934–1944, the diabetes mortality in women averaged 145/million, and the crude fiber content of wheat flour was 1.8 g/kg; between 1942 and 1946, the average mortality fell to 123/million, while the crude fiber content of the wheat was 5.0 g/kg. In the two periods cited, fat intake was similar (18–20 kg/person per year); sugar intake averaged 39 kg/person per year in 1939–1941 and 34 kg/person per year in the period 1942–1946.

Stone and Connor (1963) conducted a study in which two groups of diabetics were fed the 2200-calorie diet approved by the American Diabetic Association. The control diet contained 17% of calories as protein, 41% of calories as carbohydrate, and 42% of calories as fat. The experimental diet contained 64% of calories as carbohydrate and 20% of calories as fat. There are no data relating to the type of carbohydrate that was fed. In a one-year study, the control patients showed practically no change in serum lipid levels. In the experimental group, cholesterol levels fell from 250 to 198 mg/dl ($p <$ 0.001) and triglycerides from 150 to 124 mg/dl ($p < 0.1$).

Recently, Kiehm et al. (1976) have conducted a similar study. In this experiment, the patients were their own controls and were switched from a 2200-calorie diet that contained 234 g of carbohydrate (43% of calories) to one containing 414 g of carbohydrate (75% of calories). The ratio of starch to simple sugars in the first diet was 1.15, and in the second it was 2.63. The control diet contained 4.7 g of crude fiber and the test diet, 14.2 g. Although there was no weight change, average cholesterol levels fell from 198 to 151 mg/dl ($p < 0.01$), triglyceride levels from 165 to 140 mg/dl, and blood glucose from 183 to 136 mg/dl ($p < 0.05$). The patients on oral hypoglycemic therapy or on low insulin maintenance were able to suspend therapy.

Grande (1974) summarized results from 12 studies in which starch was substituted for sucrose on an isocaloric basis. The percentage of calories exchanged ranged from 16 to 40 and the duration of the studies from two to six weeks. In every case there was a reduction in serum cholesterol levels (average, 13 mg/dl). The foods used for substitution were bread, legumes, rice, potatoes, and fruits. Although no data on fiber content are available, clearly the exchange diets were richer in this component.

7. Overconsumption of Fiber

Dietary fiber is not an unmixed blessing. In populations subsisting on diets very high in fiber, cases of sigmoid volvulus can be observed (Sutcliffe, 1968). This condition is due to increased colonic content of volatile gases that may lead to twisting of the sigmoid colon.

Various types of fiber, especially those with a high content of phytate, bind divalent cations and render them unavailable for absorption. Thus, in populations ingesting a suboptimal diet, there is the danger of a deficiency in zinc, iron, calcium, phosphorus, and magnesium. Reinhold *et al.* (1974) found that the availability of zinc (for rats) was severalfold greater from leavened than from unleavened bread. This group has also tested the effects of different breads upon trace metal (Ca^{2+}, Mg^{2+}) absorption in humans (Haghshenass *et al.*, 1972; Reinhold *et al.*, 1973, 1976) and found lower absorption in subjects eating whole-grain breads. When two men were fed wheat in whole-meal bread, they were shown to develop negative balances of calcium, magnesium, phosphorus, and zinc. Fecal losses of these elements were correlated with fecal dry matter, which was, in turn, proportional to fecal fiber excretion. The same subjects were able to utilize these elements when fed white bread. The level of negative balance was on the order of 10% of intake. The absorption of iron in subjects eating whole-meal bread has been shown to be significantly lower than that in the same subjects eating white bread (Dobbs and Baird, 1977).

It has also been reported recently that fiber, especially wheat bran, may contain trypsin and chymotrypsin inhibitors that may be detrimental to the health of populations subsisting on a marginal diet (Mistuanaga, 1974; Schneeman, 1977).

8. Summary

Fiber is the polysaccharide and lignin component of plant structures. It consists of carbohydrate polymers, cellulose, hemicellulose, pectin, and gums and of the hydroxylated phenylpropane polymer called lignin. The exact composition of plant fiber depends upon the source and age of the plant.

Fiber analysis has progressed from treatment of foods with acid, base, and solvents to sophisticated schemes that can provide an exact analysis of components.

Fiber has the capacity to adsorb water and to bind metal ions and bile acids. It has yet to be determined which of these properties underlies its mechanism(s) of metabolic action.

For many years, dietary fiber was considered necessary for laxation and little else. Beginning with the observations of Cleave (1956, 1974; Cleave and Campbell, 1966), there has been interest in the role of high- or low-fiber diets in the development of various disease conditions. The major recent impetus to the study of dietary fiber has come from the observations of Burkitt (1969) and Trowell (1973).

Beneficial effects from dietary fiber have been claimed for a wide spectrum of diseases. The two conditions that have received the greatest amount of attention are heart disease and colon cancer. There are no firm data relating to these hypotheses, and they remain to be put to a rigorous test.

Indirect tests, such as examining the effects of fiber on serum cholesterol levels, have revealed the individual properties of fiber. Thus, bran is ineffective in reducing cholesterolemia in humans or experimental animals, whereas pectin has been shown to be hypocholesterolemic in humans and to reduce atherogenesis in chickens. In the field of colon cancer, the experimental data are sparse. Many experiments are in progress, but few results have been reported.

It is not surprising to find controversy when one dietary component, fiber, is proposed as the most important single variable in diseases of multiple etiology. Dietary fiber is not a panacea, but it may play a role in alleviating symptoms in some patients. If this is so, the furor over fiber has been worthwhile.

ACKNOWLEDGMENT. This work was supported, in part, by grants (HL-03209; AG-0076) and a Research Career Award (HL-0734) from the National Institutes of Health.

9. References

Aries, V., Crowther, J. S., Drasar, B. S., Hill, M. J., and Williams, R. E. O., 1969, Bacteria and the aetiology of cancer of the large bowel, *Gut* **10**:334.

Aries, V. C., Crowther, J. S., Drasar, B. S., Hill, M. J., and Ellis, F. R., 1971, The effect of a strict vegetarian diet on the faecal flora and faecal steroid concentration, *J. Pathol.* **103**:54.

Baird, I. McL., Walters, R. L., Davies, P. S., Hill, M. J., Drasar, B. S., and Southgate, D. A. T., 1977, The effects of two dietary fiber supplements on gastrointestinal transit, stool weight and frequency, and bacterial flora, and fecal bile acids in normal subjects, *Metabolism* **26**:117.

Balmer, J., and Zilversmit, D. B., 1974, Effects of dietary roughage on cholesterol absorption, cholesterol turnover and steroid excretion in the rat, *J. Nutr.* **104**:1319.

Berenson, L. M., Bhandaru, R. R., Radhakrishnamurthy, B., Srinivasan, S. B., and Berenson, G. S., 1975, The effect of dietary pectin on serum lipoprotein cholesterol in rabbits, *Live Sci.* **16**:1533.

Birkner, H. J., and Kern, F. Jr., 1974, *In vitro* adsorption of bile salts to food residues, salicyclazosulfapyridine and hemicellulose, *Gastroenterology* **67**:237.

Blot, W. J., Fraumeni, F. J., Jr., Stone, B. J., and McKay, F. W., 1976, Geographic patterns of large bowel cancer in the United States, *J. Natl. Cancer Inst.* **57**:1225.

Burkitt, D. P., 1969, Related disease—related cause, *Lancet* **2**:1229.

Burkitt, D. P., 1971, Epidemiology of cancer of the colon and rectum, *Cancer* **28**:3.

Burkitt, D. P., 1973, Epidemiology of large bowel disease: The role of fibre, *Proc. Nutr. Soc.* **32**:145.

Burkitt, D. P., and Trowell, H. C. (eds.), 1975, *Refined Carbohydrate Foods and Disease,* Academic Press, London.

Burkitt, D. P., Walker, A. R. P., and Painter, N. S., 1972, Effect of dietary fibre on stools and transit times and its role in the causation of disease, *Lancet* **2**:1408.

Burkitt, D. P., Walker, A. R. P., and Painter, N. S., 1974, Dietary fiber and disease, *J. Am. Med. Assoc.* **229**:1068.

Cleave, T. L., 1956, The neglect of natural principles in current medical practice, *J. R. Nav. Med. Serv.* **42**:55.

Cleave, T. L., 1974, *The Saccharine Disease,* John Wright and Sons, Bristol, England.

Cleave, T. L., and Campbell, G. D., 1966, *Diabetes, Coronary Thrombosis and the Saccharine Diseases,* John Wright and Sons, Bristol, England.

Connell, A. M., and Smith, C. L., 1974, The effect of dietary fibre on transit time, in: *Proceedings of the Fourth International Symposium on Gastrointestinal Motility,* pp. 365–369, Mitchell Press, Vancouver.

Cookson, F. B., Altschul, R., and Fedoroff, S., 1967, The effects of alfalfa on serum cholesterol and in modifying or preventing cholesterol-induced atherosclerosis in rabbits, *J. Atheroscler. Res.* **7**:69.

Cowgill, G. R., and Anderson, W. E., 1932, Laxative effect of wheat bran and washed bran in healthy man, *J. Am. Med. Assoc.* **98**:1886.

Cowgill, G. R., and Sullivan, A. J., 1933, Further studies on the use of wheat bran as a laxative, *J. Am. Med. Assoc.* **100**:795.

Cummings, J. H., 1976, What is fiber? in: *Fiber in Human Nutrition* (G. A. Spiller and R. J. Amen, eds.), pp. 1–30, Plenum Press, New York.

Cummings, J. H., Hill, M. J., Jenkins, D. J. A., Pearson, J. R., and Wiggins, H. S., 1976, Changes in fecal composition and colonic function due to cereal fiber, *Am. J. Clin. Nutr.* **29**:1468.

Dobbs, R. J., and Baird, I. McL., 1977, Effect of wholemeal and white bread on iron absorption in normal people, *Br. Med. J.* **1**:1641.

Durrington, P. N., Manning, A. P., Bolton, C. H., and Hartog, M., 1976, Effect of pectin on serum lipids and lipoproteins, whole gut transit time and stool weight, *Lancet* **2**:394.

Eastwood, M. A., and Boyd, G. S., 1967, The distribution of bile salts along the small intestine of rats, *Biochim, Biophys. Acta* **137**:393.

Eastwood, M. A., and Hamilton, D., 1968, Studies on the adsorption of bile salts to non-absorbed components of diet, *Biochim. Biophys. Acta* **152**:165.

Eastwood, M. A., Kirkpatrick, J. R., Mitchell, W. D., Bone, A., and Hamilton, T., 1973, Effects of dietary supplements of wheat bran on faeces and bowel function, *Br. Med. J.* **4**:392.

Eastwood, M. A., Anderson, R., Mitchell, W. D., Robertson, J., and Pocock, S., 1976, A method to measure the adsorption of bile salts to vegetable fiber of differing water holding capacity, *J. Nutr.* **106**:1429.

Enstrom, J. E., 1975, Colorectal cancer and consumption of beef and fat, *Br. J. Cancer* **32**:432.

Fahrenbach, M. J., Riccardi, B. A., Saunders, J. C., Lourie, I. N., and Heider, J. G., 1965, Comparative effects of guar gum and pectin on human serum cholesterol levels, *Circulation* **32** (Suppl. II):11.

Findlay, J. M., Smith, A. N., Mitchell, W. D., Anderson, A. J. B., and Eastwood, M. A., 1974, Effects of unprocessed bran on colon function in normal subjects and in diverticular disease, *Lancet* **1**:146.

Fisher, H., Soller, W. G., and Griminger, P., 1966, The retardation by pectin of cholesterol-induced atherosclerosis in the fowl, *J. Atheroscler. Res.* **6**:292.

Goering, H. K., and Van Soest, P. J., 1970, *Forage Fiber Analyses,* U.S. Dep. Agric., Agric. Handb. No. 379.

Grande, F., 1974, Sugars in cardiovascular disease, in: *Sugars in Nutrition* (H. L. Sipple and K. W. McNutt, eds.), pp. 401–437, Academic Press, New York.

Gregor, O., Toman, R., and Prusova, F., 1969, Gastrointestinal cancer and nutrition, *Gut* **10**:1031.

Haghshenass, M., Mahlondji, M., Reinhold, J. G., and Mohammadi, N., 1972, Iron deficiency anemia in an Iranian population associated with high intakes of iron, *Am. J. Clin. Nutr.* **25**:1143.

Hamilton, R. M. G., and Carroll, K. K., 1976, Plasma cholesterol levels in rabbits fed low fat, low cholesterol diets: Effect of dietary proteins, carbohydrates and fibre from different sources, *Atherosclerosis* **24**:47.

Hardinge, M. G., and Stare, F. J., 1954, Nutritional studies of vegetarians. II. Dietary and serum levels of cholesterol, *Am. J. Clin. Nutr.* **2**:83.

Hardinge, M. G., Chambers, A. C., Crooks, H., and Stare, F. J., 1958, Nutritional studies of vegetarians. III. Dietary levels of fiber, *Am. J. Clin. Nutr.* **6**:523.

Harvey, R. F., Pomare, E. W., and Heatson, K. W., 1973, The effects of increased dietary fibre on intestinal transit, *Lancet* **1**:1278.

Hill, M. J., 1976, Fecal steroids in the etiology of large bowel cancer, in: *The Bile Acids,* Vol. 3 (P. P. Nair and D. Kritchevsky, eds.), pp. 169–200, Plenum Press, New York.

Horlick, L., Cookson, F. B., and Fedoroff, S., 1967, Effect of alfalfa feeding on the excretion of fecal neutral sterols in the rabbit, *Circulation* **36** (Suppl. II):18.

Horwitz, W., 1970, Crude fiber, in: *Official Methods of Analysis of the Association of Official Analytical Chemists, U.S.A.,* 11th ed., pp. 129–133, Association of Official Analytical Chemists, Washington, D.C.

Jenkins, D. J. A., Hill, M. S., and Cummings, J. H., 1975a, Effect of wheat fiber on blood lipids, fecal steroid excretion and serum iron, *Am. J. Clin. Nutr.* **28**:1408.

Jenkins, D. J. A., Leeds, A. R., Newton, C., and Cummings, J. H., 1975b, Effect of pectin, guar gum and wheat fibre on serum cholesterol, *Lancet* **1**:1116.

Judd, P. A., Kay, R. M., and Truswell, A. S., 1976, Cholesterol lowering effect of lignin in rats, *Proc. Nutr. Soc.* **35**:71A.

Kay, R. M., and Truswell, A. S., 1977, Effect of citrus pectin on blood lipids and fecal steroid excretion in man, *Am. J. Clin. Nutr.* **30**:171.

Keys, A., Grande, F., and Anderson, J. T., 1961, Fiber and pectin in the diet and serum cholesterol concentration in man, *Proc. Soc. Exp. Biol. Med.* **106**:555.

Kiehm, T. G., Anderson, J. W., and Ward, K., 1976, Beneficial effects of a high carbohydrate, high fiber diet on hyperglycemic diabetic men, *Am. J. Clin. Nutr.* **29**:895.

Kiriyama, S., Okazaki, Y., and Yoshida, A., 1969, Hypocholesterolemic effect of polysaccharides and polysaccharide-rich foodstuffs in cholesterol-fed rats, *J. Nutr.* **97**:382.

Kirwan, W. O., Smith, A. N., McConnell, A. A., and Eastwood, M. A., 1974, Action of different bran preparations on colonic function, *Br. Med. J.* **4**:187.

Kritchevsky, D., 1964, Experimental atherosclerosis in rabbits fed cholesterol-free diets, *J. Atheroscler. Res.* **4**:103.

Kritchevsky, D., and Story, J. A., 1974, Binding of bile salts *in vitro* by non-nutritive fiber, *J. Nutr.* **104**:458.

Kritchevsky, D., and Story, J. A., 1975, *In vitro* binding of bile acids and bile salts, *Am. J. Clin. Nutr.* **28**:305.

Kritchevsky, D., and Tepper, S. A., 1965, Factors affecting atherosclerosis in rabbits fed cholesterol-free diets, *Life Sci.* **4**:1467.

Kritchevsky, D., and Tepper, S. A., 1968, Experimental atherosclerosis in rabbits fed cholesterol-free diets: Influence of chow components, *J. Atheroscler. Res.* **8**:357.

Kritchevsky, D., Casey, R. P., and Tepper, S. A., 1973, Isocaloric, isogravic diets in rats. II. Effect on cholesterol absorption and excretion, *Nutr. Rep. Int.* **7**:61.

Kritchevsky, D., Davidson, L. M., Shapiro, I. L., Kim, H. K., Kitagawa, M., Malhotra, S., Nair, P. P., Clarkson, T. B., Bersohn, I., and Winter, P. A. D., 1974a, Lipid metabolism and experimental atherosclerosis in baboons: Influence of cholesterol-free, semi-synthetic diets, *Am. J. Clin. Nutr.* **27**:29.

Kritchevsky, D., Tepper, S. A., and Story, J. A., 1974b, Isocaloric, isogravic diets in rats. III. Effects of non-nutritive fiber (alfalfa or cellulose) on cholesterol metabolism, *Nutr. Rep. Int.* **9**:301.

Kritchevsky, D., Story, J. A., and Walker, A. R. P., 1976, Binding of sodium taurocholate by cereal products, *S. Afr. Med. J.* **50**:1831.

Kritchevsky, D., Davidson, L. M., Kim, H. K., Krendel, D. A., Malhotra, S., Vander Watt, J. J., du Plessis, J. P., Winter, P. A. D., Ipp, T., Mendelsohn, D., and Bersohn, I., 1977a, Influence of semi-purified diets on atherosclerosis in African green monkeys, *Exp. Mol. Pathol.* **26**:28.

Kritchevsky, D., Tepper, S. A., Williams, D. E., and Story, J. A., 1977b, Experimental atherosclerosis in rabbits fed cholesterol-free diets. Part 7. Interaction of animal or vegetable protein with fiber, *Atherosclerosis* **26**:397.

Lambert, G. F., Miller, J. P., Olsen, R. T., and Frost, D. V., 1958, Hypercholesteremia and atherosclerosis induced in rabbits by purified high fat rations devoid of cholesterol, *Proc. Soc. Exp. Biol. Med.* **97**:544.

Leveille, G. A., and Sauberlich, H. E., 1966, Mechanism of the cholesterol-depressing effect of pectin in the cholesterol-fed rat, *J. Nutr.* **88**:209.

Lin, T. M., Kim, K. S., Karvinen, E., and Ivy, A. C., 1957, Effect of dietary pectin, "proto-pectin," and gum arabic on cholesterol excretion in rats, *Am. J. Physiol.* **188**:66.

Malmros, H., and Wigand, G., 1959, Atherosclerosis and deficiency of essential fatty acids, *Lancet* 2:749.

Mangold, D. E., 1934, The digestion and utilisation of crude fibre, *Nutr. Abstr. Rev.* 3:647.

McCance, R. A., and Widdowson, E. M., 1955, Old thoughts and new work on breads white and brown, *Lancet* 2:205.

McConnell, A. A., and Eastwood, M. A., 1974, A comparison of methods of measuring fibre in vegetable material, *J. Sci. Food Agric.* 25:1.

McConnell, A. A., Eastwood, M. A., and Mitchell, W. D., 1974, Physical characteristics of vegetable foodstuffs that could influence bowel function, *J. Sci. Food Agric.* 25:1457.

Meeker, D. R., and Kesten, H. D., 1940, Experimental atherosclerosis and high protein diets, *Proc. Soc. Exp. Biol. Med.* 45:543.

Mistuanaga, T., 1974, Some properties of protease inhibitors in wheat grain, *J. Nutr. Sci. Vitaminol.* 20:153.

Moore, J. H., 1967, The effect of the type of roughage in the diet on plasma cholesterol levels and aortic atherosis in rabbits, *Br. J. Nutr.* 21:207.

Moore, W. E. C., Cato, E. P., and Holdeman, L. V., 1969, Anaerobic bacteria of the gastrointestinal flora and their occurrence in clinical infections, *J. Infect. Dis.* 119:641.

Painter, N. S., 1969, *Diseases of the Colon, Rectum, and Anus,* Heineman, London.

Painter, N. S., and Burkitt, D. P., 1971, Diverticular disease of the colon: A deficiency disease of Western civilization, *Br. Med. J.* 2:450.

Painter, N. S., Almeida, A. Z., and Colebourne, K. W., 1972, Unprocessed bran in treatment of diverticular disease of the colon, *Br. Med. J.* 1:137.

Palmer, G. H., and Dixon, D. G., 1966, Effect of pectin dose on serum cholesterol levels, *Am. J. Clin. Nutr.* 18:437.

Parks, T. G., 1974, The effects of low and high residue diets on the rate of transit and composition of the faeces, in: *Proceedings of the Fourth International Symposium on Gastrointestinal Motility* (E. Daniel, J. A. L. Gilbert, B. Schofield, T. K. Schnitke, and G. Scott, eds.), pp. 369–374, Mitchell Press, Vancouver.

Parsons, D. S., 1973, Dietary fiber, stool output and transit time, *Lancet* 1:152 (Letter).

Payler, D. K., Pomare, E. W., Heaton, K. W., and Harvey, R. F., 1975, The effect of wheat bran on intestinal transit, *Gut* 16:209.

Plumley, P. F., and Francis, B., 1973, Dietary management of diverticular disease, *J. Am. Diet. Assoc.* 63:527.

Pomare, E. W., and Heaton, K. W., 1973, Alteration of bile salt metabolism by dietary fibre (bran), *Br. Med. J.* 4:262.

Pomare, E. W. Heaton, K. W., Low-Beer, T. S., and Espiner, H. J., 1976, The effect of wheat bran upon bile salt metabolism and upon the lipid composition of bile in gallstone patients, *Am. J. Dig. Dis.* 21:521.

Portman, O. W., 1960, Nutritional influences on the metabolism of bile acids, *Am. J. Clin. Nutr.* 8:462.

Price, A. B., 1975, Diverticular disease (pathology), in: *Fiber Deficiency and Colonic Disorders* (R. W. Reilly and J. B. Kirsner, eds.), pp. 101–108, Plenum Press, New York.

Reddy, B. S., and Wynder, E. L., 1973, Large bowel carcinogenesis: Fecal constituents of populations with diverse incidence rates of colon cancer, *J. Natl. Cancer Inst.* 50:1437.

Reinhold, J. G., Nasr, K., Lahimgarzadeh, A., and Hedayati, H., 1973, Effects of purified phytate and phytate-rich bread upon metabolism of zinc, calcium, phosphorus and nitrogen in man, *Lancet* 1:28.

Reinhold, J. G., Parsa, A., Kariman, N., Hammick, J. W., and Ismail-Beigi, F., 1974, Availability of zinc in leavened and unleavened wholemeal wheaten breads as measured by solubility and uptake by rat intestine *in vitro, J. Nutr.* 104:976.

Reinhold, J. G., Faradji, B., Abadi, P., and Ismail-Beigi, F., 1976, Decreased absorption of calcium, magnesium, zinc and phosphorus by humans due to increased fiber and phosphorus consumption as wheat bread, *J. Nutr.* 106:493.

Scala, J., 1975, The physiological effects of dietary fiber, in: *Physiological Effects of Food Carbohydrates* (A. Jeanes and J. Hodge, eds.), pp. 325–335, American Chemical Society, Washington, D.C.

Schneeman, B. O., 1977, The effect of plant fiber on trypsin and chymotrypsin activity *in vitro,* *Fed. Proc. Fed. Am. Soc. Exp. Biol.* **36:**1118.

Southgate, D. A. T., 1969a, Determination of carbohydrates in foods. I. Available carbohydrates, *J. Sci. Food Agric.* **20:**326.

Southgate, D. A. T., 1969b, Determination of carbohydrates in food. II. Unavailable carbohydrates, *J. Sci. Food Agric.* **20:**331.

Southgate, D. A. T., 1976a, The analysis of dietary fiber, in: *Fiber in Human Nutrition* (G. A. Spiller and R. J. Amen, eds.), pp. 73–107, Plenum Press, New York.

Southgate, D. A. T., 1976b, The chemistry of dietary fiber, in: *Fiber in Human Nutrition* (G. A. Spiller and R. J. Amen, eds.), pp. 31–72, Plenum Press, New York.

Southgate, D. A. T., Bailey, B., Collinson, E., and Walker, A. F., 1976, A guide to calculating intakes of dietary fibre, *J. Hum. Nutr.* **30:**303.

Spiller, G. A., and Amen, R. J. (eds.), 1976, *Fiber in Human Nutrition,* Plenum Press, New York.

Spriggs, E. I., and Marper, O. A., 1927, Multiple diverticula of the colon, *Lancet* **1:**1067.

Stone, D. B., and Connor, W. E., 1963, The prolonged effects of a low cholesterol, high carbohydrate diet upon the serum lipids in diabetic patients, *Diabetes* **12:**27.

Story, J. A., and Kritchevsky, D., 1975, Binding of sodium taurocholate by various foodstuffs, *Nutr. Rep. Int.* **11:**161.

Story, J. A., and Kritchevsky, D., 1976, Comparison of the binding of various bile acids and bile salts *in vitro* by several types of fiber, *J. Nutr.* **106:**1292.

Story, J. A., Czarnecki, S. K., Baldino, A., and Kritchevsky, D., 1977, Effect of components of fiber on dietary cholesterol in the rat, *Fed. Proc. Fed. Am. Soc. Exp. Biol.* **36:**1134.

Sutcliffe, M. M. L., 1968, Volvulus of the sigmoid colon, *Br. J. Surg.* **55:**903.

Trowell, H., 1972a, Crude fibre, dietary fibre and atherosclerosis, *Atherosclerosis* **16:**138.

Trowell, H., 1972b, Ischemic heart disease and dietary fiber, *Am. J. Clin. Nutr.* **25:**926.

Trowell, H., 1973, Dietary fibre, ischaemic heart disease and diabetes mellitus, *Proc. Nutr. Soc.* **32:**151.

Trowell, H., 1974, Diabetes mellitus death rates in England and Wales 1920–1970 and food supplies, *Lancet* **2:**998.

Trowell, H., 1975a, Coronary heart disease and dietary fiber, *Am. J. Clin. Nutr.* **28:**798.

Trowell, H., 1975b, Obesity in the western world, *Plant Foods for Man* **1:**157.

Truswell, A. S., and Kay, R. M., 1976, Bran and blood lipids, *Lancet* **1:**367.

Tsai, A. C., Elias, J., Kelly, J. J., Lin, R. S. C., and Robson, J. R. K., 1976, Influence of certain dietary fibers on serum and tissue cholesterol levels in rats, *J. Nutr.* **106:**118.

Van Soest, P. J., 1963, Use of detergents in the analysis of fibrous feeds. I. Preparation of fiber residues of low nitrogen content, *J. Assoc. Off. Agric. Chem.* **46:**825.

Van Soest, P. J., 1973, The uniformity and nutritive availability of cellulose, *Fed. Proc. Fed. Am. Soc. Exp. Biol.* **32:**1804.

Van Soest, P. J., and McQueen, R. W., 1973, The chemistry and estimation of fibre, *Proc. Nutr. Soc.* **32:**123.

Walker, A. R. P., 1961, Crude fiber, bowel motility and pattern of diet, *S. Afr. Med. J.* **35:**114

Walker, A. R. P., 1976, Gastrointestinal diseases and fiber intake with special reference to South African populations, in: *Fiber in Human Nutrition* (G. A. Spiller and R. J. Amen, eds.), pp. 241–261, Plenum Press, New York.

Walker, A. R. P., and Arvidson, U. B., 1954, Fat intake, serum cholesterol concentration and atherosclerosis in the South African Bantu. I. Low fat intake and age trend of serum cholesterol concentrations in the South African Bantu, *J. Clin. Invest.* **33:**1366.

Walker, A. R. P., and Walker, B. F., 1969, Bowel motility and colonic cancer, *Br. Med. J.* **3:**238.

Walker, A. R. P., Mortimer, K. L., Kloppers, P. J., Botha, D., Grusin, H., and Seftel, H. C., 1961, Coronary heart disease in South African poor whites and white prisoners habituated to a Bantu type of diet, *Am. J. Clin. Nutr.* **9:**643.

Walker, A. R. P., Walker, B. F., and Richardson, B. D., 1970, Bowel transit times in Bantu population, *Br. Med. J.* **3:**48.

Wells. A. F., and Ershoff, B. H., 1961, Beneficial effects of pectin in prevention of hypercholesterolemia and increase in liver cholesterol in cholesterol-fed rats, *J. Nutr.* **74:**87.

Williams, R. D., and Olmstead, W. H., 1936, The manner in which food controls the bulk of the feces, *Ann. Intern. Med.* **10:**717.

Wyman, J. B., Heaton, K. W., Manning, A. P., and Wicks, A. C. B., 1976, The effect on intestinal transit and the feces of raw and cooked bran in different doses, *Am. J. Clin. Nutr.* **29:**1474.

Appendix

Table 1. Food and Nutrition Board, National Academy of Sciences—National

| | | Weight | | Height | | | Fat-soluble vitamins | | | Water-soluble vitamins | |
	Age (yr)	(kg)	(lbs)	(cm)	(in.)	Protein (g)	Vitamin A (μg R.E.)[c]	Vitamin D (μg)[d]	Vitamin E (mg αT.E.)[e]	Vitamin C (mg)	Thia-mine (mg)
Infants	0.0–0.5	6	13	60	24	kg × 2.2	420	10	3	35	0.3
	0.5–1.0	9	20	71	28	kg × 2.0	400	10	4	35	0.5
Children	1–3	13	29	90	35	23	400	10	5	45	0.7
	4–6	20	44	112	44	30	500	10	6	45	0.9
	7–10	28	62	132	52	34	700	10	7	45	1.2
Males	11–14	45	99	157	62	45	1000	10	8	50	1.4
	15–18	66	145	176	69	56	1000	10	10	60	1.4
	19–22	70	154	177	70	56	1000	7.5	10	60	1.5
	23–50	70	154	178	70	56	1000	5	10	60	1.4
	51+	70	154	178	70	56	1000	5	10	60	1.2
Females	11–14	46	101	157	62	46	800	10	8	50	1.1
	15–18	55	120	163	64	46	800	10	8	60	1.1
	19–22	55	120	163	64	44	800	7.5	8	60	1.1
	23–50	55	120	163	64	44	800	5	8	60	1.0
	51+	55	120	163	64	44	800	5	8	60	1.0
Pregnant						+30	+200	+5	+2	+20	+0.4
Lactating						+20	+400	+5	+3	+40	+0.5

[a] Designed for the maintenance of good nutrition of practically all healthy people in the U.S.
[b] The allowances are intended to provide for individual variations among most normal persons as they live in the United States under usual environmental stresses. Diets should be based on a variety of common foods in order to provide other nutrients for which human requirements have been less well defined.
[c] Retinol equivalents. 1 R.E. = 1 μg retinol or 6 μg carotene.
[d] As cholecalciferol. 10 μg cholecalciferol = 400 IU vitamin D.
[e] α-Tocopherol equivalents. 1 αT.E. = 1 mg d-α-tocopherol.
[f] Niacin equivalents. 1 N.E. = 1 mg niacin or 60 mg dietary tryptophan.
[g] The folacin allowances refer to dietary sources as determined by *Lactobacillus casei* assay after treatment with enzymes ("conjugases") to make polyglutamyl forms of the vitamin available to the test organism.

Table 2. Estimated Safe and Adequate Daily Dietary Intakes

| | | Vitamins | | | Trace elements[c] | |
	Age (yr)	Vitamin K (μg)	Biotin (μg)	Pantothenic acid (mg)	Copper (mg)	Manganese (mg)
Infants	0–0.5	12	35	2	0.5–0.7	0.5–0.7
	0.5–1	10–20	50	3	0.7–1.0	0.7–1.0
Children and	1–3	15–30	65	3	1.0–1.5	1.0–1.5
adolescents	4–6	20–40	85	3–4	1.5–2.0	1.5–2.0
	7–10	30–60	120	4–5	2.0–2.5	2.0–3.0
	11+	50–100	100–200	4–7	2.0–3.0	2.5–5.0
Adults		70–140	100–200	4–7	2.0–3.0	2.5–5.0

[a] From Recommended Dietary Allowances, revised 1979. Food and Nutrition Board, National Academy of Sciences—National Research Council, Washington, D.C.
[b] Because there is less information on which to base allowances, these figures are not given in the main table of the RDA and are provided here in the form of ranges of recommended intakes.

Research Council Recommended Daily Dietary Allowances (Revised 1979)[a,b,j]

Water-soluble vitamins					Minerals					
Ribo-flavin (mg)	Niacin (mg N.E.)[f]	Vitamin B_6 (mg)	Folacin[g] (μg)	Vitamin B_{12} (μg)	Cal-cium (mg)	Phos-phorus (mg)	Mag-nesium (mg)	Iron (mg)	Zinc (mg)	Iodine (μg)
0.4	6	0.3	30	0.5[h]	360	240	50	10	3	40
0.6	8	0.6	45	1.5	540	360	70	15	5	50
0.8	9	0.9	100	2.0	800	800	150	15	10	70
1.0	11	1.3	200	2.5	800	800	200	10	10	90
1.4	16	1.6	300	3.0	800	800	250	10	10	120
1.6	18	1.8	400	3.0	1200	1200	350	18	15	150
1.7	18	2.0	400	3.0	1200	1200	400	18	15	150
1.7	19	2.2	400	3.0	800	800	350	10	15	150
1.6	18	2.2	400	3.0	800	800	350	10	15	150
1.4	16	2.2	400	3.0	800	800	350	10	15	150
1.3	15	1.8	400	3.0	1200	1200	300	18	15	150
1.3	14	2.0	400	3.0	1200	1200	300	18	15	150
1.3	14	2.0	400	3.0	800	800	300	18	15	150
1.2	13	2.0	400	3.0	800	800	300	18	15	150
1.2	13	2.0	400	3.0	800	800	300	10	15	150
+0.3	+2	+0.6	+400	+1.0	+400	+400	+150	[i]	+ 5	+25
+0.5	+5	+0.5	+100	+1.0	+400	+400	+150	[i]	+10	+50

[h] The RDA for vitamin B_{12} in infants is based on average concentration of the vitamin in human milk. The allowances after weaning are based on energy intake (as recommended by the American Academy of Pediatrics) and consideration of other factors such as intestinal absorption.

[i] The increased requirement during pregnancy cannot be met by the iron content of habitual American diets nor by the existing iron stores of many women; therefore, the use of 30–60 mg of supplemental iron is recommended. Iron needs during lactation are not substantially different from those of nonpregnant women, but continued supplementation of the mother for 2–3 months after parturition is advisable in order to replenish stores depleted by pregnancy.

[j] Reproduced with the permission of the National Academy of Sciences, Washington D.C.

of Additional Selected Vitamins and Minerals[a,b]

Trace elements[c]				Electrolytes		
Fluoride (mg)	Chromium (mg)	Selenium (mg)	Molybdenum (mg)	Sodium (mg)	Potassium (mg)	Chloride (mg)
0.1–0.5	0.01–0.04	0.01–0.04	0.03–0.06	115–350	350–925	275–700
0.2–1.0	0.02–0.06	0.02–0.06	0.04–0.08	250–750	425–1275	400–1200
0.5–1.5	0.02–0.08	0.02–0.08	0.05–0.1	325–975	550–1650	500–1500
1.0–2.5	0.03–0.12	0.03–0.12	0.06–0.15	450–1350	775–2325	700–2100
1.5–2.5	0.05–0.2	0.05–0.2	0.1 –0.3	600–1800	1000–3000	925–2775
1.5–2.5	0.05–0.2	0.05–0.2	0.15–0.5	900–2700	1525–4575	1400–4200
1.5–4.0	0.05–0.2	0.05–0.2	0.15–0.5	1100–3300	1875–5625	1700–5100

[c] Since the toxic levels for many trace elements may be only several times usual intakes, the upper levels for the trace elements given in this table should not be habitually exceeded.

Table 3. Mean Heights and Weights and Recommended Energy Intake[a,b]

Category	Age (yr)	Weight (kg)	Weight (lb)	Height (cm)	Height (in.)	Energy needs (with range) (kcal)	Energy needs (with range) (MJ)
Infants	0.0–0.5	6	13	60	24	kg × 115 (95–145)	kg × .48
	0.5–1.0	9	20	71	28	kg × 105 (80–135)	kg × .44
Children	1–3	13	29	90	35	1300 (900–1800)	5.5
	4–6	20	44	112	44	1700 (1300–2300)	7.1
	7–10	28	62	132	52	2400 (1650–3300)	10.1
Males	11–14	45	99	157	62	2700 (2000–3700)	11.3
	15–18	66	145	176	69	2800 (2100–3900)	11.8
	19–22	70	154	177	70	2900 (2500–3300)	12.2
	23–50	70	154	178	70	2700 (2300–3100)	11.3
	51–75	70	154	178	70	2400 (2000–2800)	10.1
	76+	70	154	178	70	2050 (1650–2450)	8.6
Females	11–14	46	101	157	62	2200 (1500–3000)	9.2
	15–18	55	120	163	64	2100 (1200–3000)	8.8
	19–22	55	120	163	64	2100 (1700–2500)	8.8
	23–50	55	120	163	64	2000 (1600–2400)	8.4
	51–75	55	120	163	64	1800 (1400–2200)	7.6
	76+	55	120	163	64	1600 (1200–2000)	6.7
Pregnant						+ 300	
Lactating						+ 500	

[a] From Recommended Dietary Allowances, revised 1979. Food and Nutrition Board, National Academy of Sciences—National Research Council, Washington, D.C.

[b] The data in this table have been assembled from the observed median heights and weights of children shown in Table 1, together with desirable weights for adults given in Table 2 for the mean heights of men (70 inches) and women (64 inches) between the ages of 18 and 34 years as surveyed in the U.S. population (HEW/NCHS data).

The energy allowances for the young adults are for men and women doing light work. The allowances for the two older age groups represent mean energy needs over these age spans, allowing for a 2% decrease in basal (resting) metabolic rate per decade and a reduction in activity of 200 kcal/day for men and women between 51 and 75 years, 500 kcal for men over 75 years, and 400 kcal for women over 75. The customary range of daily energy output is shown for adults in parentheses and is based on a variation in energy needs of ± 400 kcal at any one age, emphasizing the wide range of energy intakes appropriate for any group of people.

Energy allowances for children through age 18 are based on median energy intakes of children these ages followed in longitudinal growth studies. The values in parentheses are 10th and 90th percentiles of energy intake, to indicate the range of energy consumption among children of these ages.

Index